Critical Assessment and Strategies for Increased Student Retention

Ruth Claire Black
Namseoul University, South Korea

A volume in the Advances in Higher Education and Professional Development (AHEPD) Book Series

Published in the United States of America by
IGI Global
Information Science Reference (an imprint of IGI Global)
701 E. Chocolate Avenue
Hershey PA, USA 17033
Tel: 717-533-8845
Fax: 717-533-8661
E-mail: cust@igi-global.com
Web site: http://www.igi-global.com

Library of Congress Cataloging-in-Publication Data

Names: Black, Ruth Claire, 1964- editor.
Title: Critical assessment and strategies for increased student retention / Ruth Claire Black, editor.
Description: Hershey, PA : Information Science Reference, 2018. | Includes bibliographical references.
Identifiers: LCCN 2017012052| ISBN 9781522529989 (hardcover) | ISBN 9781522529996 (ebook)
Subjects: LCSH: College dropouts--Prevention. | College attendance. | Minority college students. | First-generation college students. | Limited English-proficient students. | College students with disabilities.
Classification: LCC LC148.15 .C75 2018 | DDC 378.1/6913--dc23 LC record available at https://lccn.loc.gov/2017012052

This book is published in the IGI Global book series Advances in Higher Education and Professional Development (AHEPD) (ISSN: 2327-6983; eISSN: 2327-6991)

British Cataloguing in Publication Data
A Cataloguing in Publication record for this book is available from the British Library.

For electronic access to this publication, please contact: eresources@igi-global.com.

Advances in Higher Education and Professional Development (AHEPD) Book Series

Jared Keengwe
University of North Dakota, USA

ISSN:2327-6983
EISSN:2327-6991

Mission

As world economies continue to shift and change in response to global financial situations, job markets have begun to demand a more highly-skilled workforce. In many industries a college degree is the minimum requirement and further educational development is expected to advance. With these current trends in mind, the **Advances in Higher Education & Professional Development (AHEPD) Book Series** provides an outlet for researchers and academics to publish their research in these areas and to distribute these works to practitioners and other researchers.

AHEPD encompasses all research dealing with higher education pedagogy, development, and curriculum design, as well as all areas of professional development, regardless of focus.

Coverage

- Adult education
- Assessment in Higher Education
- Career Training
- Coaching and Mentoring
- Continuing Professional Development
- Governance in Higher Education
- Higher Education Policy
- Pedagogy of Teaching Higher Education
- Vocational Education

IGI Global is currently accepting manuscripts for publication within this series. To submit a proposal for a volume in this series, please contact our Acquisition Editors at Acquisitions@igi-global.com or visit: http://www.igi-global.com/publish/.

The Advances in Higher Education and Professional Development (AHEPD) Book Series (ISSN 2327-6983) is published by IGI Global, 701 E. Chocolate Avenue, Hershey, PA 17033-1240, USA, www.igi-global.com. This series is composed of titles available for purchase individually; each title is edited to be contextually exclusive from any other title within the series. For pricing and ordering information please visit http://www.igi-global.com/book-series/advances-higher-education-professional-development/73681. Postmaster: Send all address changes to above address.

Titles in this Series

For a list of additional titles in this series, please visit: www.igi-global.com/book-series

Fostering Positive Civic Engagement Among Millennials Emerging Research and Opportunities
Darrell Hucks (Keene State College, USA) Tanya Sturtz (Keene State College, USA) and Katherine Tirabassi (Keene State College, USA)
Information Science Reference • copyright 2018 • 98pp • H/C (ISBN: 9781522524526) • US $135.00 (our price)

Fostering Effective Student Communication in Online Graduate Courses
Abigail G. Scheg (Western Governors University, USA) and Melanie Shaw (Northcentral University, USA)
Information Science Reference • copyright 2018 • 387pp • H/C (ISBN: 9781522526827) • US $195.00 (our price)

Fostering Reflective Teaching Practice in Pre-Service Education
Zineb Djoub (Abdelhamid Ibn Badis University of Mostaganem, Algeria)
Information Science Reference • copyright 2018 • 366pp • H/C (ISBN: 9781522529637) • US $215.00 (our price)

Violence Prevention and Safety Promotion in Higher Education Settings
Mihaela Badea (Petroleum-Gas University of Ploiesti, Romania) and Mihaela Suditu (Petroleum-Gas University of Ploiesti, Romania)
Information Science Reference • copyright 2018 • 266pp • H/C (ISBN: 9781522529606) • US $205.00 (our price)

Personalized Professional Learning for Educators Emerging Research and Opportunities
Kristi Meeuwse (Drayton Hall Elementary School, USA) and Diane Mason (Lamar University, USA)
Information Science Reference • copyright 2018 • 156pp • H/C (ISBN: 9781522526858) • US $135.00 (our price)

Ideological Function of Deming Theory in Higher Education Emerging Research and Opportunities
Thanos Kriemadis (University of Peloponnese, Greece) Ioanna Thomopoulou (University of Peloponnese, Greece) and Anastasia Sioutou (University of Peloponnese, Greece)
Information Science Reference • copyright 2018 • 176pp • H/C (ISBN: 9781522522652) • US $135.00 (our price)

Examining the Changing Role of Supervision in Doctoral Research Projects Emerging Research and Opportunities
Annette Lerine Steenkamp (Lawrence Technological University, USA)
Information Science Reference • copyright 2018 • 176pp • H/C (ISBN: 9781522526100) • US $145.00 (our price)

IGI Global
DISSEMINATOR OF KNOWLEDGE

701 East Chocolate Avenue, Hershey, PA 17033, USA
Tel: 717-533-8845 x100 • Fax: 717-533-8661
E-Mail: cust@igi-global.com • www.igi-global.com

Editorial Advisory Board

Table of Contents

Detailed Table of Contents

Noncognitive skills are a rather popular topic in higher education today. Factors such as "grit," "character," and "growth mindset" are an increasing part of many student success efforts. This chapter will discuss several issues related to noncognitive factors, seeking to bridge the gap from research to practice. First, we will review various ways in which the noncognitive space has been defined. Second, we will review research supporting the relevance of noncognitive variables in higher education, including their espoused importance, empirical relationships with student outcomes, and finally, evidence that such factors can serve a somewhat compensatory role for many students. Third, we will provide an overview of methods of assessing noncognitive tools, particularly within a context of student retention and success. Finally, we will review common practices to implementing noncognitive factors in student success strategies, including holistic placement, enhanced student advising, and other interventions focused directly on the development of noncognitive skills.

This chapter discusses the positive correlations between study abroad and student success broadly, as well as the ways in which study abroad programs operate from a student success model. Both may help influence home-campus practitioners in working with all students. First, this chapter will first focus on the history of study abroad within higher education to contextualize it specifically relating to retention and student success. Second, the chapter will focus on the positive, and not often discussed, positive relationship between student retention and graduation rates correlated to study abroad. Third will be a

In integrating the two theories, the researcher was able to predict or explain the study outcomes and link them to the existing body of knowledge. This research presented the theoretical framework of the study on factors that affected retention among freshmen students at Historically Black Colleges and Universities. The research problem in this chapter was factors that affected retention among freshmen students at HBCUs. This research problem anchored the entire study and formed the basis of constructing the study's theoretical framework. The key variables in this study were the freshmen students, theories, retention rates, and the factors affecting retention.

Student engagement, retention, and success are not issues limited to traditional academic settings. Corporations and non-traditional business entities allocate significant resources to identify and develop viable solutions that positively impact employee engagement, retention and success. While most corporate training and employee development programs are driven by corporate mandates to improve efficiency and cost containment, the most dynamic retention and student success elements within these programs deserve further study and exploration. Indeed, the possible transfer of these effective retention and student success elements from the corporate training and development space to other learning, training and student development settings, can yield significant benefits to any institution or organization that seeks to improve its initiatives focused on training and education.

This paper explores anonymous online learning as a tool to overcome specific teaching and learning issues within Korean post-secondary institutions. The chapter utilizes a survey of a small group of ESL students at a single Korean university to better understand student preferences and opinions about non-traditional learning options and opportunities in Korea. While many students in Korea have not been exposed to online learning, the students surveyed expressed interest in learning online and they were especially interested in collaborative learning opportunities. As more online classes and online learning opportunities start to become available in South Korea, this study explores anonymous online learning as an effective tool to overcome some significant and distinct teaching and learning challenges at Korean post-secondary institutions. The anonymous online learning suggestions and approaches in the paper can be implemented within fully online courses and blended classes but they can also be used as stand-alone online components of traditional face to face and ESL courses.

This Chapter explores the reasons for and provides a range of ways for faculty members, staff and administrators to build a bridge that will encourage and help a higher number of at-risk, underrepresented and female students matriculate directly into the MBA program upon completion of an undergraduate degree. First, the chapter will provide data driven reasons why it is in the best interest of at-risk, underrepresented minority and female students to pursue a graduate degree. Second, it will provide concrete and proven examples of how to help these student's move directly from undergraduate degrees into the MBA program. The examples of how to encourage this direct pathway fall into three categories: faculty focus, curriculum design and university-wide systems.

South Korea is one of the highest achieving education countries in the world. However, there are several factors that inhibit South Koreans from realizing their full potential. South Korean students exhibit low scores on a number of learning and success measures, including low satisfaction with education and career prospects, low numbers of positive feelings and high numbers of negative feelings on a range of measures such as, poor family connections and poor social connections. These factors contribute to high depression rates and high rates of suicide among young adults. The depressive states that are common among young South Koreans can be addressed using a Hope Theory based professional coaching intervention to generate and then support more positive feelings about education and personal and professional goals. The combined approach can significantly and positively impact depressive issues and address student confidence, motivation, and learning ability. Thereby, leading to improved metrics on a wide range of life and educational satisfaction indexes.

While there is no direct causal link between academic advising and increased student persistence, the role of the academic advisor can be key to an institution's success. This chapter examines one university's approach to redesigning the academic advising model from the ground up and committing to a philosophy of continuous improvement in academic advising, retention and student success. A decade in the making, the tools and approaches created through this process now play an important part in the institution's success and can be a road-map for other institutions to follow as they aspire to revise and improve their academic advising models and to improve student success.

Student retention is becoming critical as the budgets for higher education are shrinking and the economies are becoming more and more knowledge driven. Student retention if not addressed in a holistic manner would mean loss not only to the institution but also to the development of human resources and equitable economic development. This chapter intends to review the causes of student dropout, the current approaches adopted for retaining students with focus on master level programmes in management discipline at a Namibian University. In doing so it also addresses the socio-cultural aspects of student retention and intends to provide a holistic framework for enhancing the rate of student retention.

This case study is the result of a reconnoitre style approach to observation, analysis and subsequent planning. This approach comprised several (on-going) extensive informal interviews with staff across departments responsible for quality, administration, teaching and management. This has been supported with the analysis and observations of the University's staff and student culture, coupled with benchmarking these insights with other HEIs and current retention academic literature. Before presenting the case, it is necessary to ground the reader in current understanding of student retention in HEI's. The literature reviewed is drawn from both UK and international scholarship and following the presentation of current literature on student retention and the University's case study, further empirical review is integrated within the case study where appropriate, to contextualise, corroborate or reflect.

This chapter uses the literature on student retention in online programs as a foundation to present a case study illustrating how a well-designed, affordable and high-quality online program substantially increased access, retention rates, and progression to graduation. The case study covers how, by using the principles of designing with the end in mind. The University of Texas at Arlington College of Nursing and Health Innovation (CONHI) developed and implemented a highly successful, award-winning online Bachelor of Science in Nursing (BSN) completion program for registered nurses (RN) who had entered the profession through an associate degree program. This RN-to-BSN program was specifically designed to overcome issues known to impact enrollment and completion in nursing programs by leveraging technology and the best practices for online education. The case study also explores how best practices in online education were incorporated into the development of the RN-to-BSN program.

Online learning has shown persistent and unrelenting growth over the past few years and serves a wide variety of modern educational needs. However, attrition can be a troublesome phenomenon unless course designers and instructors develop methods of engaging students through use of the Four Pillars discussed in this chapter. The strategies and tactics associated with and springing from these Pillars will go a long way in retaining students in online programs. Face-to-face instructional methods simply will not work in the online environment, and the innovative methods discussed in this chapter will yield student engagement and completion of their educational goals.

Research literature in public education often categorizes self-efficacy in order for practitioners of K-12 academic settings to understand how to relate to students. They often view self-efficacy in students' perception, intrinsic motivation and behavior in academic settings. This chapter utilizes the research literature from public education as a bridge to discover the importance of self-actualization of efficacy in developing reading specialists. Too often, students in public education who have been referred to a reading specialist due to struggles with reading, low academic performance or a high dropout risk are missing a level of self-efficacy. In that same context, the authors in this study looked to a specific program which prepares teachers of those students and incorporated rigorous self-efficacy scales to inform and shape practice. This chapter asserts that teachers are paramount in the successful development of an at-risk learner. This chapter provides empirical research centered around two core programs, undergraduate and graduate with specialization in urban education and reading.

This chapter extrapolates information processing theory and advances in research in memory formation and effective learning practices on to the first-year student. The authors will lay a context for what FYE is, and why it matters, survey advances in memory research, and explore the study practices that students actually use. Finally, the authors posit a structure for organizing FYE courses around the best practices in cognitive and educational psychology so as to change student study behavior and help them increase their academic performance.

Foreword

Critical Assessment and Strategies for Increased Student Retention is an important and pivotal reference resource that will advance retention and student success efforts at a wide range of academic institutions in the US and abroad. It is also long overdue that we have access to a book which illustrates a range of best practices and approaches to improving student success and retention. Ruth Claire Black, after her time leading Cal State Online and previously starting Brandman University's online efforts, is the right person to assemble this work.

Higher Education is experiencing the convergence of a number of long-term trends which may ultimately cause significant changes in how future students are educated. Many institutions are already at the crossroads of how to respond to these trends. The most significant trend is the decrease in the number of college enrollments. While recently the number of high school graduates has increased, National Student Clearinghouse Research Center data suggest that college enrollment rates and the number of enrollments have actually decreased for the fifth straight year. For the most recent year the decrease in enrollments was 1.4%.

The emergence of non-traditional educational providers is creating lower cost pathways for high school graduates. Coding academies started by providing accelerated, focused coding training with assured job placement for graduates. Some are now expanding offerings by leveraging the same business model for areas such as marketing, business, data analytics, project management, and computer assisted design.

Other providers are utilizing massive open online courses (MOOC's) or low cost/no cost online classes to reduce the number of courses a student must take when they do enroll in their degree granting institution. It is also becoming more acceptable for high school graduates to opt for trades that provide middle class income while avoiding the time, expense and debt of a traditional institution.

These trends have resulted in significant competition for the decreasing number of college enrollments, especially of high school graduates who opt to take the traditional pathway from high school to college. At the same time, the selection criteria that students consider when choosing an institution are shifting. In the past, inputs were important in evaluating an institution; now, outcomes have assumed a more important role in the process of college selection.

The exploding student debt crisis is an overriding concern of students. In the past ten years median salaries have stagnated while during the same period the increase in education costs have far exceeded inflation. It is no wonder that student debt has exploded to over $1.4 trillion. Borrowing to fund an education is a great investment if one graduates in a major that has good job prospects; however, it is a disaster if one incurs debt and fails to graduate, which is why institutions with low graduation rates have high default rates. This trend is why graduation rates are becoming a more important factor considered by prospective students.

Currently, the federal government, through the Integrated Postsecondary Education Data System (IPEDS), published each institution's graduation rate for first-time, full-time freshmen who start in the fall of each year. This graduation rate is not relevant for the majority of students who now graduate from the second or third institution that they attend. For the past two years IPEDS has been collecting non-first-time graduation rate information and this year, is expected to publish both first-time and non-first-time graduation rates. The availability of the non-first-time graduation rates will, for the first time, provide transfer students with useful information on how to compare institutions and select an institution where they are more likely to graduate. It will also be useful in developing university rankings for institutions who serve transfer students.

As we see the growing importance of graduation rates, the importance of retention comes into focus. Evaluating retention rates each year will facilitate continuous improvements that will ultimately lead to higher graduation rates. While retention rates may differ a bit in meaning in traditional and non-traditional institutions, they provide an annual result that will be used to set goals and measure progress.

In an environment where institutions are competing over a decreasing number of enrollments, it will become more expensive to enroll new students. This will tilt the economics to make it more favorable to expend additional resources to retain current students and this is why this book on the best thinking in the field of retention is so timely, useful and important.

As you read Critical Assessment and Strategies for Increased Student Retention, you will see that incremental improvement in retention can be achieved by focusing on improving existing systems and by using sophisticated data tools and predictive analytics. You will also note that greater improvements can be achieved by focusing on adopting best practices. While a cross departmental retention committee can be helpful, like anything else, executive leadership of a university wide effort will produce the best results.

As the leader of Brandman University, when we transitioned from being Chapman University College, we had the opportunity to change everything, and that is what we did based upon our research of best practices for retention and student success. Changes included:

- All students receive success coaching
- Creating a duel advising model where each student has a dedicated adviser
- Creating one-stop student services
- Designing and implementing an online writing and math community
- Transitioning all on ground classes to the blended class format
- Ensuring that students can take classes on ground or online or both
- Six 8 week sessions per school year to reduce time to degree
- Focus on relevant curriculum with DQP and AAC&U LEAP competencies embedded

Brandman serves a non-traditional student body that is 40% from underrepresented groups, 46% of whom receive Pell grants, 49% first generation and 24% active military and veterans. Yet, a few months ago when we reported to IPEDS non-first-time, full-time graduation rates for our 2008 cohort our six-year graduation rate was 82% and our eight-year rate was 83%, with a 4% cohort default rate.

When a student begins or restarts their educational journey, coaching may enable them to successfully make the transition. During busy times at home or work, being able to shift to online classes might enable a student to persist. Accelerated 8 week sessions are beneficial to many students and having a dedicated adviser that gets to know the student and their needs could prove to be the difference.

There is not a magic bullet to retention. High retention and graduation rates are the result of adopting best practices that are the most relevant to the students you serve. Critical Assessment and Strategies for Increased Student Retention is an essential tool in identifying and implementing the best practices that will be most effective in enhancing your student's success.

Gary Brahm
Brandman University, USA

Preface

I wanted to produce this book for well over a decade and I have notes referring to ideas and possible content areas for the book dating all the way back to 1997. However, over the last seven years, the need for a comprehensive and critical look at retention and student success became more intense and clear.

For me, the scrutiny of retention rates and student success in online, blended, and face-to-face programs steadily intensified after the U.S. Department of Education's 2010 report confirmed what I had believed for years, that "students in online conditions performed modestly better, on average, than those learning the same material through traditional face-to-face instruction" and that "online learning is superior as a medium" (p. xvi, xviii, original emphasis). In addition, in 2010 President Barack Obama delivered an important speech on education during which he said, "over a third of America's college students and over half of our minority students don't earn a degree, even after six years. So, we don't just need to open the doors of college to more Americans; we need to make sure they stick with it through graduation. That is critical" (President Obama, 2010, para. 34).

We simply need to do better at retaining our college students through to graduation. From my own teaching and administrative experiences, principally in online education at both public and private non-profit institutions, I believed that the 2010 U.S. Department of Education's report on online education was correct. Yet, why was retention in online programs continuing to significantly lag face-to-face programs; why were self-reported measures of student success lower in online programs and non-traditional student communities; and why, given all of the attention and efforts at improvement, wasn't the gap narrowing?

The moment when I finally committed to delivering this book came in early 2015 when I first read the proposed regulatory package that came to be known as the Gainful Employment Act by the U.S. Department of Education. It was then that I realized that efforts to regulate, shape and limit education opportunities were squarely focused on retention rates. Retention rates were the tipping point for determining a host of important education and societal issues, such as financial aid, university rankings, professional job opportunities, community participation and social impact, to name just a few. It has taken almost three years since that moment of total commitment to produce this book; but, it has been time well spent.

Various reviews of the literature over the years confirmed the significant retention and student success gap between online and face-to-face programs (Boston, Diaz, Gibson, Ice, Richardson & Swan, 2009). The literature also confirmed that non-traditional and non-first-time students, especially those enrolled in online programs, have retention rates that are significantly lower than those of first time freshmen (Hill, 2010). Further, retention has been and continues to be a difficult problem to assess due to the retention gap between first time freshman and non-first-time students, and between programs; face-to-face, blended and online. The gap in retention has not appreciably narrowed over the last twenty years (Allen

& Seaman, 2015). As demonstrated in the literature, improving retention rates requires comprehensive institution-wide strategies that systemically tackle extensive and pervasive problems (Hill, 2010).

Since the release of the Spellings Report in 2006, U.S. based post-secondary institutions have been under pressure to become more transparent about retention and graduation rates (United States Department of Education, 2006). Two years after the Spellings Report, the federal government amended the Higher Education Act of 1965 to regulate and mandate disclosure and annual reporting of graduation and retention data as well as cost of attendance (Kelly & Schneider, 2012; United States Department of Education, 2009). Government and accreditation agency pressures to improve outcomes and raise graduation rates have been steadily building for the last decade. Universities are increasingly being evaluated in relation to their retention rates and critical aspects of education such as financial aid and institutional rankings now focus heavily on retention rates. Institutions need some practical approaches, best practices and new ideas to support them in meeting these rising expectations.

We simply must do a better job in retaining our college students to graduation. The data and knowledge about how to do better was out there. It just wasn't curated, organized, or structured in a practical and sensible way that would help scholars, politicians, institutions, and other education stakeholders understand the enormous and diverse issues, challenges and obstacles to retention that exist. We needed a "call to arms" on retention and student success that also provided a road map of how to move forward from whatever place a program or institution finds itself in when they commit to improving their retention rates.

What was needed was a way to inspire positive and practical action on retention and student success in a systematic, data driven, and process oriented way. From my experience, there were five key components to addressing the challenges: 1) a comprehensive set of studies devoted to "how to" assess the issues, challenges and obstacles, 2) a range of strategies to help close the retention gap that could be relevant and meaningful to a wide range of programs and institutions, 3) a comprehensive and critical assessment of retention and student success from the perspective of a wide range of university stakeholders, 4) the discussion needed to include faculty members, university administrators and staff members, as well as, governments, regulators, accreditation agencies, and technology and student support vendors, and 5) the discussion also needed to include a range of academic, teaching and learning institutions, organizations and corporations working at the post-secondary level, both in the U.S. and abroad.

This book collects case chapters of scholarly work, along with the specific road maps that could have practical and impactful results and success for a wide range of readers and audiences, in order to motivate, encourage, and inspire positive action within programs, institutions, and organizations. It is designed to communicate the issues, challenges, and obstacles to retention and student support work in diverse settings while setting a framework for approaches, strategies, processes and solutions to improve retention for many types of academic, teaching and training institutions and organizations.

The contributors offer a critical, yet practical perspective that is constructive and productive, rooted in how to understand and assess, and then, how to systematically address and improve retention and student success. The emphasis on practical research does not mean that the studies are isolated or removed from the academic literature and seminal studies in the areas of retention and student success. All of the chapters include literature review sections that contextualize their retention and student success work within the discipline. Each of the authors in the book describe plans for future research in their chapters. I expect that all of them will continue to periodically publish their subsequent findings on these important issues over the next several decades.

The contributors come from a wide range of perspectives, program levels and institution types, and deal with a broad range of retention issues, challenges and obstacles. In this collection, the reader is offered a multifaceted range of critical perspectives that are helpful, valuable and insightful no matter how uncomfortable some of the material and some of the data may prove for some readers. While contributions are mainly from North America, important contributions from Asia, Africa and Europe are also included.

The critical perspective and focus of the book is to emphasize that institutions must look closely at their program and institution retention rates, and then, carefully and strategically invest in resources, programs and processes to evaluate and improve them. The results of the research detailed in this book clearly illustrate that meaningful improvements in retention rates are possible, and that the longer the program or institution remains systematically committed to critical assessment and continuous improvement, the better. Improving retention requires a serious long term and comprehensive institution-wide commitment.

The book does not promote the legitimacy of specific retention approaches or theories, nor does it advocate for the acceptance or integration of specific retention rates and student success statistics into institutional assessment processes or ranking metrics. The chapter contributors were selected because of their attempts to move the field of retention and student success forward. They offer their best wisdom, recommendations, and solutions as a way to develop and then advance more effective ways of addressing the issues inevitably encountered in student retention, while maintaining a commitment to quality, data driven research. The authors are also mindful of their particular accreditation issues and their national education agendas. A critical foundation is either implied or overt in each chapter as each author attempts to reveal diverse retention issues of underlying concern that suggest the need to do things differently. Yet a common framework of dialog is apparent in the commitment to providing a better experience for students, faculty, and all the stakeholders involved in the post-secondary education process.

I am personally and professionally pleased to be the editor of this collection. I have connected with many wonderful scholars and practitioners while developing the content and learned much about this field in which I have so much invested. I am confident that the reader will gain valuable insight by reading these excellent case chapters. The chapters are divided into four sections: 1) Retention Perspectives and Approaches, 2) Supporting and Encouraging Student Success, 3) Assessing the Impact of Retention and Student Success Programs, and 4) Pedagogy and Program Design to Support Retention and Student Success.

In Section 1, "Retention Perspectives and Approaches," the chapters focus on the way that different approaches to retention in various teaching, learning and assessment contexts can be instructive. Markle and Rikoon's (1) chapter introduces us to non-cognitive factors in retention and explores individual characteristics and behaviors that increase the likelihood of a student persisting until graduation. It also provides approaches and tools for how to measure and assess these important retention factors. Canny's (2) essay focuses on the approach to retention and student success utilized in study aboard programs and suggests ways that programs and institutions can leverage these approaches to impact and improve retention on the home campus. Savas's (3) article provides a statistical model for understanding gender and race as critical retention factors and provides insights into the evaluation of this data for improved retention. Little, Gaier and Spoutz's (4) contribution emphasizes the importance of values, beliefs and culture in retention.

Section 2, "Supporting and Encouraging Student Success," starts with a theoretical analysis by Hinton (5) of the factors that influence retention of freshmen students at historically black colleges and universities (HBCUs). The author details how to support retention and improve student success, as well as, institutional growth in the unique teaching and learning setting of the HBCU. Wilcox and Black (6) look beyond the academic institution for retention and student success lessons and solutions through an analysis of a multi-level marketing distributor operating in fifty countries. Cromlish (7) then introduces the concept of anonymous online learning as a student success tool. Eckmann (8) writes about a process and approach that helps at-risk, underrepresented and female students matriculate into an MBA program directly upon completion of the bachelor's degree. To complete Section 2, Cromlish and Black (9) suggest a specialized coaching intervention to address systemic social and student success issues within the South Korean post-secondary education system.

In Section 3, "Assessing the Impact of Retention and Success Programs," Nemeth (10) offers university administrators a process and a framework to assess and then improve academic advising. Baporikar (11) provides a holistic framework for assessing and then enhancing the rate of student retention using a master's level program at a public Namibian University as a case study. Armstrong (12) then provides a comprehensive and systematic approach to assessing and then improving retention and student success using a private non-profit London based university as a case study.

In Section 4, "Pedagogy and Program Design to Support Retention and Student Success," the contributors describe pedagogy and best practices for retention focused course design. Mancini, Cipher and Ganji (13) provide an approach to designing and developing retention-focused online instruction. The chapter is a case study based on their curriculum design experience with a fully online nursing program at a flagship public university system. Galloway (14) also focuses on designing online courses and programs with the "end in mind" and offers "Four Pillars" of engaged course design as a model. Watson, Moreland and Smith (15) describe a reading and literacy program as a case study of engaged, self-efficacious and impactful teaching and learning at a university with strong ties to the Cherokee Tribal Nation. Korstange and Krahenbuhl (16) study the mind and how learning is processed and stored and how to leverage this for retention and student success generally and within first-year experience courses specifically. To complete the section and the book, Estes (17) looks at the role of communication in teaching and learning and how best practices in communication can facilitate improved retention and student success.

I am confident that you will find the case chapters highly readable, informative, and provocative. The time spent reading this collection of practical studies and interconnecting experiences will be well-rewarded with valuable insights that are readily applicable in your professional activities.

Ruth Claire Black
Namseoul University, South Korea

REFERENCES

Allen, I. E., & Seaman, J. (2015, February). *Grade level: Tracking online education in the United States 2015.* Retrieved from http://www.onlinelearningsurvey.com/reports/ gradechange.pdf

Boston, W., Diaz, S., Gibson, A., Ice, D., Richardson, J., & Swan, K. (2009). An exploration of the relationship between indicators of the community inquiry framework and retention in online programs. *Journal of Asynchronous Learning Networks*, *13*(3), 67–83.

Hill, C. (2010). *Faculty focus special report: Strategies for increasing online student retention and satisfaction.* Retrieved from http://www.facultyfocus.com/free-reports/strategies-for-increasing-online-student-retention-and-satisfaction/

Kelly, A. P., & Schneider, M. (2012). *Getting to graduation: The completion agenda in higher education.* Baltimore, MD: John Hopkins University Press.

Obama, B. (2010). *Remarks by the President on Higher Education and the Economy at the University of Texas at Austin.* Retrieved from https://obamawhitehouse.archives.gov/the-press-office/2010/08/09/remarks-president-higher-education-and-economy-university-texas-austin

United States Department of Education. (2006, September). *A Test of Leadership: Charting the Future of U.S. Higher Education: A Report of the Commission Appointed by Secretary of Education Margaret Spellings.* Retrieved from https://www2.ed.gov/about/bdscomm/ list/hiedfuture/reports/pre-pub-report.pdf

United States Department of Education. (2009, October 27). *Institutional eligibility under the Higher Education Act of 1965, as amended, and the secretary's recognition of accrediting agencies; final rule* (Federal Register, Vol. 7, No. 206, Doc E9-25186). Retrieved from https://www2.ed.gov/legislation/FedRegister/finrule/2009-4/102709a.html

United States Department of Education. (2012, February). *Pathways to success, integrating learning with life and work to increase national college completion.* Retrieved from http://www2.ed.gov/about/bdscomm/list/acsfa/ptsreport2.pdf

Acknowledgment

First and foremost, I wish to express my utmost appreciation to my former colleague and long-time friend, Vicki DeVore of the California State University System, Office of the Chancellor, who reviewed many chapter drafts and gave me unflagging and bountiful support and encouragement during the entire book development process. I must also express my gratitude to Jim Mathis for his steadfast friendship, his instrumental advice on life and career options and for his enduring enthusiasm and support of this book project as well as for his good counsel on the Preface. Thank you also to Myron Orleans for his friendship and professional collaborations over the years and for involving me in his 2014 book, Cases on Critical and Qualitative Perspectives in Online Higher Education, and then encouraging me to develop this work as a companion title to his 2014 book. I am also thankful to Myron for his advice on the Preface, though most certainly, the content and views expressed throughout the Preface are entirely my own and my responsibility.

Much of the success of the chapters is attributable to the thoughtful and detailed reviews during the double-blind evaluation process that were done by the members of the book's Editorial Advisory Board: Kiley Adolph, Kaye Bragg, Russell Brodie, Daniel Chatham, Alan Cromlish, Brad DeVore, Vicki DeVore, Myron Orleans, Kurt Takamine, Matthew Weaver and Alexandra McDermott Wilcox. Thank you also to the following contributing authors for participating in the peer review process: Neeta Baporikar, Eric Canny, Diasha Cipher, Alan Cromlish, Judi Estes and Alexandra McDermott Wilcox.

I want to thank Gary Brahm, Chancellor of Brandman University, not only for his gracious foreword, but for all of his encouragement over the many years of our friendship and professional relationship. I wish to thank Courtney Tychinski of IGI, our publisher's representative, who kindly and professionally guided me through the details and specifics of arranging for all the component parts of this collection and who was a warm and wonderful colleague on whom I relied mightily throughout the book's development process.

Thank you also to Sean Shin, Nari Kim, Alan Cromlish and others throughout Seoul, South Korea for their help with and support of my writing projects during my tenure as a visiting scholar at Namseoul University. I must thank Wendy Bearns for hosting me as I wrote these acknowledgements and the Preface and took care of the final details of the book from her beautiful loggia in the Comune of Monterchi, in Tuscany, Italy. Finally, I am eternally grateful to Diane Edwards-LiPera, Marilyn G. Geller, Marie G. Schrup, Helen Eckmann, Laura Galloway and Arthur Ruzzano for all of their help and support with this project and for their important and enduring friendships.

Ruth Claire Black
Namseoul University, South Korea

Section 1

Retention Perspectives and Approaches

Chapter 1
Noncognitive Skills and Student Success:
Relevance, Assessment, and Strategies

Ross Markle
Educational Testing Service, USA

Samuel H. Rikoon
Educational Testing Service, USA

ABSTRACT

Noncognitive skills are a rather popular topic in higher education today. Factors such as "grit," "character," and "growth mindset" are an increasing part of many student success efforts. This chapter will discuss several issues related to noncognitive factors, seeking to bridge the gap from research to practice. First, we will review various ways in which the noncognitive space has been defined. Second, we will review research supporting the relevance of noncognitive variables in higher education, including their espoused importance, empirical relationships with student outcomes, and finally, evidence that such factors can serve a somewhat compensatory role for many students. Third, we will provide an overview of methods of assessing noncognitive tools, particularly within a context of student retention and success. Finally, we will review common practices to implementing noncognitive factors in student success strategies, including holistic placement, enhanced student advising, and other interventions focused directly on the development of noncognitive skills.

INTRODUCTION

Historically, a great deal of research on student retention and success has focused on identifying the characteristics of those who succeed versus those who do not. Much of this research has focused on demographic, academic, contextual, and socioeconomic factors associated with student success. For example, students from certain minority groups, first-generation college students, and students with lower levels of academic preparation have tended to experience lower rates of success than do their peers (Burrus et al., 2013; Kuh, Kinzie, Buckley, Bridges, & Hayek, 2006). Eaton and Bean (1995) referred to

DOI: 10.4018/978-1-5225-2998-9.ch001

this avenue of research as taking a sociological approach to understanding student success. While this research has identified many traditionally underserved populations, Eaton and Bean pointed out that "we know less about the characteristics of individuals within such a group that increase the likelihood of their remaining in school until graduation" (p. 617).

One strand of more recent studies of "noncognitive" (aka "soft skills," or "character") factors and their relationships with student success has approached the topic from a psychological rather than a sociological perspective. That is, they have focused on key behaviors, skills, attitudes, and mindsets that successful students employ (e.g., Poropat, 2009; Richardson, Abraham, & Bond, 2012; Robbins et al., 2004). This work has also sought to reveal how to best identify and assess such factors, as well as examining methods of fostering them in students to help them navigate the challenges they will face in higher education. The field has established a group of factors related to both academic and persistence-related outcomes, further demonstrating that these factors are predictive of academic success even when controlling for many traditional markers of academic preparation.

Turning an explicit focus on noncognitive factors may represent a change to the daily practice of institutions focused on student retention and success. Most faculty and staff are not accustomed to assessing and teaching skills such as perseverance, self-efficacy, or help-seeking behavior directly. Of course, these important elements have been an implicit focus of many curricular and co-curricular interventions. These and other similar factors are also identified as critically important by advisors, instructors, and administrators. By transitioning from implicit to explicit attention to noncognitive factors, colleges and universities may gain the ability to better understand and address these vital behaviors with their students as integral parts of existing efforts to improve retention rates.

In this chapter, we discuss several issues related to understanding noncognitive factors as they relate to student success in higher education, summarizing important research findings and offering practical recommendations. First, we review several ways in which the noncognitive construct space has been defined in the field, discussing frameworks articulating an array of skills, behaviors, and mindsets. Second, we review research supporting the relevance and importance of noncognitive variables to higher education, including their empirical relationships with student outcomes and evidence of their potential to serve in a compensatory capacity for some students, helping to overcome other challenges students may bring to college. Third, we provide an overview of methods for assessing noncognitive constructs, with particular attention to the context of student retention and success efforts. Finally, we review common practices aimed at integrating attention to noncognitive factors within student success strategies. These include holistic placement, enhanced student advising, and other interventions focused directly on the development of noncognitive skills.

REVIEW OF THE LITERATURE

The term "noncognitive" has increased in prevalence over the last several years - in higher education as well as other sectors - but there appears to be little consensus as to what factors define the noncognitive space. The foundational meta-analyses that demonstrated the value of such variables in both workforce (Bowles, Gintis, & Osborne, 2001) and educational (Heckman & Rubenstein, 2001) sectors generally referred to "noncognitive" factors as those being outside the targeted scope of traditional measures of cognitive ability. Many educational practitioners question the use of the term "noncognitive," pointing out that nearly everything we do in our daily lives (apart from involuntary physical responses) involves a cognitive component (e.g., Fickel, 2015). The important distinction here is that (in the context of education and assessment) "noncognitive" is used to contrast the array of attitudes and social-emotional inter- and intrapersonal skills relevant to academic success from traditionally defined cognitive ability or intelligence (i.e., in this case, "noncognitive" is not intended to imply that no cognition is involved).

However, beyond example lists, the above meta-analyses did not specifically identify frameworks or taxonomies to help organize the field's thinking around noncognitive variables. For example, Heckman and Rubinstein (2001) referred to "persistence, reliability, and self-discipline," "persistence, self-esteem, optimism, and time preference," and "motivation and self-discipline." Building upon their work, there have since been several efforts to outline and define key noncognitive factors, and to acknowledge their importance alongside more traditional academic outcomes.

Several recent frameworks organizing higher education student learning outcomes have expanded the consideration of such measures beyond traditional definitions of domains such as critical thinking, written and oral communication, and quantitative fluency. These frameworks recognize additional complexity present in defining and assessing traditional cognitive outcomes and explicitly identify the expression of noncognitive factors as a critical expectation of college graduates. One example is Liberal Education and America's Promise (LEAP), a framework by the American Association of Colleges & Universities (AAC&U), outlining key student learning domains including teamwork and problem solving, civic knowledge and engagement, intercultural knowledge and competence, ethical reasoning and action, and "foundations and skills for lifelong learning" (AAC&U; 2011). Another example, by the Council for the Advancement of Standards in Higher Education (Strayhorn, 2006), noted a broad array of learning areas, including self-esteem, leadership development, social responsibility, and spiritual awareness. A recent review of student learning outcomes frameworks in higher education (Markle, Brenneman, Jackson, Burrus, & Robbins, 2013), found that areas such as creativity, teamwork, and civic engagement – among other noncognitive factors – were common across several similar efforts (see also Oliveri & Markle, 2017).

The benefits of a postsecondary education include more than just the academic knowledge and skills comprising one's major and general education courses. The framework efforts discussed above have contributed to the field by providing language and organizational rubrics encompassing a diverse set of skills. These are helpful for articulating the full scope of characteristics ideally observed among those who persisted in obtaining a college degree (vs. those who did not). Yet, while these expected *outcomes* of higher education suggest a need to consider and support student noncognitive skill development throughout their postsecondary careers, the specific skills, behaviors, and mindsets helpful for students to succeed and persist *during* college should be expected to vary across both individuals and institutional contexts. Another body of research has sought to identify such key predictors of academic success and persistence, highlighting the need for improved assessment of noncognitive factors and drawing connections to the classic Five Factor Model of personality (Kyllonen, Lipnevich, Burrus, & Roberts, 2014; Poropat, 2009; Richardson, Abraham, & Bond, 2012).

In developing ETS' *SuccessNavigator* assessment, Markle, Olivera-Aguilar, Jackson, Noeth, and Robbins (2013) organized students' noncognitive strengths and challenges in four general areas. *Academic Skills* refer to behavioral strategies useful for fostering success in the classroom. These include key classroom factors such as attendance, timeliness, and engagement (operationalized as "meeting class expectations") as well as key organizational behaviors outside the classroom (studying, planning, etc.). *Commitment*, deals with the ways students approach their goals, including overall drive toward a college degree, as well as institutional commitment –a significant component of Tinto's seminal work on retention (e.g., Tinto, 1993). *Self-Management* refers to regulatory strategies students' may employ to help them navigate or cope with the challenges they encounter during college including their tendency to feel and cope with stress, and their academic self-efficacy. Finally, *Social Support* refers to students' connections with others and the extent to which that connectivity supports or hinders their academic success in college. Social support incorporates students' sense of belonging, help-seeking behaviors, and the support (or lack thereof) provided by external factors such as one's family, work, and finances. *SuccessNavigator* is discussed in further detail later in this chapter.

Through several large-scale studies, the work of Robbins and colleagues has not only demonstrated the value of noncognitive or psychosocial skills, but also articulated some of the ways via which they may influence student success. This group has also emphasized assessment-based strategies for using noncognitive data to address retention and student success issues and challenges across a range of institutions and organizations (Habley, Bloom, & Robbins, 2012; Markle, Olivera-Aguilar, et al., 2013; Porchea, Allen, Robbins, & Phelps, 2010; Robbins et al., 2009; Robbins, Allen, Casillas, Peterson, & Le, 2006; Robbins et al., 2004; Robbins, Oh, Le & Button, 2009).

Although their work has considered multiple specific constructs, one contribution of Robbins et al. has been to articulate the noncognitive space in higher education[1] as an organized array of skills, behaviors, and mindsets. Several of the studies cited above were conducted during the development of ACT's Engage® assessment, the model for which includes three domains each inclusive of multiple underlying scales (ACT, 2013). Motivation includes both goal-related behaviors (effort toward a goal, persistence) and study skills. Social engagement includes both social engagement behaviors and feelings of belonging. Self-regulation refers to the regulation of stress and emotions to facilitate academic success.

Developed based on similar literatures, *SuccessNavigator* and Engage® are notable in their identification and targeting of a diverse array of skills, behaviors, and mindsets related to student success. The scope of these assessments can be contrasted with prominent psychological research focused on particular constructs, such as that of Duckworth and colleagues on grit and self-control (Duckworth, 2016; Duckworth, Gendler, & Gross, 2016; Duckworth, Peterson, Matthews, & Kelly, 2007; Duckworth, White, Matteucci, Shearer, & Gross, 2016) and Dweck and colleagues on growth mindset (Dweck, 2008; Yeager & Dweck, 2012; Yeager et al., 2016).

Grit is defined as "perseverance and passion for long-term goals" (Duckworth et al., 2007; p. 1087). In her writings, talks, and work with schools across the educational spectrum, Duckworth discusses the importance of effort, perseverance, and self-control in academic achievement, noting that students who experience structural or other challenges (e.g., low socioeconomic status, previous academic failure) may have a higher likelihood of future success if they demonstrate such characteristics. This message has resonated in both academia and popular culture (e.g. Tough, 2013), with a growing body of work supporting grit and related motivational constructs as important characteristics for student success.

Another noted effort in this space has been that work around growth mindset. A growth mindset is one that views a given trait or skill (e.g. intelligence, academic ability, personality) as malleable and thus able to be developed with effort over time (Dweck, 2008). The opposite of a growth mindset is a fixed mindset, in which one assumes such a characteristic is immutable. The terms "fixed" versus "growth" mindset are analogues of Dweck's earlier work on "entity" versus "incremental" implicit theories of personal characteristics, respectively (Yeager & Dweck, 2012). The implications of student mindsets are significant because those with a growth mindset are more likely to work hard to overcome challenges and succeed, whereas those adopting a fixed mindset are more likely to give up and attribute failure to their own lack of ability.

Though others continue to make important contributions in the noncognitive skills space with a particular focus on higher education (e.g., Fong, Davis, Kim, Kim, Marriott, & Kim, 2017), the two groups of researchers highlighted above have been particularly effective in shining a national spotlight on the role noncognitive, psychosocial, and specifically motivational factors play in student success. Many of these factors and the ways in which they are typically organized in research can be linked back to the Five Factor Model of personality (Goldberg, 1990, 1992). For example, conscientiousness tends to encompass several motivational and behavioral facets closely aligned with grit (MacCann, Duckworth, & Roberts, 2009; Rikoon et al., 2016). Conscientiousness is also the Big-5 factor most strongly and consistently related to academic success (Poropat, 2009). Indeed, a recent meta-analysis confirmed grit's strong relationship to its parent domain conscientiousness (Credé, Tynan, & Harms, 2016), an association noted by proponents of grit since the construct's modern emergence in the literature (Duckworth et al, 2007; Eskreis-Winkler, Duckworth, Shulman, & Beal, 2014).

The close association of facet-level characteristics (e.g., grit) to their broad domain-level traits (i.e., conscientiousness) is an empirical fact. However, it does not speak on its own to the value of emphasizing (or avoiding) discussions of either factor in institutional practice, communication, interventions, pedagogy, or policy discussions. Each context (defined by a given educational environment, student-level characteristics, and their interaction) within each institution comes with its own challenges, and thus the breadth of focus and variety of factors considered to meet those challenges should be expected to vary with each new type of situation encountered in the field.

Beyond acknowledging the value of noncognitive factors, institutions of higher education can benefit from existing frameworks through articulating students' strengths and challenges and developing a "theory of action" (Bennett, 2010) around how to best support student noncognitive skill development. There are several examples of such frameworks specifically targeting issues in both higher education (Conley, 2015; Kyllonen et al., 2014; Markle, Olivera-Aguilar, et al., 2013) and the K12 sectors (Farrington et al., 2012; Kautz, Heckman, Diris, Ter Weel, & Borghans, 2014; National Research Council, 2012). We highlight both types of frameworks here because each can provide value to colleges and universities via their organization of both terminology and research evidence supporting the relevance and malleability of noncognitive skills (i.e., their susceptibility to intervention).

At a basic level, recognizing the connections (both theoretical and empirical) between different types of noncognitive skills across a student's enrollment is helpful to organize our thinking, even though an institution may only seek to prevent challenges with (or intervene to support) a small number of areas at a given time. These types of connections also show the utility of assessing a wide array of student noncognitive skills, which permits institutions to take a holistic approach to connecting students with supports tailored to their specific challenges. Frameworks such as those referenced above have generated common language, rubrics, and an evidence base for us to articulate the specific strengths and challenges students face, facilitating our ability to connect them with appropriate supportive resources.

THE IMPORTANCE OF NONCOGNITIVE FACTORS

As evidenced by the mission statements of colleges and universities (Oswald, Schmitt, Kim, Ramsay, & Gillespie, 2004), it could be argued that noncognitive factors are naturally appealing as outcomes criteria for institutions of higher education, since success in one's work and life are so obviously dependent on both inter- and intrapersonal skills (in addition to cognitive ability). Of course, there is also a substantial body of research to support their relevance to student success in higher education, particularly relating to retention. Below we briefly summarize three types of evidence in this area.

First, there is evidence for the importance of noncognitive factors that might be characterized as "face validity," mainly derived from reports showing that stakeholders in both higher education and the workforce explicitly identify the vital nature of noncognitive factors to the success of their constituent populations. Numerous organizations of such stakeholders have identified noncognitive outcomes as key competencies that students should be expected to acquire while in college (Adelman, Ewell, Gaston, & Schneider, 2014; AAC&U, 2011; Council for the Advancement of Standards in Higher Education [CAS], 2011). A similar approach was taken by Schmitt and colleagues (Oswald et al., 2004; Schmitt et al., 2011) who examined the mission statements of 35 colleges and universities to determine their publicly stated goals. They found that mission statements often referred to noncognitive factors – leadership, engagement, life-long learning – rather than strictly academic or traditional "cognitive" domains.

There is also evidence that employers place a premium on the ability of college graduates to demonstrate noncognitive skills. A survey conducted by The Conference Board (in collaboration with other organizations) found that teamwork, professionalism, and work ethic were ranked as the most important skills, with more than 90% of responding employers saying those factors were "very important" for college graduates to have. This level of importance was similar to more traditional, "cognitive" factors such as communication, critical thinking, and problem solving (Casner-Lotto & Benner, 2006). Using the Five Factor Model of personality as an organizing framework, Sackett and Walmsley (2014) provided a helpful summary of the relative importance of different characteristics across 22 "broad occupational families" spanning the five levels of job complexity recognized in the United States Department of Labor's Occupational Information Network (aka O*NET; Peterson, Mumford, Borman, Jeanneret, & Fleishman, 1999). In agreement with much of the literature on personality and noncognitive factors more generally, their review found conscientiousness and agreeableness to be the two broad personality factors most consistently rated as important to workplace success across job levels.

In addition to the stated importance of noncognitive skills, there is also evidence that these types of factors are predictive of student outcomes. While a vast research literature in psychology and education supports an array of specific noncognitive variables (e.g., goal setting, self-efficacy, study skills), here we highlight meta-analytic studies integrating findings across research areas. These studies make use of large effective sample sizes helpful for obtaining a sense of the state of the field's knowledge on a given topic – in this case, supporting a positive relationship between noncognitive factors and measures of student academic achievement.

Several meta-analyses have shown that higher levels of noncognitive factors tend to be associated with higher levels of academic achievement (e.g., Crede, Roch, & Kieszczynka, 2010; O'Connor & Paunonen, 2007; Richardson et al., 2012). Two such studies warrant further mention here because their findings shed light on our understanding of student success in higher education. First is a meta-analysis by Poropat (2009) of relationships between the Big-5 personality factors and intelligence with academic achievement at the primary, secondary, and post-secondary levels. Poropat found that, in primary education (kindergarten through 8th grade), intelligence tended to be the strongest predictor of success, with other personality factors having moderate relationships with student achievement. In high school and college, however, these relationships shifted such that the association between intelligence and achievement was somewhat weaker and of nearly the same magnitude as that between achievement and conscientiousness (in college it was essentially equivalent). These findings suggest that cognitive ability and behaviors indicative of conscientiousness (e.g., perseverance, organization) may deserve similar levels of attention when it comes to encouraging students' success in college.

Second is a meta-analysis by Robbins et al. (2004), which showed that not all achievement-related outcomes are the same in terms of their relationships with noncognitive factors. Whereas Poropat primarily operationalized achievement as defined by student grades, Robbins et al. (2004) looked at academic success in terms of earned grades/GPA as well as retention. They found that, when predicting collegiate grades, focusing on academic preparation as a key indicator was reasonable, as standardized test scores and high school grades tended to demonstrate the strongest relationships versus noncognitive factors. The addition of noncognitive variables significantly increased the predictive efficacy of their model, but academic indicators still accounted for the largest amount of variance in college GPA. However, when predicting retention, a different story emerged. In that case, adding noncognitive factors nearly doubled the predictive efficacy of the model with several of these school-related noncognitive factors (e.g., academic goals, institutional commitment) outperforming standardized admissions tests (but not high school GPA) as predictors of retention. In their models predicting both college GPA and retention, Robbins et al. found that noncognitive factors showed significant incremental prediction when controlling for socioeconomic status, test scores, and high school grades.

Beyond the statistical prediction of student success, another important point about noncognitive factors is that, to some extent at least, they appear to help students compensate for other challenges they may encounter during college. That is, students demonstrating relatively low levels of academic preparation may be able to overcome such a deficit through sustained and focused effort, managing their work effectively, or connecting with appropriate support systems. Several small-scale studies have examined this issue in different demographic groups, finding that noncognitive factors were stronger predictors of success than traditional academic markers, such as standardized test scores (e.g., Dennis, Phinney, & Chuateco, 2005; Engle, & Tinto, 2008; Pascarella, Pierson, Wolniak, & Terenzini, 2004; Tracey & Sedlacek, 1982).

One study that demonstrates this possible compensatory effect was conducted by Li et al. (2013). Looking at success in developmental mathematics courses in a large community college system, they examined students' preparation (as measured by academic placement test scores) and engagement in the course, the latter measured by faculty rating of each student's attendance, timeliness, and classroom engagement. Looking at students with low, moderate, and high levels of preparation, they found a positive relationship between engagement and course completion in all three groups. Two other notable findings emerged as well. First, students with high levels of engagement tended to pass their mathematics course regardless of their level of preparation at enrollment. Even at the lowest level of placement test score, highly engaged students first enrolling in the Fall semester passed their mathematics course 74% of the time (64% of those first enrolling in the Spring semester passed as well. Second, the gap in developmental mathematics course success rates between students with low versus high levels of engagement was greater in the subgroups of students with lower and moderate levels of preparation (relative to the highly prepared group). In other words, within-class behaviors were important for all students, but appeared to be associated with greater differences in success rates for students entering college with academic challenges. This result might be expected given higher base rates of course success among those entering college with higher levels of preparation, but nonetheless speaks to the potential value of effort-related behaviors among less prepared students.

ASSESSING NONCOGNITIVE FACTORS VIA SELF-REPORT MEASURES

Once colleges and universities have identified key noncognitive factors of interest, acquiring the right tools for assessing such characteristics becomes a critical component of informing student success efforts. After all, simply understanding the importance of noncognitive skills does little to inform institutional strategy. Assessments, and more importantly the data and information they provide, can help institutions understand the extent to which their students express or demonstrate noncognitive factors and thus, which academic or co-curricular resources might be best suited to support their success.

By far the most common noncognitive assessment method used in both research and practice is student self-report ratings of individual statements (aka Likert-type items). Students provide responses on a predefined scale (e.g. five points ranging from Strongly Disagree to Strongly Agree) to statements such as "I show up to class," or "I ask for help when I have a problem." Such responses are then analyzed statistically to determine which are appropriate to group together, generate scores for, and interpret as a single scale. In most cases, the inferences drawn in applied psychological research reporting on student noncognitive skill levels are based on this type of "observed" scale score. That is, a student's score on a given noncognitive factor is a combination of their responses to all of the assessment items targeting that factor. This type of assessment is distinguished from traditional surveys, which tend to be more focused on summarizing responses to individual items. For example, a survey might report on the percentage of students who indicated that they attended class regularly, versus aggregating student responses to multiple items (one or more of which might have to do with class attendance) to generate a scaled score estimating their level of course engagement.

The above indirect method of assessing noncognitive factors differs from direct self-report (Pauhlus & Vazire, 2007), which asks students to self-report a global assessment of a skill or trait level. For example, a direct assessment of time management skills might ask, "How good are you at managing your time?" This method may appear to have the advantages of being straightforward and efficient, but indirect self-reports based on multiple items are preferable from a measurement perspective since responses to any one item may exhibit random variation that can be minimized through analyzing multiple item responses in combination. Interested readers should see Raykov & Marcoulides (2011) for a more thorough treatment of the above and related measurement issues.

While it remains true that the clear majority of our understanding of student noncognitive factors has been derived using indirect self-report assessments, these types of tools are subject to limitations surrounding their validity for certain purposes. Under low- or no-stakes conditions (i.e., when the student understands that there will be no significant consequences or decisions made on the basis of their scores), student self-reports of their skills, behaviors, and mindsets can provide an understanding of the extent to which they see themselves exhibiting such characteristics relative to their perceived peer group. That said, these types of measures are also susceptible to phenomena such as intentional response distortion or "faking" in which students may engage in impression management by responding to an assessment on the basis of how they think their answers are likely to be interpreted instead of how they actually feel (Dilchert, Ones, Viswesvaran, & Deller, 2006; Ziegler, MacCann, & Roberts, 2011). Another issue is that student responses may reflect unintentional positive or negative biases about themselves (Alliger & Dwight, 2000; Pauhlus & Vazire, 2007). A major source of inaccurate self-perception in noncognitive skills assessment is reference group bias, in which students respond differently to assessment items not because they actually differ in their expressed trait levels, but instead because the groups they have compared themselves to in generating a response naturally differ (Duckworth & Yeager, 2015). Self-report measures should not be used on their own to make high-stakes decisions such as who will be admitted to a given institution since the context of those decisions naturally raise the likelihood of students altering their responses to appear more favorable to reviewers. In low- or no-stakes settings, however, such as an advising context where student responses will be used in combination with expert in-person guidance to help direct them to appropriate support services, the use of self-report measures to inform student skill levels is more appropriate (Markle, Wang, Sullivan, & Russell, 2015).

The discussion above has focused on self-report measures due to their prevalence in both research and practical applications. Nonetheless, several other assessment formats are available to evaluate personality or other noncognitive factors. Kyllonen et al. (2014) provide a helpful review as well as advantages and disadvantages of several such formats including situational judgment tests (Weekley & Ployhart, 2013), forced-choice (Drasgow, Chernyshenko, & Stark, 2010) biographical data, the use of transcript and interview data, and third-party ratings (among others). While it is beyond our current scope to review each type of assessment, just as assessing multiple constructs is valuable to gaining a holistic understanding of students, using multiple assessment methods offers institutions the opportunity to study whether different types of data are differentially helpful in predicting student success or suggesting areas in which intervention may help bolster their skills.

STRATEGIES AND INTERVENTIONS FOCUSING ON NONCOGNITIVE SKILLS

While data on noncognitive factors might help us better understand and predict student retention, without identifying effective related interventions, colleges and universities will be unable to leverage this information to improve student outcomes. A large body of research in primary and secondary education supports social and emotional interventions, finding that their impact on academic achievement meets and often exceeds academically focused interventions (e.g., Catalano, Berglund, Ryan, Lonczak, & Hawkins, 2002; Durlak, Weissberg, Dymnicki, Taylor, & Schellinger, 2011). Yet research in postsecondary settings is less prevalent, so conversations about noncognitive skills in higher education may still boil down to "We know noncognitive skills are important, but how do we help students develop them? Moreover, if we are able to do so, will that improve retention and graduation rates?"

The topic of higher education interventions targeting noncognitive skills is critical to moving the field forward in its goal of supporting student development. In their extensive review of research on persistence and retention, Habley et al. (2012) both emphasize the critical nature of student noncognitive skills and focus on four "proven" practices to improve student success rates: assessment and course placement, developmental education initiatives, academic advising, and first-year training programs (a type of co-curricular support). Below we discuss elements of the latter two practices directly related to noncognitive factors, continuing to outline recent work describing the conditions and contexts within which these and similar interventions might take place.

HOLISTIC ADVISING

One way that student noncognitive skills can be encouraged via postsecondary intervention is through the academic advising process (Drake, Jordan, & Miller, 2013). After all, the primary objective of academic advisors (which could include faculty, counselors, and coaches as well, depending on your institution's terminology and structures) is to support student on their path to graduation. While this may have traditionally been conceptualized in terms of helping students identify a major, navigate confusing institutional bureaucracy, and ultimately complete curricular requirements (Deil-Amen & Rosenbaum, 2003), it must be acknowledged that advising can also help students overcome other types of nonacademic challenges (e.g., establishing goal orientation or time management strategies, developing a sense of belonging; Drake et al., 2013). In these instances, incorporating an explicit discussion of noncognitive skills with students can be an important and meaningful aspect of a holistic advising relationship.

In our work with the *SuccessNavigator* assessment at both the collegiate and high school levels (Markle, Olivera-Aguilar, et al., 2013; Rikoon, Liebtag, Olivera-Aguilar, Steinberg, & Robbins, 2015), for example, the provision of explicit support for student noncognitive skills development serves a key function. Students and their advisors are provided with student noncognitive skill score levels accompanied by written feedback to help them interpret each score. A resource guide is also made available to help advisors provide students with activities relevant to each assessed domain and spur a discussion of other campus resources that may meet the student's needs. This strategy facilitates connecting students to both existing institutional supports and exercises that can be done within the advising context. Implemented in this way, the advising experience itself could be considered an intervention. While there are several studies showing a positive impact on student retention after implementing a centralized advising strategy or particular advising or coaching intervention (e.g., Bettinger & Baker, 2011; Deil-Amen & Rosenbaum, 2003; Kot, 2014; Ryan, 2013), not all such efforts have been successful (e.g., Schwebel, Walburn, Klyce, & Jerrolds, 2012). Although rigorous experiments comparing different approaches to college advising are difficult to implement, future research to determine which specific types of advising are likely to have the greatest impact on student retention rates and academic achievement – particularly across different types of institutions and students - would provide a valuable service to the field. Regardless of the exact mechanism through which advising is implemented, however, attending to a holistic set of student needs incorporating both academic and noncognitive factors remains an important goal to help students overcome challenges and persist in their education.

CO-CURRICULAR SUPPORTS

Beyond advising services, every college and university offers an array of co-curricular supports to facilitate student engagement and success (e.g., first-year experience programs, tutoring, interest-focused clubs, recreational or public service activities, athletics, student success courses, and others). Webber and Ehrenberg (2010) found that when institutions spent more on student support services they tended to realize higher persistence and graduation rates, with larger such effects at schools enrolling greater numbers of students with relatively low test scores and higher amounts of Pell Grant funding. Yet studies of the effectiveness of student supports (e.g., student success courses, orientation programs, living-learning communities) often show mixed results, which Kuh, Kinzie, Buckley, Bridges, and Hayek (2006) suggest may due to inconsistent implementation strategies across different institutional settings.

A key recommendation of Kuh et al. (2006) was that institutions increase their capacity for using assessment and other data to inform their improvement efforts, a charge we interpret as implicitly supportive of noncognitive assessment given the authors' focus on issues such as student engagement, satisfaction, and personal development, evaluating the quality of students' educational experiences, and student learning outcomes. Assessing student noncognitive skills may be particularly helpful in aiding schools to target existing co-curricular interventions they already provide to those students whose needs are most closely aligned with each program's objectives (Olivera-Aguilar, Rikoon, & Robbins, 2017). This type of matching could help address inconsistencies in implementation by connecting students with resources catered to their identified noncognitive challenges or, alternatively, intended to augment their identified strengths. At the institutional-level, aligning resources to a noncognitive skills framework should help identify gaps or potential efficiencies to be realized in existing co-curricular support programs.

TARGETED INTERVENTIONS

A review of studies of brief interventions by Yeager and Walton (2011) looked at research exploring social-psychological strategies for improving noncognitive factors, and the impact of these types of interventions on outcomes such as academic success and persistence. Summarizing their review, Yeager and Walton identified several themes that relatively simple and cost-effective interventions had in common in producing their effects. First, they were targeted at specific, well-established psychological phenomena related to student success (e.g., a growth mindset, value affirmation, sense of belonging). Second, the interventions were interactive. That is, each study asked students to participate in brief structured activities (e.g., letter writing) designed to ingrain certain lessons or perspectives rather than simply communicating the lessons explicitly. Third, successful interventions tended to avoid directly raising students' awareness of the underlying issue. For example, students who were writing letters were not aware that the exercise was designed to drive their own perceptions of the malleability of intelligence. While it is straightforward to imagine how such targeted, stand-alone interventions might be applied within different types of existing co-curricular programs, such initiatives would need to achieve substantial buy-in, be implemented with high fidelity, and take elements of both organizational and student contexts into account to take effective root within existing higher education structures (Weiss, Bloom, & Brock, 2014).

FUTURE RESEARCH DIRECTIONS

Although a large body of research has demonstrated the value of noncognitive skills (and related assessments) for predicting academic success in higher education, evidence of the practical use of noncognitive data to connect students with interventions and improve success rates is less prevalent. Moving forward, several areas could benefit from additional research. First, rigorously implemented experiments or quasi-experiments evaluating theory-based interventions are vital for building a base of evidence to support the widespread adoption of tactics helpful for improving success rates. Lacking a basis of high-quality evidence, determinations as to which interventions are likely to be effective in a given situation are essentially guesswork. Second, there are important questions to be answered around methods of delivery for each intervention, perhaps on a school-by-school basis. We have mentioned several mechanisms via which schools might incorporate an explicit focus on noncognitive skills into existing strategies (e.g., advising, co-curricular supports), but it may be the case that a given institution determines its existing structures, strategies, and cultures are insufficient in their current forms to support meaningful interventions. That is, once effective interventions have been identified, the evidence supporting them should be considered within each institution to determine ideal setting(s) for implementation.

The above points lead to a third area of need for future research on noncognitive skills in higher education: generalizability. It is important to evaluate whether "effective" strategies appear to exert differential impacts on certain subgroups of students (e.g., defined by academic, noncognitive, demographic, or socioeconomic characteristics) versus their effects being similar across the breadth of a student population. For example, gathering data on noncognitive factors may help to articulate the strengths of and challenges faced by traditionally underserved populations, thus informing the development of interventions designed to help close academic achievement gaps. Generalizability at the institutional level is an equally important issue, since we must also determine whether, for example, a set of interventions appears to benefit community colleges more than four-year institutions (or vice versa). Similar questions could be asked of public versus private, rural versus urban, selective versus non-selective schools, and a host of other between-school characteristics. Indeed, given the natural variation within each of the above categories, great potential exists for more nuanced comparisons at the school level dependent on the questions of interest to researchers, institutions, or political stakeholders.

CONCLUSION

If colleges and universities are to leverage noncognitive factors to improve student retention, they must first develop an understanding of key skills, strategies, and mindsets based on the research findings that inform this space. The goal of this chapter has been to provide background and substance to the conversation around noncognitive skills and their meaningful role in supporting student success in higher education. By demonstrating their relevance to student retention through existing research efforts, we hope to elevate the priority given to supporting the development of such characteristics in the higher education community, particularly concerning issues of academic readiness. Utilizing assessment methods appropriate for a given intended use of noncognitive skill scores is fundamental to detecting student strengths and challenges with appropriate rigor, and permits the identification of distinct noncognitive skill profiles across student subgroups. Finally, the consistent implementation of effective interventions will be crucial to encouraging student noncognitive skills development. Combining knowledge of the field, rigorous assessment strategies, and salient interventions can help equip higher education institutions to adopt a holistic perspective toward understanding and predicting student success, as well as improving its likelihood.

REFERENCES

ACT. (2013). *ACT Engage College Domains and Subscales.* Retrieved from http://www.act.org/content/dam/act/unsecured/documents/EngageCollegeHandout.pdf

Adelman, C., Ewell, P., Gaston, P., & Schneider, C. G. (2014). *The Degree Qualifications Profile.* Lumina Foundation. Retrieved from http://www.luminafoundation.org/publications/The_Degree_Qualifications_Profile.pdf

Alliger, G. M., & Dwight, S. A. (2000). A meta-analytic investigation of the susceptibility of integrity tests to coaching and faking. *Educational and Psychological Measurement*, *60*(1), 59–72. doi:10.1177/00131640021970367

Association of American Colleges and Universities. (2011). *The LEAP vision for learning: Outcomes, practices, impact, and employers' views*. Washington, DC: Author.

Bennett, R. E. (2010). Cognitively based assessment of, for, and as learning: A preliminary theory of action for summative and formative assessment. *Measurement: Interdisciplinary Research and Perspectives*, *8*(2-3), 70–91. doi:10.1080/15366367.2010.508686

Bettinger, E., & Baker, R. (2011). *The effects of student coaching in college: An evaluation of a randomized experiment in student mentoring (No. w16881)*. National Bureau of Economic Research. doi:10.3386/w16881

Bowles, S., Gintis, H., & Osborne, M. (2001). The determinants of earnings: A behavioral approach. *Journal of Economic Literature*, *39*(4), 1137–1176. doi:10.1257/jel.39.4.1137

Burrus, J., Elliott, D., Brenneman, M., Markle, R., Carney, L., Moore, G., . . . Roberts, R. (2013). *Putting and keeping students on track: Toward a comprehensive model of college persistence and goal attainment.* (ETS Research Report No. RR-13-14). Princeton, NJ: Educational Testing Service.

Casner-Lotto, J., & Benner, L. (2006). *Are they really ready to work? Employers' perspectives on the basic knowledge and applied skills of new entrants into the 21st century workforce*. New York, NY: The Conference Board, Inc. Retrieved from http://www.p21.org/storage/documents/FINAL_REPORT_PDF09-29-06.pdf

Catalano, R. F., Berglund, M. L., Ryan, J. A., Lonczak, H. S., & Hawkins, J. D. (2002). Positive youth development in the United States: Research findings on evaluations of positive youth development programs. *Prevention & Treatment*, *5*(1), 15a.

Conley, C. S. (2015). SEL in higher education. In J. A. Durlak, C. E. Domitrovich, R. P. Weissberg, & T. P. Gullotta (Eds.), *Handbook of social and emotional learning: Research and practice* (pp. 197–212). New York: Guilford.

Council for the Advancement of Standards in Higher Education. (2011). *The role of counseling services.* Washington, DC: Author. Retrieved from http://collegecounseling.org/docs/CAS.pdf

Credé, M., Roch, S. G., & Kieszczynka, U. M. (2010). Class attendance in college a meta-analytic review of the relationship of class attendance with grades and student characteristics. *Review of Educational Research*, *80*(2), 272–295. doi:10.3102/0034654310362998

Credé, M., Tynan, M. C., & Harms, P. D. (2016). Much ado about Grit: A meta-analytic synthesis of the grit literature. *Journal of Personality and Social Psychology*. doi:10.1037/pspp0000102 PMID:27845531

Deil-Amen, R., & Rosenbaum, J. E. (2003). The social prerequisites of success: Can college structure reduce the need for social know-how? *The Annals of the American Academy of Political and Social Science*, *586*(1), 120–143. doi:10.1177/0002716202250216

Dennis, J. M., Phinney, J. S., & Chuateco, L. I. (2005). The role of motivation, parental support, and peer support in the academic success of ethnic minority first-generation college students. *Journal of College Student Development*, *46*(3), 223–236. doi:10.1353/csd.2005.0023

Dilchert, S., Ones, D. S., Viswesvaran, C., & Deller, J. (2006). Response distortion in personality measurement: Born to deceive, yet capable of providing valid self-assessments? *Psychological Science, 48*(3), 209.

Drake, J. K., Jordan, P., & Miller, M. A. (Eds.). (2013). *Academic advising approaches: Strategies that teach students to make the most of college*. John Wiley & Sons.

Drasgow, F., Chernyshenko, O. S., & Stark, S. (2010). 75 years after Likert: Thurstone was right! *Industrial and Organizational Psychology: Perspectives on Science and Practice, 3*(4), 465–476. doi:10.1111/j.1754-9434.2010.01273.x

Duckworth, A. (2016). *Grit: Passion, Perseverance, and The Science Of Success*. Scribner.

Duckworth, A. L., Gendler, T. S., & Gross, J. J. (2016). Situational strategies for self-control. *Perspectives on Psychological Science, 11*(1), 35–55. doi:10.1177/1745691615623247 PMID:26817725

Duckworth, A. L., Peterson, C., Matthews, M. D., & Kelly, D. R. (2007). Grit: Perseverance and passion for long-term goals. *Journal of Personality and Social Psychology, 92*(6), 1087–1101. doi:10.1037/0022-3514.92.6.1087 PMID:17547490

Duckworth, A. L., White, R. E., Matteucci, A. J., Shearer, A., & Gross, J. J. (2016). A stitch in time: Strategic self-control in high school and college students. *Journal of Educational Psychology, 108*(3), 329–341. doi:10.1037/edu0000062 PMID:27158155

Duckworth, A. L., & Yeager, D. S. (2015). Measurement matters: Assessing personal qualities other than cognitive ability for educational purposes. *Educational Researcher, 44*(4), 237–251. doi:10.3102/0013189X15584327 PMID:27134288

Durlak, J. A., Weissberg, R. P., Dymnicki, A. B., Taylor, R. D., & Schellinger, K. B. (2011). The impact of enhancing students social and emotional learning: A meta-analysis of school-based universal interventions. *Child Development, 82*(1), 405–432. doi:10.1111/j.1467-8624.2010.01564.x PMID:21291449

Dweck, C. S. (2008). *Mindset: The new psychology of success*. Random House Digital, Inc.

Eaton, S. B., & Bean, J. P. (1995). An approach/avoidance behavioral model of college student attrition. *Research in Higher Education, 36*(6), 617–645. doi:10.1007/BF02208248

Engle, J., & Tinto, V. (2008). *Moving beyond access: College success for low-income, first-generation college students*. The Pell Institute. Retrieved from http://files.eric.ed.gov/fulltext/ED504448.pdf

Eskreis-Winkler, L., Duckworth, A. L., Shulman, E. P., & Beal, S. (2014). The grit effect: Predicting retention in the military, the workplace, school and marriage. *Frontiers in Psychology, 5*, 36. doi:10.3389/fpsyg.2014.00036 PMID:24550863

Farrington, C. A., Roderick, M., Allensworth, E., Nagaoka, J., Keyes, T. S., Johnson, D. W., & Beechum, N. O. (2012). *Teaching adolescents to become learners. The role of noncognitive factors in shaping school performance: A critical literature review*. Chicago: University of Chicago Consortium on Chicago School Research.

Fickel, L. (2015, May 1). What's in a terrible name? *U.S. News*. Retrieved from https://www.usnews.com/opinion/knowledge-bank/2015/05/01/non-cognitive-skills-are-important-but-have-a-terrible-name

Fong, C. J., Davis, C. W., Kim, Y., Kim, Y. W., Marriott, L., & Kim, S. (2016). Psychosocial factors and community college student success a meta-analytic investigation. *Review of Educational Research*, *87*(2), 388–424.

Goldberg, L. R. (1990). An alternative description of personality: The Big-Five factor structure. *Journal of Personality and Social Psychology*, *59*(6), 1216–1229. doi:10.1037/0022-3514.59.6.1216 PMID:2283588

Goldberg, L. R. (1992). The development of markers for the Big-Five factor structure. *Psychological Assessment*, *4*(1), 26–42. doi:10.1037/1040-3590.4.1.26

Habley, W. R., Bloom, J. L., & Robbins, S. (2012). *Increasing persistence: Research-based strategies for college student success*. John Wiley & Sons.

Heckman, J. J., & Rubinstein, Y. (2001). The importance of noncognitive skills: Lessons from the GED testing program. *The American Economic Review*, *91*(2), 145–149. doi:10.1257/aer.91.2.145

Kautz, T., Heckman, J. J., Diris, R., Ter Weel, B., & Borghans, L. (2014). *Fostering and measuring skills: Improving cognitive and non-cognitive skills to promote lifetime success (No. w20749)*. National Bureau of Economic Research.

Kot, F. C. (2014). The impact of centralized advising on first-year academic performance and second-year enrollment behavior. *Research in Higher Education*, *55*(6), 527–563.

Kuh, G. D., Kinzie, J., Buckley, J. A., Bridges, B. K., & Hayek, J. C. (2006, July). What matters to student success: A review of the literature. *Commissioned report for the national symposium on postsecondary student success: spearheading a dialog on student success*. Retrieved from http://nces.ed.gov/npec/pdf/kuh_team_report.pdf

Kyllonen, P. C., Lipnevich, A. A., Burrus, J., & Roberts, R. D. (2014). *Personality, motivation, and college readiness: A prospectus for assessment and development. Educational Testing Service Research Report No: RR-14-06*. Princeton, NJ: Educational Testing Service.

Li, K., Zelenka, R., Buonaguidi, L., Beckman, R., Casillas, A., Crouse, J., & Robbins, S. et al. (2013). Readiness, behavior, and foundational mathematics course success. *Journal of Developmental Education*, *37*(1), 14.

MacCann, C., Duckworth, A. L., & Roberts, R. D. (2009). Empirical identification of the major facets of conscientiousness. *Learning and Individual Differences*, *19*(4), 451–458. doi:10.1016/j.lindif.2009.03.007

Markle, R., Brenneman, M., Jackson, T., Burrus, J., & Robbins, S. (2013). *Synthesizing frameworks of higher education student learning outcomes* (Research Report No. RR-13-22). Princeton, NJ: Educational Testing Service. doi:10.1002/j.2333-8504.2013.tb02329.x

Markle, R. E., Olivera-Aguilar, M., Jackson, T., Noeth, R., & Robbins, S. (2013). *Examining evidence of reliability, validity, and fairness for SuccessNavigator*. (ETS Research Report 13-12). Princeton, NJ: Educational Testing Service.

Markle, R. E., Wang, J. Z., Sullivan, T., & Russell, J. A. (2015, April). Myths and realities: Effort and response distortion in low-stakes, self-report assessments of noncognitive skills. *Paper presented to the annual meeting of the American Educational Research Association*, Denver, CO.

National Research Council. (2012). *Education for Life and Work: Developing Transferable Knowledge and Skills in the 21st Century*. Washington, DC: The National Academies Press.

OConnor, M. C., & Paunonen, S. V. (2007). Big Five personality predictors of post-secondary academic performance. *Personality and Individual Differences*, *43*(5), 971–990. doi:10.1016/j.paid.2007.03.017

Olivera-Aguilar, M., Rikoon, S. H., & Robbins, S. B. (2017). Noncognitive skill profiles and their relationships with early college academic outcomes. *The Journal of Higher Education*, *88*(2), 234–257.

Oliveri, M. E., & Markle, R. (2017). *Continuing a culture of evidence: Expanding skills in higher education* (Research Report No. RR-17-09). Princeton, NJ: Educational Testing Service. doi:10.1002/ets2.12137

Oswald, F. L., Schmitt, N., Kim, B. H., Ramsay, L. J., & Gillespie, M. A. (2004). Developing a biodata measure and situational judgment inventory as predictors of college student performance. *The Journal of Applied Psychology*, *89*(2), 187–207. doi:10.1037/0021-9010.89.2.187 PMID:15065969

Pascarella, E. T., Pierson, C. T., Wolniak, G. C., & Terenzini, P. T. (2004). First-generation college students: Additional evidence on college experiences and outcomes. *The Journal of Higher Education*, *75*(3), 249–283. doi:10.1353/jhe.2004.0016

Paulhus, D. L., & Vazire, S. (2007). The self-report method. In R. C. Fraley & R. F. Kruger (Eds.), *Handbook of research methods in personality psychology* (pp. 224–239). New York: Guilford.

Peterson, N. G., Mumford, M. D., Borman, W. C., Jeanneret, P. R., & Fleishman, E. A. (Eds.). (1999). *An occupational information system for the 21st century: The development of O*NET*. Washington, DC: American Psychological Association. doi:10.1037/10313-000

Porchea, S. F., Allen, J., Robbins, S., & Phelps, R. P. (2010). Predictors of long-term enrollment and degree outcomes for community college students: Integrating academic, psychosocial, socio-demographic, and situational factors. *The Journal of Higher Education*, *81*, 750–778.

Poropat, A. E. (2009). A meta-analysis of the Five-Factor Model of personality and academic performance. *Psychological Bulletin*, *135*(2), 322–338. doi:10.1037/a0014996 PMID:19254083

Raykov, T., & Marcoulides, G. A. (2011). *Introduction to psychometric theory*. Routledge.

Richardson, M., Abraham, C., & Bond, R. (2012). Psychological correlates of university students' academic performance: A systematic review and meta-analysis. *Psychological Bulletin*, *138*(2), 353–387. doi:10.1037/a0026838 PMID:22352812

Rikoon, S. H., Brenneman, M., Kim, L. E., Khorramdel, L., MacCann, C., Burrus, J., & Roberts, R. D. (2016). Facets of conscientiousness and their differential relationships with cognitive ability factors. *Journal of Research in Personality*, *61*, 22–34. doi:10.1016/j.jrp.2016.01.002

Rikoon, S. H., Liebtag, T., Olivera-Aguilar, M., Steinberg, J., & Robbins, S. B. (2015). *Anticipating College Enrollment: Adapting SuccessNavigator for High School Students. (ETS RR-15-37)*. Princeton, NJ: Educational Testing Service.

Robbins, S. B., Allen, J., Casillas, A., Akamigbo, A., Saltonstall, M., Campbell, R., & Gore, P. A. (2009). Associations of resource and service utilization, risk level, and college outcomes. *Research in Higher Education*, *50*(1), 101–118. doi:10.1007/s11162-008-9106-7

Robbins, S., Allen, J., Casillas, A., Peterson, C., & Le, H. (2006). Unraveling the differential effects of motivational and skills, social, and self-management measures from traditional predictors of college outcomes. *Journal of Educational Psychology*, *98*(3), 598–616. doi:10.1037/0022-0663.98.3.598

Robbins, S. B., Lauver, K., Le, H., Davis, D., Langley, R., & Carlstrom, A. (2004). Do psychosocial and study skill factors predict college outcomes? A meta-analysis. *Psychological Bulletin*, *130*(2), 261–288. doi:10.1037/0033-2909.130.2.261 PMID:14979772

Robbins, S. B., Oh, I. S., Le, H., & Button, C. (2009). Intervention effects on college performance and retention as mediated by motivational, emotional, and social control factors: Integrated meta-analytic path analyses. *The Journal of Applied Psychology*, *94*(5), 1163–1184. doi:10.1037/a0015738 PMID:19702363

Ryan, M. G. (2013). Improving retention and academic achievement for first-time students at a two-year college. *Community College Journal of Research and Practice*, *37*(2), 131–134. doi:10.1080/10668926.2012.715266

Sackett, P. R., & Walmsley, P. T. (2014). Which personality attributes are most important in the workplace? *Perspectives on Psychological Science*, *9*(5), 538–551. doi:10.1177/1745691614543972 PMID:26186756

Schmitt, N., Billington, A., Keeney, J., Reeder, M., Pleskac, T. J., Sinha, R., & Zorzie, M. (2011). *Development and validation of measures of noncognitive student potential* (College Board Research Report 2011-1). New York: The College Board.

Schwebel, D. C., Walburn, N. C., Klyce, K., & Jerrolds, K. L. (2012). Efficacy of advising outreach on student retention, academic progress and achievement, and frequency of advising contacts: A longitudinal randomized trial. *NACADA Journal*, *32*(2), 36–43. doi:10.12930/0271-9517-32.2.36 PMID:23270066

Strayhorn, T. L. (2006). *Frameworks for assessing learning and development outcomes*. Washington, DC: Council for the Advancement of Standards in Higher Education.

Tinto, V. (1993). *Leaving college: Rethinking the causes and cures of student attrition* (2nd ed.). Chicago: The University of Chicago Press.

Tough, P. (2013). *How children succeed*. Random House.

Tracey, T. J., & Sedlacek, W. E. (1982, March). Non-cognitive variables in predicting academic success by race. *Paper presented at the annual meeting of the American Educational Research Association*, New York.

Webber, D. A., & Ehrenberg, R. G. (2010). Do expenditures other than instructional expenditures affect graduation and persistence rates in American higher education? *Economics of Education Review*, *29*(6), 947–958. doi:10.1016/j.econedurev.2010.04.006

Weekley, J. A., & Ployhart, R. E. (Eds.). (2013). *Situational judgment tests: Theory, measurement, and application*. Psychology Press.

Weiss, M. J., Bloom, H. S., & Brock, T. (2014). A conceptual framework for studying the sources of variation in program effects. *Journal of Policy Analysis and Management*, *33*(3), 778–808. doi:10.1002/pam.21760

Yeager, D. S., & Dweck, C. S. (2012). Mindsets that promote resilience: When students believe that personal characteristics can be developed. *Educational Psychologist*, *47*(4), 302–314. doi:10.1080/00461520.2012.722805

Yeager, D. S., Romero, C., Paunesku, D., Hulleman, C. S., Schneider, B., Hinojosa, C., & Trott, J. et al. (2016). Using design thinking to improve psychological interventions: The case of the growth mindset during the transition to high school. *Journal of Educational Psychology*, *108*(3), 374–391. doi:10.1037/edu0000098 PMID:27524832

Yeager, D. S., & Walton, G. M. (2011). Social-psychological interventions in education: Theyre not magic. *Review of Educational Research*, *81*(2), 267–301. doi:10.3102/0034654311405999

Ziegler, M., MacCann, C., & Roberts, R. (Eds.). (2011). *New perspectives on faking in personality assessment*. Oxford University Press. doi:10.1093/acprof:oso/9780195387476.001.0001

KEY TERMS AND DEFINITIONS

Academic Skills: Within the chapter academic skills refer to behavioral strategies useful for fostering success in the classroom. These include key classroom factors such as attendance, timeliness, and engagement (operationalized as "meeting class expectations") as well as key organizational behaviors outside the classroom (studying, planning, etc.).

Co-Curricular Supports: Beyond advising services, every college and university offers an array of co-curricular supports to facilitate student engagement and success (e.g., first-year experience programs, tutoring, interest-focused clubs, recreational or public service activities, athletics, student success courses, and others).

Commitment: Commitment deals with the ways students approach their goals, including overall drive toward a college degree, as well as institutional commitment –a significant component of Tinto's seminal work on retention (e.g., Tinto, 1993).

Five Factor Model (FFM): The five-factor model of personality is a set of five broad trait dimensions or domains, often referred to as the "Big Five": Extraversion, Agreeableness, Conscientiousness, Neuroticism and Openness to Experience.

Fixed Mindset: A fixed mindset, in which one assumes such a characteristic is immutable. The terms "fixed" versus "growth" mindset are analogues of Dweck's earlier work on "entity" versus "incremental" implicit theories of personal characteristics, respectively (Yeager & Dweck, 2012).

Grit: The chapter utilizes Duckworth's definition of grit. Duckworth defined grit as "perseverance and passion for long-term goals" (Duckworth et al., 2007, p. 1087).

Growth Mindset: A growth mindset is one that views a given trait or skill (e.g. intelligence, academic ability, personality) as malleable and thus able to be developed with effort over time (Dweck, 2008).

Liberal Education and America's Promise (LEAP): LEAP is a framework by the American Association of Colleges & Universities (AAC&U), outlining key student learning domains including teamwork and problem solving, civic knowledge and engagement, intercultural knowledge and competence, ethical reasoning and action, and "foundations and skills for lifelong learning."

Motivation: Motivation includes both goal-related behaviors (effort toward a goal, persistence) and study skills.

Noncognitive Skills: Are often called "soft skills," or "character" factors. Noncognitive factors and their relationship with retention and student success approach the topic from a psychological rather than a sociological perspective and they focus on key behaviors, skills, attitudes, and mindsets that successful students employ and how to foster these skills to help students navigate the issues, challenges and obstacles that they will face in higher education. "Noncognitive" is used to contrast the array of attitudes and social-emotional inter- and intrapersonal skills relevant to academic success from traditionally defined cognitive ability or intelligence.

Self-Management: Self-management refers to regulatory strategies students' may employ to help them navigate or cope with the challenges they encounter during college including their tendency to feel and cope with stress, and their academic self-efficacy.

Self-Regulation: Self-regulation refers to the regulation of stress and emotions to facilitate academic success.

Social Engagement: Social engagement includes both social engagement behaviors and feelings of belonging.

Social Support: Social support refers to students' connections with others and the extent to which that connectivity supports or hinders their academic success in college. Social support incorporates students' sense of belonging, help-seeking behaviors, and the support (or lack thereof) provided by external factors such as one's family, work, and finances.

***SuccessNavigator* Assessment Tool:** The ETS SuccessNavigator is an assessment, designed to help colleges reach at-risk incoming students and improve retention and completion rates. The tool is a 30-minute, non-proctored online assessment that gives programs and institutions a holistic view of the critical factors that most greatly influence incoming student success, academic skills, commitment, self-management and social support.

ENDNOTE

1 Similar widely used frameworks in the K-12 space include those by the Collaborative for Academic, Social, and Emotional Learning (www.casel.org) and Consortium for Chicago School Research (Farrington et al., 2012).

Chapter 2
Study Abroad Effect on Retention and Student Success

Eric Canny
CEA Study Abroad, USA

ABSTRACT

This chapter discusses the positive correlations between study abroad and student success broadly, as well as the ways in which study abroad programs operate from a student success model. Both may help influence home-campus practitioners in working with all students. First, this chapter will first focus on the history of study abroad within higher education to contextualize it specifically relating to retention and student success. Second, the chapter will focus on the positive, and not often discussed, positive relationship between student retention and graduation rates correlated to study abroad. Third will be a section devoted to exploring the research findings of the high-impact nature of study abroad linked to learning outcomes, as data shows deep student learning, including self-reflection, which could be indicators of student success. Finally, the chapter will draw correlations between these topics and international on-site models of student success to support this transformational and challenging experience for students.

INTRODUCTION

This chapter will focus broadly on the history of study abroad within higher education to contextualize it specifically as it relates to retention and student success. First, the chapter will focus on the positive, but not often discussed, relationship between student retention and graduation rates correlated to study abroad. The subject of student success and retention while studying abroad is rarely discussed within traditional discourse on these topics. It has however been a dialogue within the global education field. Instead, the focus of discourse regarding study abroad within broader educational research has focused on intercultural competence and learning, global education, and language acquisition.

Though individual overseas programs may track retention, there is no central reporting of retention and completion data within the U.S. higher educational field relating to study abroad. While there are now possible exceptions to this due to the development of branch campuses where students can complete their entire academic program, traditional study abroad programs do not measure graduation rates as

DOI: 10.4018/978-1-5225-2998-9.ch002

students' time overseas ranges from a few weeks to at most one year. There is, however, data showing that participation in study abroad does improve overall retention.

Second, the chapter will be a focus on the high-impact nature of study abroad linked to learning outcomes, as data shows deep student learning, including self-reflection, which are important indicators of student success. Studies have shown that not only do students learn academically while abroad, but, they also acquire skills that correlate to student success and metacognitive reflection. Areas of student success that have been identified as positively impacted due to study abroad include self-confidence, maturity, and increased focus and interest in academics.

Finally, the chapter draws correlations between these topics and international on-site models of student success to support this transformational, and challenging experience for students. As such, study abroad programs have largely focused on ensuring that students succeed for their period abroad, regardless of length. Additionally, a student's success overseas tends to be black and white: remain or return home. There are limited options to transfer to other courses, return to the home campus to complete the term, or address long-term issues as the time-frame of the study experience is not a four-year program, with few exceptions.

Due to these factors, the pressure placed upon the individuals who administer and teach in these programs can be more intense than on traditional campuses. It is even arguable that the vernacular of "student success", which is now common on U.S. campuses but not usually associated with study abroad, has in fact infused overseas programs for decades in how administrators and faculty work with students. Indeed, study abroad as a field has been undertaking and focusing on "student success" without using this specific terminology for decades, and lessons from this part of the higher educational sector could be applied to other areas.

BACKGROUND OF STUDY ABROAD IN RELATION TO HIGHER EDUCATION

To understand study abroad today, it must be put into historical context as a component of global education and U.S. higher education that connects to student success and retention. First, there is data showing positive correlation between student success, retention, and study abroad. Second, the volume of students studying abroad is increasing, which is aligned to an expanding breadth of academic disciplines offered (i.e., STEM, internships) as well as an increase of diversity of students (i.e., First Generation, non-traditional, and higher minority participation rates.

REVIEW OF THE LITERATURE

Given the uniqueness of study abroad as a singular part of a holistic educational experience, for the purposes of this chapter student success will be defined utilizing the American Association of Colleges and Universities work on Inclusive Excellence (AAC&U, 2014). The action of making access, student success, and high-quality learning inclusive of all students, or inclusive excellence, requires that institutions discover inequities in student success, distinguish effective educational practices, and build these effective practices into their missions and institutional operations for sustained change to occur in the areas of diversity, equity, and educational quality (AAC&U, 2014). This chapter will address all of the key elements of this definition: learning inclusive of all students; discovering inequities in student success and developing remedies for them; distinguishing effective educational practices; and building these practices into their missions and institutional operations for sustained change for all students. Utilizing this definition, the relationship between study broad experiences and programs provide insights into effective strategies for student success related to knowledge acquisition, self-efficacy and degree completion within a holistic higher educational model.

The literature outlines wide-ranging trends for the continually shifting field of global education, and for study abroad in particular (ACE, 2011, 2014; Knight, 1997, 2003, 2004; McCabe, 2001; NAFSA, 2010; Olsen, Green, & Hill, 2005, 2006; Olson, Evans, & Shoenberg, 2007; Olson, Green, & Hill, 2005, 2006; Rudzki, 2000; van der Wende, 2007; Wächter, 2003; Woolf, 2002). The shifts cited include complexity, speed of development, integration into core curricular areas, and the globalization of higher education itself. These factors are applicable to study abroad in particular: the increase of programs, types of program, and the volume and diversity of offerings student can select from.

The speed at which global education affects higher education is accelerating along with the breadth of activities that embody globalization in higher education. Study abroad, however, is not a new phenomenon. That there were activities prior to World War II is not in doubt. As such, while not historically documented from a student success lens, there is a history of working with students overseas that spans at least 60 years.

It was post-WWII when several factors aligned to influence globalization in higher education, including study abroad. First, study abroad sites reopened and expanded after the War ended (Desruisseaux, 1999, 1999a, 1999b). Data continues to show expansion and growth in study abroad sites, with students studying in more locations than before. Second, there was a relationship of increased competition of global knowledge as a result of the Cold War, and Sputnik in particular (Desruisseaux, 1999, 1999a, 1999b, 2000; Goodman, 1999). Again, this is a trend that continues today in relation to national security linked to scholarships and language acquisition, and ensuring a work force which through study abroad has knowledge and skills that enable them to compete in a truly globalized economy. Third, a shifting within institutions around the philosophy of area studies, linked to the nationalistic needs stemming from the Cold War and funding, was also a factor (Tarrow & Hall, 1998). These factors have continued to influence study abroad in terms of: expansion of academic programs from the traditional language and humanities to now include business and STEM fields; rationale for increasing study abroad participation linked to an educated work force able to compete in both the current and future globalized economy; and a constant expansion of study abroad options from short to long-term, as well as multi-country.

There is debate about when globalization and higher education accelerated as the phenomenon spans decades of interchanging activities and geographic diversity. However, the consensus is that the mid to late 1990's was a period of rapid development, with new models of activities developing in a broader range of countries, coupled with preliminary references to and definitions of the term globalization (Altbach & Knight, 2007; Carr, 1999; Desruisseaux, 1999, 1999a, 1999b, 2000; Hudzik, 2011; Knight, 1993 & 2004; Yang, 2002).

Higher education institutions, indeed, are bound to internationalize in a response to globalization forces (Coughlin, 1994; Lipsman, 2000; Sassen, 2001; Yang, 2002). An early definition of internationalization is "the process of integrating an international, intercultural or global dimension into the purpose, functions or delivery of postsecondary education (Knight, 2004, p. 11)." Higher education has been and continues to alter and evolve practices to integrate global components into academic and co-curricular programs, research, and operations; in all three areas in the realms of direct services, management, and strategy and policy (NAFSA, 2015). There is great variance, based upon institutional type and other factors, of how globalization affects an institution (Yang, 2002). A tendency toward business-related practices, including privatization and entrepreneurialism coupled with an increase in the breadth of the type of programs offered is often cited (Altbach & Knight, 2007; Mazzarol & Soutar, 2002; Stromquist, 2002). Another example of the evolution of the field is a recent professional document that highlights five broad areas of global education: comprehensive internationalization; education abroad; international enrollment management; international student and scholar services; and cross-cutting competencies (NAFSA, 2015).

Internationalization of institutional activities centers around two factors, educational and economic (Raby, 2009). First, institutions internationalize because they must provide an education relevant to a changing world (Armstrong, 2007). This is for both educational as well as business reasons in terms of attracting students with a curriculum perceived as applicable and relevant. Second, on the business side, pressures in the form of increased competition from other institutions' internationalization efforts coupled with unchanging or declining student enrollment force institutions to undertake new initiatives (Armstrong, 2007; Mazzarol & Soutar, 2002; Stromquist, 2002). Additional factors leading to the proliferation of internationalization activities include peace building, national security, reputational advancement, human capital development, expanded tertiary access, demographic growth, and research collaboration (Altbach & Knight, 2007; Edwards, 2007; Hser, 2005; Knight, 2004).

STUDY ABROAD AND RETENTION

While the learning benefits of study abroad discussed in the next section are often cited as reasons for students to participate, research has found participation in study abroad can both: decrease the time to graduation post participation as well as improve overall student retention. The study abroad experience can, additionally: increase discipline specific knowledge, and both academic skills and self-efficacy broadly. All of these factors positively correlate to degree completion and they also have a positive impact on the institution's rate of student retention.

Research conducted at Michigan State University by Ingraham and Peterson found that "[d]ata on time to graduation and number-of-terms-enrolled show that, on average, study abroad participants enroll for more terms before graduating *but* that they take less time to graduate (2004, p. 98)." This echoes research undertaken by Malmgren and Galvin (2008) at the University of Minnesota. Their research found that the difference in graduation rates between students who participated in study abroad and those who did not was statistically significant. Overall, participants had higher graduation rates, and participation in study abroad did not delay graduation in their study pool. The study also showed that students who studied abroad have 10.0% higher odds of graduating in four years (2008).

The research done by Malmgren and Galvin (2008) also highlighted two key findings relating to students falling into groups that traditionally have lower retention and student success rates. First, for students of color, there were strong positive correlations between those who had studied abroad and graduation rates. Second, the findings indicated that at-risk students who participated in study abroad had higher graduation rates than other at-risk students who did not partake in study abroad.

The largest research undertaken to date on study abroad and retention is from the Georgia Learning Outcomes of Students Studying Abroad Research Initiative (GLOSSARI). This research initiative is a comparison of several outcomes between study abroad participants and non-participants attending sixteen varied public institutions within the University System of Georgia. Initial phases of the GLOSSARI initiative, published in 2004, were focused on learning outcomes. Additionally, findings published in 2010 showed a direct correlation between student participation in study abroad and graduation rates compared to students who did not undertake this activity. First, overall four-year graduation rates for study abroad participants were 17.8% higher than non-participants. Second, while the percentage decreased for study-abroad participants, for five-year graduation rates, those who studied abroad graduated at a rate 10.6% higher that their non-study abroad peers, and for six-year graduation rates, the percentage continued to remain positive at 6.4% (GLOSSARI, 2010).

When looking at specific student demographics and student abroad participation, similar positive correlations relating to graduation rates were found. First, males who studied abroad graduated at rates 6-12% higher than their non-study abroad peers. Second, females who studied abroad graduated at rates 6-19% higher than non-study abroad females. Third, graduation rates for non-white students who participated in study abroad were 7-18% higher than those who did not participate. Finally, examination of African-American student data showed that participation in study abroad resulted in graduation rates that were 13-31% higher (GLOSSARI, 2010).

The data presented above is frequently discussed within the field of international education. However, within the broader literature on student success and retention, there is little mention of how study abroad as part of a student's collegiate experience can lead to improved retention. Study abroad is often presented as a superfluity activity rather than a core means of deep learning and skill acquisition in public discourse surrounding the cost of education and student retention. In fact, while institutions are attempting to increase the number of students studying abroad, they are cutting the financial resources which enable them to do so. Rather than funding study abroad programs institutions are focusing their limited resources on the home campus and other activities engage students on their home campuses by developing independent research opportunities and off campus programs that are domestically based such as Washington DC internship programs.

STUDY ABROAD AND LEARNING OUTCOMES

Study abroad is increasingly part of a student's higher educational career. Study abroad programs and now a mainstream part of campus recruitment efforts. Study abroad coursework options are now also included as a part of the campus academic advising program to ensure that coursework during the study abroad semester is a meaningful part of the student's degree program. Study abroad was, and can still be, presented as a superfluous activity rather than as a core method of deep learning and meaningful life skills acquisition. Research, as outlined in this chapter, is evidence and support to counter this incorrect but regularly held viewpoint. Data shows a positive relation between knowledge acquisition and a study abroad experience. Data also supports the correlation between study abroad and student success. This is a critical point regarding knowledge and skills that is not often examined within the larger higher education research context. As such, focusing on the topic of success and retention, both historically and as an emerging institutional structure and approach, is a central element for the larger higher education community to understand and consider.

In the literature of student success, there is a wide-ranging discussion over which variables and data are used to measure or identify success. While some accreditation and university organizations focus on graduation rates and time to degree completion, others include measures of program learning outcomes and employment. This chapter illustrates the correlation between study abroad and these measures of student success. However, there are also limitations with the data depending upon the measures an entity chooses to use within its definition of student success. For example, measuring study abroad program learning outcomes can be identified and correlate directly to measuring student success, however, while data measuring graduation rates and time to degree completion is positive there could be other intervening factors. As such, it is possible lessons learned from study abroad programs that are related to this debate may contribute to student success and positively impact retention rates, as study abroad programs are broadly seen as resources for enhancing the academic, and campus experience of students.

This section will focus on the learning outcomes related to study abroad and their correlation to student success and retention. As the terms "retention" and "student success" are not commonly associated with study abroad, the following descriptions will be used to ground study abroad within the commonly used higher educational definitions.

- Successful completion of students' academic goals of degree attainment (Levitz, 2001).
- Students meeting clearly defined educational goals whether they are course credits, career advancement, or achievement of new skills (Tinto, 1993).
- Students' successful academic and social integration into the college community, marked by the feeling that one fits at the institution and positive educational attitudes and experiences (Bean, 1980).
- The match between students' motivation and academic ability and their academic and social characteristics (Cabrera, Castaneda, Nora, Hengstler, 1992).

Research shows that study abroad is one of, if not the most life changing experience a student can have during their higher education career (Dwyer &Peters 2004). The table below reflects the percentage of students who studied abroad that indicated a positive correlation between the experience and the associated item. Few, if any other experiences in college show such high positive correlation rates.

Table 1.

These direct and positive retention benefits are a result of the many deep changes that a student undergoes while abroad. They included mental, emotion, social, cultural, and academic transformations all of which are often purposefully integrated into the study abroad experience to immerse the student into the new culture in as many ways as possible. Correspondingly, these changes and transformations can be highly stressful for the student by design. As such, the entire study abroad experience results in profound changes to the student. Study abroad has utilized the cultural adaptation model as a core instrument for developmental strategies for working with students as they move through the stages while overseas. Indeed, many study abroad programs share this model with students to increase self-awareness of the stages and possible consequences. In doing so, the programs reinforce self-responsibility, de-stigmatize issues like mental health, and encourage students to engage with local staff from the outset.

The research by Dwyer and Peters shows a positive correlation between study abroad and all definitions of student success. First, there is demonstrated successful completion of students' academic goals of degree attainment, higher than students who do not study abroad (Levitz, 2001). Second, data demonstrates that students are meeting clearly defined educational goals in the areas of course credit attainment, career advancement, or achievement of new skill proficiencies (Tinto, 1993). Third, linked to the areas of intercultural development, students demonstrate successful academic and social inte-

Table 1. Percentage of U.S. students correlating study abroad to personal change

Survey Item	% Full Year	% Fall Semester	% Spring	% Semester Summer	% Total
Personal Development					
Increased self-confidence.	98%	95%	96%	97%	96%
Served as a catalyst for increased maturity.	98%	97%	97%	95%	97%
Has had a lasting impact on world view.	97%	95%	94%	92%	95%
Academic Commitment					
Enhanced interest in academic study.	81%	80%	79%	84%	80%
Influenced subsequent educational experiences.	91%	85%	86%	84%	87%
Reinforced commitment to foreign language study.	88%	83%	85%	90%	86%
Intercultural Development					
Helped me better understand my own cultural values and biases.	99%	97%	97%	95%	98%
Influenced me to seek out a greater diversity of friends.	94%	88%	89%	86%	90%
Continues to influence interactions with people from different cultures.	97%	93%	92%	92%	94%
Career development					
Acquired skill sets that influenced career path	82%	73%	74%	71%	76%
Ignited an interest in a career direction pursued after the experience	70%	57%	59%	59%	62%

Dwyer &Peters 2004

gration into the college community, whether that be their home institution or a third-party, marked by the feeling of fitting at the institution and positive educational attitudes and experiences (Bean, 1980). Fourth, linked to both personal development and academic development, research shows an increase in students' motivation and academic ability and their academic and social characteristics (Cabrera, Castaneda, Nora, Hengstler, 1992).

There is a growing emphasis within the literature on assessing high-impact educational practices that help students to achieve essential learning outcomes within, as well as outside of, the traditional classroom setting (AAC&U, 2014; Brownell & Swaner, 2012; Kuh, 2012; Kuh & O'Donnell, 2013). Examples of high-impact educational practices include: Common Intellectual Experiences, Learning Communities, Writing-Intensive Courses, Collaborative Assignments and Projects, Undergraduate Research, Diversity/Global Learning, Service Learning, Community-Based Learning, Internships, and Capstone Courses and Projects (AAC&U, 2014). As seen in Tables 2 and 3, data shows the positive correlation between high-impact practices and student learning. Additionally, while only study abroad is clearly international, based upon the AAC&U data, study abroad can be imbedded in all the other high-impact activities (i.e., intercultural learning communities, internships with a global focus, service learning overseas, and research with a global focus).

Table 2. Relationships between selected high-impact activities, deep learning, and self-reported gains

	Deep Learning	Gains: General	Gains: Personal	Gains: Practical
	First Year			
Learning Communities	+++	++	++	++
Service Learning	+++	++	+++	+++
	Senior			
Study Abroad	++	+	+	++
Student-Faculty Research	+++	++	++	++
Internships	++	++	++	++
Service Learning	+++	++	+++	+++
Senior Culminating Experience	+++	++	++	++

+ $p<0.001$, ++ $p<0.001$ & Unstd B > 0.10, +++ $p<0.001$ & Unstd B > 0.30
AAC&U 2013

Table 3. Relationships between selected high-impact activities and clusters of effective educational practices

	Level of Academic Challenge	Active and Collaborative Learning	Student-Faculty Interaction	Supportive Campus Environment
	First Year			
Learning Communities	+++	+++	+++	++
Service Learning	+++	+++	+++	+++
	Senior			
Study Abroad	++	++	++	++
Student-Faculty Research	+++	+++	+++	++
Internships	++	+++	+++	++
Service Learning	+++	+++	+++	+++
Senior Culminating Experience	++	+++	+++	++

+ $p<0.001$, ++ $p<0.001$ & Unstd B > 0.10, +++ $p<0.001$ & Unstd B > 0.30
AAC&U 2014

When correlating the AAC&U data to that of Dwyer and Peters, it is clear that study abroad not only results in deep academic learning, but also causes increases in the areas of student maturity, world view, increased focus on academics, and increased student interest and career progression skills. The field of study abroad is also shifting to incorporate other high-impact practices into the study abroad experience, in particular internships, research, and service learning. While additional research is needed, based upon the breadth of study abroad programs, initial data shows that the study abroad experience undoubtedly has positive effects on students in both academic learning outcomes as well as core skills related to student success.

STUDENT SUCCESS MODELS IN STUDY ABROAD CHALLENGES AND OPPORTUNITIES

While much public discourse around study abroad focuses on crisis management and safety, overseas programs have for decades deployed a student success and wellness model to ensure students succeed while abroad. As it has on U.S. campuses, this model has evolved over time, with emphasis of late being placed on holistic student success. The sophisticated measurement and identification of obstacles students encounter in the learning experience, whether during study abroad programs or on the home campuses, are similar to those being identified in the current research of U.S. campus based programs. Despite a broad lack of specific theories related to overseas program management, it is commonplace in the study abroad field to triangulate student life with academic progress and success at a level arguably only started on U.S. campuses within the past decade. Finally, U.S. institutions are attempting to bridge student success with their overseas counterparts in a coordinated integration of reflection during the entire study abroad experience. As such, important lessons could be learned from examining the research and literature of students abroad in this area, including work that bridges U.S. campuses to their overseas counterparts.

Signs of student issues can be a display of cultural adaptation issues, academic struggle, or mental or physical health issues because of being overseas. As such signals can be exhibited in classrooms, housing, or on excursions tracking the information and sharing it across traditional academic silos (e.g. housing, academic, co-curricular) is critical. While such triangulation is becoming more common on U.S. campuses, the sharing of such information has been common within the study abroad field from a student success perspective for decades. While maintaining confidentiality, overseas programs monitor and share information on their students at a level rarely seen on U.S. campuses, quickly deploying resources to aid students and assessing who and what resources can be deployed to reach the core issues. In addition, as study abroad programs can enroll students from many institutions, the field is accustomed to securing and sharing information with the students' home campuses and families. This not only complicates the student assistance process, but usually also happens at high speed with the additional complication of time zones, adding a level of complexity to the student success efforts employed by study abroad staff. As noted later, how this is accomplished, including role clarity, processes, and ensuring confidentiality could be a model for home campuses to emulate.

As the prior sections and data highlighted, students experience emotional and intellectual challenges during study abroad because of direct cultural encounters. One means of ensuring these experiences result in positive learning outcomes, both academic as well as personal, is through guided reflection. This process includes metacognitive reflection on experiences, sharing and comparing with fellow students, and input from educators and study abroad staff (Engle & Engle, 2003). When deployed successfully, guided reflection in study abroad offers students ways to apply their knowledge within and outside the classroom in ways that further resilience and student success (Smith-Paríolá & Gòkè-Paríolá, 2006; Vande Berg, 2007). As Braskamp, Braskamp, and Merrill state, "experience integrated with rigorous and critical, constructive, and creative thinking may help students learn and develop even more holistically" (Braskamp, Braskamp, and Merrill, 2009, p. 113). This work is being undertaken by most study abroad programs.

Study abroad is not isolated, but is deeply engrained in the U.S. higher educational student demographic. As such, overseas programs work with students with mental and physical health issues, learning and physical disabilities, and low-grade point averages. Like U.S. based institutions, study abroad programs accept and put students on disciplinary or academic probation for a wide range of issues. While there are usually robust mechanisms for students to self-identify these issues and to articulate their needs for support, as in the U.S., many students elect not to self-disclose. This places tremendous pressure on staff to respond to student needs which are exacerbated by the cultural adaptation the students are undergoing and to attempt resolution within the program before these needs become a true emergency. In part due to local limitations of services that are common place on U.S. campuses (e.g., counseling centers, writing or tutorial labs, accommodations for students), the study abroad field has developed internal mechanisms to address these issues quickly out of necessity. While external resources are available in some locations, such resources can come with a different cultural lens and usually require much more advance time to engage than on a traditional U.S. campus. As such, there are lessons to learn from how student success is achieved in an overseas setting with limited resources and staff.

One way in which U.S. institutions are attempting to bridge student success with their overseas counterparts is a coordinated integration of reflection during the entire study abroad experience. How guided reflection is integrated into programming varies by program model. However, study abroad offices are collaborating with international staff to design holistic student experiences that force reflection, even upon return to the U.S. First, this work includes exploring why and where a student wants to study abroad and the student's academic and career goals. Second, during the student's departure preparation, the challenges in preparation for going abroad (e.g., securing a visa, research on the country, preparing for culture shock) are articulated. Third, while abroad, more active work is done by on-site staff, linked to the prior work of selecting a program and preparing to go abroad, to set goals and move students through purposeful reflection processing their experience overseas. Finally, on-site staff increasingly are conducting re-entry workshops which align with continued workshops and activities implemented by the home campus to assist the student in processing the experience, as well as engaging in activities that will continue to reinforce their acquired knowledge and skills.

The study abroad field is comprised of myriad types of programs that can complicate efforts to ensure student retention and success, again forcing on-site staff to work though multifaceted institutional relationships. These programs include the following:

- Short and long-term programs where institutions send their own students overseas.
- Short and long-term programs where institutions send their own students overseas and additionally open their program to other students.
- Short and long-term programs run by private entities that send students overseas from any institution.
- Short and long-term programs run by institutions and private entities that send students overseas from any institution, coupling in-face with on-line learning.
- Hybrids of all the above including exchange institutions and partner programs while attending the "base" program.

Overseas staff have learned to navigate these complex organizational structures with differing laws around privacy, concepts of what "education" is, different attitudes around mental health, and cultural differences in the interpretation and existence of the U.S. concept of student success. As the percentage of U.S. students studying abroad is expected to increase, and the demographic will shift to include more difficult to retain student groups, discourse between those with expertise overseas and in the U.S. on this topic likely will become part of the public discourse around student success and retention.

FUTURE RESEARCH DIRECTIONS

The literature on global education, study abroad, retention and student success highlights the continually shifting nature of all these areas as well as their intersection. Study abroad program's contribution to student retention, academic knowledge and skill development and self-efficacy provide insights into specific retention, teaching and services that increase degree completion. Research areas should investigate the transfer of these study abroad strategies to campus programs, integration of campus and study abroad learning into a single undergraduate curriculum, and integration of services that are available across campus to study abroad programs. When coupled with the limitations of existing research, five areas of possible research are seen.

First is the expansion of research like those cited, that might expand the scope of correlation between study abroad and retention at multiple institutions of different types, and on different student demographics than has been undertaken. Such work could lead to more comprehensive ways to ascertain how on-campus practitioners can work with their international colleagues to ensure student success.

Second is to expand research around assessment of students studying abroad directly exploring student's metacognitive skills relating to self-efficacy and other non-academic factors that could correlate to retention and student success. Such research could be of benefit to the field of assessment, study abroad, and working with student retention broadly in factors that could be replicated on home campuses.

Third is the need to conduct longitudinal research given the shifting nature of the fields of global education, study abroad, retention and student success. Such research could go beyond a particular organizational culture and model, highlighting how key needs shift over time and must be addressed and could provide insight into global education, study abroad, retention and student success.

Fourth is conducting a comparative analysis across US and foreign institutions in study abroad, retention and student success. The data from this research could provide insight into where there is divergence as well as alignment in finding how study abroad correlates to retention and student success across different cultures and higher educational models. Given broad alignment of the internationalization of higher education and the public discourse around retention and student success, this data could prove useful to multiple areas of educational research and policy in the increasingly interconnected world.

Finally, longitudinal research should track students to determine whether studying abroad correlates to their employment over time. The University of Minnesota has undertaken such research, and others should follow their lead to improve data in this realm. Given the public discourse around return on investment of the cost of education, and research tracking students post-graduation for other reasons by career and alumni offices, such research could explore the correlation between study abroad, to jobs attained and salaries earned. The resulting data may correlate to a return on investment in new ways which could in turn provide additional specifics in the discourse around global education, study abroad, retention and student success.

CONCLUSION

The literature outlines broad trends for the continually shifting fields of global education, study abroad, retention and student success. Higher education continues to alter and evolve practices to integrate global components into academic and co-curricular programs, research, and operations in the realms of direct services, management, and strategy and policy. Based on institutional type and other factors, there is great variance in how globalization affects and is implemented at an institution. Yet, as research from this chapter shows, there is data and trends from which U.S. campuses could learn how to leverage study abroad to achieve greater student success and retention, and practitioners could learn from their overseas counterparts about how to best support students going through personal and academic development.

Student learning outcomes define students' institutionally expected knowledge, skills, and attitudes developed as a result of a learning experience. Institutions must determine whether they offer sufficient opportunities to produce the intended learning, select assessment measures (e.g., rubrics), and then assess if evidence of SLOs data matched the institutional expectations. If the SLOs are not being met, this finding should result in review and improvement of existing programs as well as the possible creation of new programs and new opportunities as needed.

Student learning assessment goes beyond the definition of traditional classroom learning. For global education experience, assessment requires gathering evidence that demonstrates student achievement of specific competencies as a result of educational experiences abroad as well as on the campus and similar measures in the co-curricular realms regardless of location. As highlighted in this chapter, global education is knowledge, both in academic as well as life skills.

Higher education institutions have been challenged not only by a shift toward increased assessment, but also by the need to systemically measure increasingly complex learning outcomes. As Robert Barr and John Tagg (1995) argued, the mission, vision, culture, and structure of higher education institutions must undergo a paradigm shift from providing instruction to producing learning and knowledge in students. Their framework, like the AAC&U model, was broad, encompassed all graduates, and called for learning assessment to be done without regard to a particular curriculum or educational experience. As the frameworks also focus on learning and knowledge rather than teaching, accreditation bodies along with other stakeholders increasingly require institutions to reliably assess learning outcomes across all graduates and the institution as a whole. As infusion of global education occurs throughout an entire institution, not just study abroad, lessons in retention and student success can be learned from study abroad. Furthermore, study abroad should actively be advocated for as a means to retain students and increase their core skills to succeed in college. This advocacy for study abroad, has arguably, been increased by recent reports and changes to demands for the future workforce.

Reflecting the international need to continually explore the intersection of student learning, skills, and global education, three reports were released in 2016. First, in January, the World Economic Forum put forth a model for the Fourth Industrial Revolution and linked this to a set of new job skills for 2020 (World Economic Forum, 2016). Within these skills and linked to the new global industrial revolution are many aspects of study abroad including both skills as well as knowledge. It is imperative that home institutions focus on these skills. While these skills can be secured overseas they can also be achieved through other high-impact practices in the U.S. There is a shift in the newly prioritized skills and knowledge around qualitative reasoning, self-awareness as well as the "other", and creativity. Arguably, all three as seen from data previously shown, correlate to student success and retention.

Second, in February, the Organisation for Economic Co-operation and Development (OECD) called for international standards of SLOs assessment (Andreas-Schleicher, 2016). As with the World Economic Forum model, this discussion should be central to U.S. institutions discussion of all student learning outcomes. Based on this OECD report, SLOs should be directly linked to intercultural competence and global understanding. Both of which correlate to student success and retention while studying abroad, and could be strengthened on U.S. campuses through other means, thus having a higher impression on a broader range of students.

Third, in May, the OECD proposed the Programme for International Student Assessment including measures of global learning and intercultural competence in 2018. These conversations are the latest in a global public discourse on higher education outcomes that are measurable and ensure an educated as well as skilled student in a globalized world. These mirror similar conversations in the U.S. around the need to increase student retention and success, as well as ensure a return on the entire educational experience that is measurable. The discourse likely will not diminish but rather continue to increase in complexity.

ACKNOWLEDGMENT

The author would first like to thank Dr. Elizabeth G. Chan for her continual support, both throughout the writing of this chapter as well as her long-term commitment to intercultural education and ensuring there is open access to education globally. She has been a collaborator without equal. Second, thank you to Dr. Cathy Krop for her constant support as both a mentor as well as fellow scholar relating to higher education and policy. Finally, but not least, to CEA Study Abroad, my employer for their support and interest in this work and true commitment to ensuring that students have meaningful life-changing experiences grounded in academic rigor with cultural integration.

REFERENCES

Altback, P., & Knight, J. (2007). The internationalization of higher education: Motivations and Realities. *Journal of Studies in International Education*, *11*(3-4), 290–306. doi:10.1177/1028315307303542

American Council on Education. (2011). Report of the blue-ribbon panel on global engagement strength through global leadership and engagement: U.S. higher education in the 21st Century. Campus internationalization. Retrieved from http://www.acenet.edu/news-room/Documents/2011-CIGE-BRPReport.pdf

American Council on Education. (2014). *Campus internationalization*. Retrieved from http://www.acenet.edu/higher-education/topics/Pages/Campus-Internationalization.aspx

Armstrong, L. (2007). Competing in the global higher education marketplace: Outsourcing, twinning, and franchising. *New Directions for Higher Education*, *140*(140), 131–138. doi:10.1002/he.287

Association of American Colleges and Universities. (2014). Essential learning outcomes. Retrieved from https://www.aacu.org/leap/essential-learning-outcomes

Astin, A. W. (1984). Student involvement: A developmental theory for higher education. *Journal of College Student Personnel*, *25*(2), 297–308.

Bean, J. P. (1980). Dropouts and turnover: The synthesis and test of a casual model of student attrition. *Research in Higher Education*, *12*(2), 155–187. doi:10.1007/BF00976194

Brownell, J., & Swaner, L. (2012). Five High-Impact Practices: Research on Learning Outcomes, Completion, and Quality. *Peer Review*, *14*(3), 29.

Cabrera, A. F., Castaneda, M. B., Nora, A., & Hengstler, D. (1992). The convergence between two theories of college persistence. *The Journal of Higher Education*, *63*(2), 143–164. doi:10.2307/1982157

Carr, S. (1999). *Distance-Learning group blends offerings of 2- and 4-Year colleges*. The Chronicle of Higher Education. Retrieved from: www.chonicle.com

Coughlin, E. K. (1994). *The emergence of the 'global city'*. The Chronicle of Higher Education. Retrieved from: www.chonicle.com

Desruisseaux, P. (1991). *Congress approves program to support overseas study*. The Chronicle of Higher Education. Retrieved from www.chonicle.com

Desruisseaux, P. (1999). *What parents should know about study-abroad programs*. The Chronicle of Higher Education. Retrieved from www.chonicle.com

Desruisseaux, P. (1999a). *15% rise in American students abroad shows popularity of non-European destinations*. The Chronicle of Higher Education. Retrieved from www.chonicle.com

Desruisseaux, P. (1999b, December 10). *Foreign Students Continue to Flock to the U.S.* The Chronicle of Higher Education. Retrieved from www.chonicle.com

Desruisseaux, P. (2000,). *As exchanges lose a political rationale, their role is debated*. The Chronicle of Higher Education. Retrieved from www.chonicle.com

Dwyer, M. M., & Peters, C. K. (2004). The benefits of study abroad. *Transitions abroad, 37*(5), 56-58.

Edwards, J. (2007). Challenges and opportunities for the internationalization of higher education in the coming decade: Planned and opportunistic initiatives in American Institutions. *Journal of Studies in International Education*, *11*(3/4), 373–381. doi:10.1177/1028315307303920

Engle, L., & Engle, J. (2003). Study abroad levels: Toward a classification of program types. *Frontiers: The Interdisciplinary Journal of Study Abroad*, *9*, 1–20.

Georgia Learning Outcomes of Students Studying Abroad Research Initiative. (2010.) Retrieved from: http://glossari.uga.edu/datasets/pdfs/FINAL.pdf

Goodman, A. E. (1999). *America is devaluing international exchanges for students and scholars*. The Chronicle of Higher Education. Retrieved from www.chonicle.com

Hser, M. P. (2005). Campus internationalization: A study of American universities' internationalization efforts. *International Education*, *35*(1), 35–48.

Hudzik, J. (2011). *Comprehensive internationalization: From concept to action.* NAFSA: Association of International Educators. Retrieved from http://www.nafsa.org

Ingraham, E. C., & Peterson, D. L. (2004). Assessing the impact of study abroad on student learning at Michigan State University. *Frontiers: The Interdisciplinary Journal of Study Abroad*, *10*, 83–100.

Knight, J. (1993). Internationalization management strategies and issues. *International Education Magazine*, *9*(6), 21–22.

Knight, J. (1997). Internationalization of higher education: A conceptual framework.

Knight, J. (2001). Internationalization. *Journal of Studies in International Education*, *5*(3), 228–243. doi:10.1177/102831530153004

Knight, J. (2003). Updating the definition of internationalisation. *Industry and Higher Education*, *33*, 2.

Knight, J. (2004). Internationalization remodeled: Definition, approaches, and rationales. *Journal of Studies in International Education*, *8*(1), 5–31. doi:10.1177/1028315303260832

Kuh, G. (2012). High-impact educational practices: What they are, who has access to them, and why they matter. *Peer Review*, *14*(3), 29.

Kuh, G., & O'Donnell, K. (2013). Ensuring quality and taking high-impact practices to scale. *Peer Review*, *15*(2), 32.

Levitz, R. (2001). What's Working Right Now in Student Retention! Building Blocks for Retention Success – the Basis for Recruiting Success. *Prepared for the National Conference for Student Retention*.

Lipsman, M. (2000). *Globalization isn't good or inevitable*. The Chronicle of Higher Education. Retrieved from www.chonicle.com

Malmgren, J., & Galvin, J. (2008). Effects of study abroad participation on student graduation rates: A study of three incoming freshman cohorts at the University of Minnesota, Twin Cities. *NACADA Journal*, *28*(1), 29–42. doi:10.12930/0271-9517-28.1.29

Mazzarol, T., & Soutar, G. N. (2002). Push-pull factors influencing international student destination choice. *International Journal of Educational Management*, *16*(2/3), 82–90. doi:10.1108/09513540210418403

McCabe, L. (2001). Globalization and internationalization: The impact on education abroad programs. *Journal of Studies in International Education*, *5*(2), 138–145. doi:10.1177/102831530152004

Merrill, K. C., Braskamp, D. C., & Braskamp, L. A. (2012). Assessing individual's global perspective. *Journal of College Student Development*, *53*(2), 356–360. doi:10.1353/csd.2012.0034

NAFSA. Association of International Educators. (2010). Members of the Teaching, Learning, and Scholarship Knowledge Community 2009 Task Force on Assessment and Evaluation. Retrieved from http://www.nafsa.org/uploadedFiles/NAFSA_Home/Resource_Library_Assets/Publications_Library/Assess%20and%20Eval%20in%20IE.pdf

NAFSA. Association of International Educators. (2015). International Education Professional Competencies. Retrieved from http://www.nafsa.org/Professional_Resources/Career_Center/IE_Competencies/International_Education_Professional_Competencies/

Noel, L., Levitz, R., & Saluri, D. et al. (1985). *Increasing Student Retention*. San Francisco, CA: Jossey-Bass.

Olsen, C. L., Green, M. F., & Hill, B. A. (2005). *Building a strategic framework for comprehensive internationalization*. Washington, DC: American Council on Education.

Olson, C. L., Evans, R., & Shoenberg, R. F. (2007). *At home in the world: Bridging the gap between internationalization and multicultural education*. Washington, DC: American Council on Education.

Olson, C. L., Green, M. F., & Hill, B. A. (2005). *Building a strategic framework for comprehensive internationalization*. Washington, DC: American Council on Education.

Olson, C. L., Green, M. F., & Hill, B. A. (2006). *A handbook for advancing comprehensive internationalization: What institutions can do and what students should learn*. Washington, DC: American Council on Education.

Porter, M. C. (2011). Assessing alternative breaks: Moving beyond sleeping on floors and pass-the-candle reflection. *About Campus: Enriching the Student Learning Experience*, *16*(5), 21–24. doi:10.1002/abc.20077

Raby, R. L. (2009). Globalization and community college model development. In R.L. Raby & E.J. Valeau (Eds.). Community college models: Globalization and high education reform (pp. 21-38). Amsterdam: Springer Press. doi:10.1007/978-1-4020-9477-4_2

Rudzki, R. (2000). Implementing internationalisation: The practical application of the fractal process model. *Journal of Studies in International Education*, *4*(2), 77–90. doi:10.1177/102831530000400208

Sassen, S. (2001). *The global city: New York, London, Tokyo*. Princeton, NJ: Princeton University Press. doi:10.1515/9781400847488

Smith-Paríolá, J., & Gòkè-Paríolá, A. (2006). Expanding the parameters of service learning: A case study. *Journal of Studies in International Education*, *10*(1), 71–86. doi:10.1177/1028315305283307

Stebleton, M. J., Soria, K. M., & Cherney, B. (2013). The high impact of education abroad: College students' engagement in international experiences and the development of intercultural competencies. *Frontiers: The Interdisciplinary Journal of Study Abroad*, 22.

Stromquist, N. P. (2002). *Education in a globalized world: the connectivity of economic power, technology and knowledge*. Lanham, MD: Rowman & Littlefield.

Sutton, R., & Rubin, D. (2004). The GLOSSARI project: Initial findings from a system-wide research initiative on study abroad learning out- comes. *Frontiers: The Interdisciplinary Journal of Study Abroad*, *10*, 65–82.

Tarrow, S., & Hall, P. A. (1998). *Globalization and area studies: When is too broad too narrow?* The Chronicle of Higher Education. Retrieved from http://chronicle.com/article/GlobalizationArea/99332/

Tinto, V. (1993). *Leaving College: Rethinking the Causes and Cures of Student Attrition* (2nd ed.). Chicago, IL: University of Chicago Press.

van der Wende, M. (2007). Internationalization of higher education in the OECD countries: Challenges and opportunities for the coming decade. *Journal of Studies in International Education*, *11*(3-4), 274–289. doi:10.1177/1028315307303543

Vande Berg, M. (2007). Intervening in the learning of U.S. students abroad. *Journal of Studies in International Education*, *11*(3-4), 392–399. doi:10.1177/1028315307303924

Wächter, B. (2003). An introduction: Internationalisation at home in context. *Journal of Studies in International Education*, *7*(1), 5–11. doi:10.1177/1028315302250176

Woolf, M. (2002). Harmony and dissonance in international education: The limits of globalisation. *Journal of Studies in International Education*, *6*(1), 5–15. doi:10.1177/102831530200600102

Yang, R. (2002). University internationalization: Its meanings, rationales and implications. *Intercultural Education*, *13*(1), 81–95. doi:10.1080/14675980120112968

KEY TERMS AND DEFINITIONS

Globalization: Changes within higher education in relation to external forces which place pressure upon institutions to implement international business-related practices, including privatization, as well entrepreneurialism coupled with an increase in breadth of the type of programs both overseas as well as on the home campus.

High-Impact Practices: Such practices include particular attributes, including: require considerable time and effort; expedite learning outside of the classroom as well as link to in-class experiences; necessitate significant interactions with faculty and students (and in the case of study abroad locals); support partnership with diverse others; and include repeated and substantive feedback.

Internationalization: "The process of integrating an international, intercultural or global dimension into the purpose, functions or delivery of postsecondary education (Knight, 2004, p. 11)."

Study Abroad: A number of possibilities by which students complete part of their degree program through educational activities outside the United States, in the case of this chapter, data is referencing those programs that last at least one whole semester.

Chapter 3
The Predictive Role of Gender and Race on Student Retention

Gokhan Savas
Social Sciences University of Ankara, Turkey

ABSTRACT

Student retention is an important issue in American higher education, and has major impacts on students' access to employment and earning potential. Furthermore, it significantly influences the finances of colleges and universities. This chapter looks at the predictive role of gender and race on students' college retention, and analyzes the Education Longitudinal Study of 2002 (ELS, 2002) that follows a nationally representative cohort of students from 2002, when they were high school sophomores, through their postsecondary education. The analytic sample of this research includes high school graduates who remained in the study from 2002-2012, and reported "any known degree attained as of June 2013." Findings indicate that male students are more likely to drop out of college compared to female students, and this gender effect does not change even after controlling for several other variables. Similar to gender, race is also found to be a significant predictor of student retention.

INTRODUCTION

Over a third of America's college students and over half of our minority students don't earn a degree, even after six years. So, we don't just need to open the doors of college to more Americans; we need to make sure they stick with it through graduation. That is critical. (Obama, 2010)

Given the role of educational credentials in American society in which over half of all workers currently have at least bachelor's degree (U.S. Department of Labor, 2015a), higher education attainment is the admission ticket to most professional occupations. Higher education is highly valued by American society and most people seek higher education to improve their job prospects and social status. Those with a bachelor's degree are more likely to find lucrative professional jobs compared to those with a high school diploma (Baum et al., 2010). The earning power of workers with a bachelor's degree is 1.7 times higher than the average earning by those with only a high school diploma, and 2.3 times higher than the earnings of high school dropouts (U.S. Department of Labor, 2015b).

DOI: 10.4018/978-1-5225-2998-9.ch003

Today the transition from high school to college is an important milestone for American youth in the pathway to adulthood. The transition to adulthood in the United States has changed in recent decades and no longer depends on the traditional milestones that mark adulthood, such as leaving home, securing full time employment, establishing a home and forming a family (Berlin et al., 2010; McLanahan et al., 2010; Arnett, 2000). In the contemporary United States, "young adults take far longer to reach economic and social maturity than their contemporaries did five or six decades ago. In large part, this shift is attributable to the expansion of higher education beginning in the late 1960s" (Berlin et al., 2010, p. 3). In other words, more and more young adults are choosing to go to college.

REVIEW OF THE LITERATURE

Compared to the past, current research observes that young adults are marrying at a later age and that they also stay in school for longer periods of time (Settersten & Ray, 2010). This is more salient for women because the median age of marriage and childbearing for women has dramatically increased, which parallels the increasing rate of college attendance among women. In other words, the priorities of women in society have changed over time; and, they are now delaying marriage and childbearing in favor of attending college. In 1960, the median age for a first marriage for a woman was 20 years old. Today the median age for a woman's first marriage is 26 years old (U.S. Bureau of the Census, 2010). A college degree for women is recognized as a form of insurance against poverty and income deprivation (Buchman & DiPirete, 2006). Buchmann (2006) explains this situation as follows:

In the 1960s and 70s, girls were getting better grades, but many young women were not going to college, or they were dropping out of college to get married. Now the benefits of a college education are growing faster for women than for men, and women are taking advantage... The generation of women who were born in the 1960s were the first to see their mothers getting divorced and having few options in the labor market. Many of these women were likely thinking, they wanted to avoid that situation by getting a college degree. (as cited in Grabmeier, 2006)

In relation to the new transition to adulthood, it is important to understand what is going on with young people after they graduate from high school. Graduating from high school is traditionally seen as a first step into adulthood. Historically, immediately after high school graduation, young adults left home, and began to look for full-time work. However, today graduating from high school is no longer considered to be an important enough achievement to ensure reasonable job and career prospects or opportunities. Therefore, many young adults seek to obtain at least some postsecondary education after graduating from high school (Berlin et al. 2010). This new longer transition to adulthood is also supported through family perception that early transition to work and marriage is no longer optimal (Berlin et al. 2010). The majority of parents expect their children to graduate from high school and complete at least some postsecondary education. In 2012, about 65% of parents with students in grades six through 12 expected that their child would attain a bachelor's degree or higher (Child Trends Data Bank, 2015).

The data illustrates that enrollment of underrepresented students has steadily increased over time. Specifically, women have outnumbered men on college campuses since the late 1980s (Snyder & Dillow, 2011). However, the proportion of both men and women enrolling in college has increased over the years, with the increase in the number of female students being more substantial. The college enrollment rate for men increased by 36% between 1999 and 2009, but the college enrollment rate for women increased by 63% (Snyder & Dillow, 2011).

Current statistics also estimate that the gender gap in college attendance will continue through 2020, when women are expected to make up 59% of undergraduate enrollment (Snyder & Dillow, 2011). The gender pattern in undergraduate enrollment and how it has changed over years is depicted in Figure 1. Current descriptive statistics also estimate that not only the gender gap but also the racial/ethnic gap in college attendance will continue through 2020. As seen in Figure 2, the racial/ethnic gap is not expected to change over years. Although racial and ethnic minorities are projected to have increasing enrollment rates, the gap between white and racial and ethnical minority group enrollments is expected to stay constant.

Unfortunately, college enrollment does not necessarily mean college graduation. In the United States, student retention is an important issue because of the consistently low rate of college graduation. In the United States, the four-year graduation rate for first-time freshman females (43.9%) from a four-year postsecondary institution is higher than for males (34.8%). Also, Asian (44.70%) and white (43.70%) students have much higher graduation rates compared to African American (21.40%) and Latino (30.40%) students (see Figure 3). These rates significantly increase for the six-year graduation rate, but the trend does not change. This means that females (61.9%) have higher graduation rate than males (56.5%); and Asians (70.5%) and whites (62.9%) have much higher rates than African Americans (40.8%) and Latinos (52.5%) (NCES, 2016).

The American Institute for Research (2010) found that there was a significant cost for students who dropped out during their first year of college. The average debt for individual students is about $29,000 and most of them are not able to pay it back given the combination of low socio-economic background

Figure 1. Number of actual and projected college enrollment by gender
Source: U.S. Department of Education, National Center for Education Statistics, Higher Education General Information Survey (HEGIS), "Fall Enrollment in Colleges and Universities" surveys, 1970 through 1985; 1990 through 2009 Integrated Postsecondary Education Data System, "Fall Enrollment Survey" (IPEDS-EF:90–99), Spring 2001 through Spring 2010; and Enrollment in Degree-Granting Institutions Model, 1980–2009.

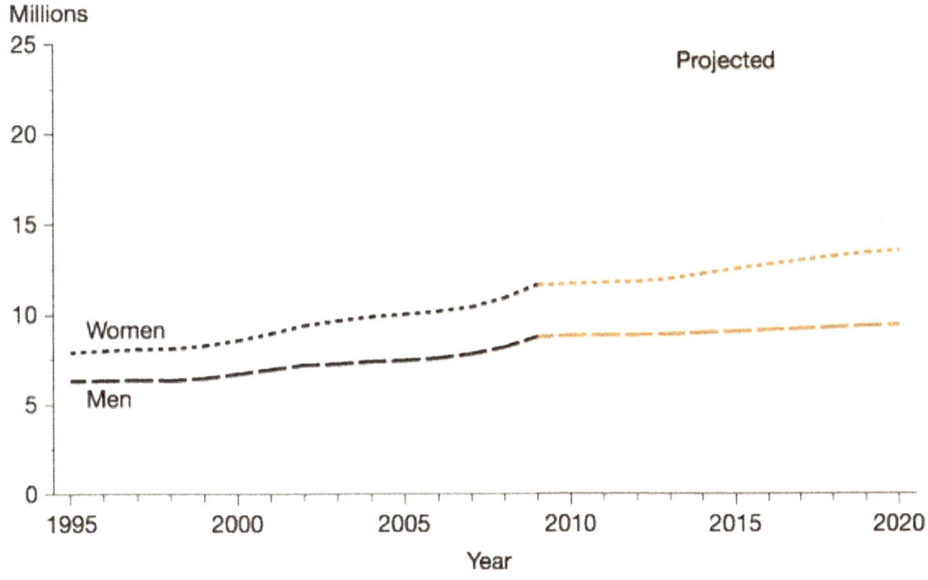

Figure 2. Number of actual and projected college enrollment by race

Source: U.S. Department of Education, National Center for Education Statistics, Integrated Postsecondary Education Data System, "Fall Enrollment Survey" (IPEDS-EF, pp. 95–99), and Spring 2001 through Spring 2010; and Enrollment in Degree-Granting Institutions by Race/Ethnicity Model, 1980–2009.

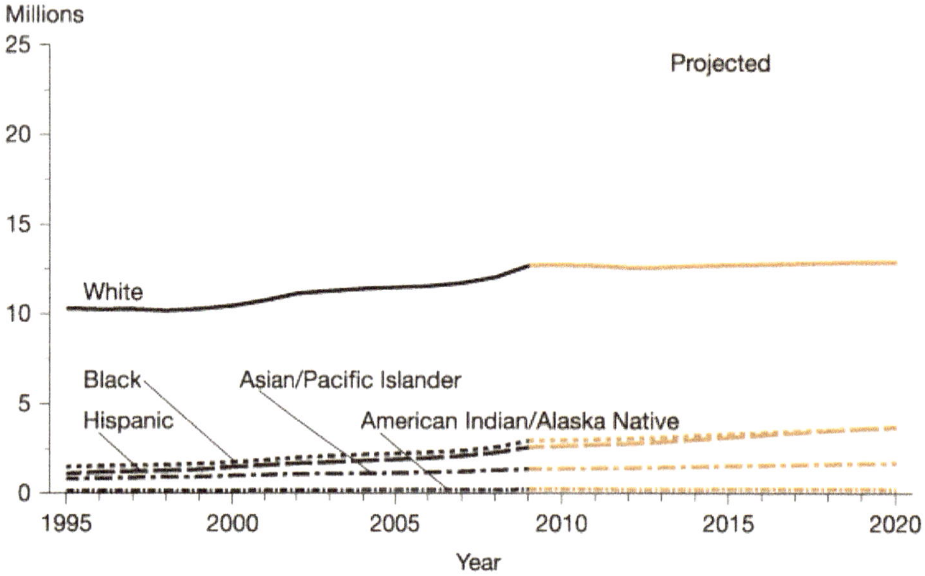

Figure 3. Graduation rate from first institution attended for first-time, full-time bachelor's degree-seeking students at 4-year postsecondary institutions within 4 years after start, by gender and race/ethnicity

Source: U.S. Department of Education, National Center for Education Statistics, Integrated Postsecondary Education Data System (IPEDS), Fall 2002 and Spring 2007 through Spring 2015, Graduation Rates component; and IPEDS Fall 2008, Institutional Characteristics component.

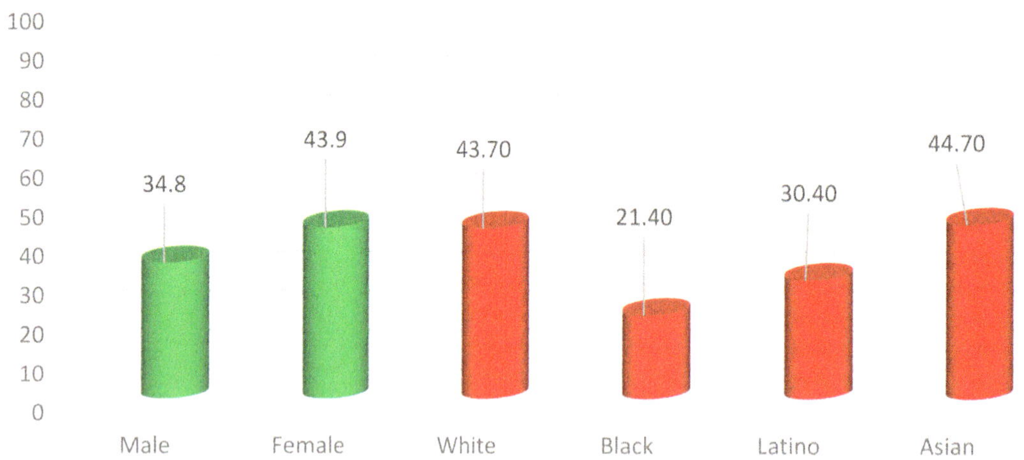

and minimal education. There are many important longitudinal studies on college persistence (Tinto, 1975, 1994, 2004; Astin, 1984, 1993; Noura, 1987; Bennet and Okinaka, 1990; Berger, 2001; Thomas, 2002; Pascarella and Terenzini, 2005; Hu, 2011; Sciarra et al., 2016). and Colleges and universities have started to pay close attention to this issue.

As one of the prominent scholars in the study of college retention, Tinto (1994, 2004) found that college persistence was linked to several factors, such as students' academic and social integration, separation from the home environment and culture, transition from high school to college, and incorporation into college life. Therefore, different groups of students need different retention programs and different types of institutional support. Bennet and Okinaka (1990) echo this by finding that feelings of social alienation and dissatisfaction influence student persistence differently for different racial/ethnic groups. Involvement in academic and social activities is also an important factor positively affecting student retention (Astin, 1993; Pascarella & Terenzini, 2005) though its impact has been found to be minimal for different groups such as Chicano students (Nora 1987). Research indicated that not only individual factors but also organizational factors such as institutional habits and the patterns of organizational behavior within colleges and universities have significant influences on student retention (Thomas 2002; Berger 2001).

RESEARCH DESIGN AND STUDY METHODS

The chapter analyzes the Education Longitudinal Study of 2002 (ELS, 2002) that follows a nationally representative cohort of students from 2002, when they were high school sophomores, through their postsecondary education. The researchers collected information from the same students in 2004 when they were high school seniors and 2006 when the majority of the students were in college. One additional follow-up was conducted in 2012, six years after the second follow-up or eight years after high school graduation. The ELS includes 16.197 10th grade students in 752 schools, which represent 3.4 million students as of 2002 in the United States.

The analytic sample of this research includes high school graduates (including the ones who reported completing a GED) who remained in the study from 2002-2012, and reported "any known degree attained as of June 2013." Therefore, there are 9.827 respondents in this research. The choice of data is important for this research since ELS, 2002 is the most recent longitudinal dataset that spans the period from high school to postsecondary enrollment through to after college graduation. Also, ELS monitored the critical transitions made by a cohort of 2002 high school sophomores through college and into their adult careers, with a special emphasis on college access and choice. The study also obtained information about student's educational motivations, experiences, and achievement, as well as school and family characteristics.

ELS is a sample and the entire population was not surveyed. It uses a complex sampling design rather than simple random sampling (SRS). Thus, not all schools and students had an equal probability of selection, as well as, not all schools and students participated in ELS; therefore, ELS weights its samples to make sure it maintains a plausible level of representativeness. In other words, each sample member has a design weight, which is the number of people in the target population represented by that sample member when selected for the study. To secure the generalizability of the research findings, this study includes the panel weight that adjusts for sample member nonresponses to maintain representativeness of the analysis sample.

As seen below in Figure 4, compared to the parallel datasets with similar designs, the ELS provides more recent data than the National Education Longitudinal Study (NELS) and High School and Beyond (HSB).

To measure students' college retention, I used a variable in the ELS study to indicate whether the student received a known degree from any postsecondary institution as of June 2013.

PRIMARY INDEPENDENT, INTERVENING, AND CONTROL VARIABLES

Gender and Race

The primary independent variables of interest were gender and race. Gender included males and females. Race included whites, African Americans, Latinos, Asians, and other.

Parental Expectation

Parental expectation includes both the mother's and the father's expectations of their children. The study also measured parental expectation as a dummy variable to indicate whether parents have any expectations of their children to earn a bachelor's degree or higher (BA degree = 1, else = 0).

Figure 4. Research design for the NCES high school cohorts
Source: NCES,

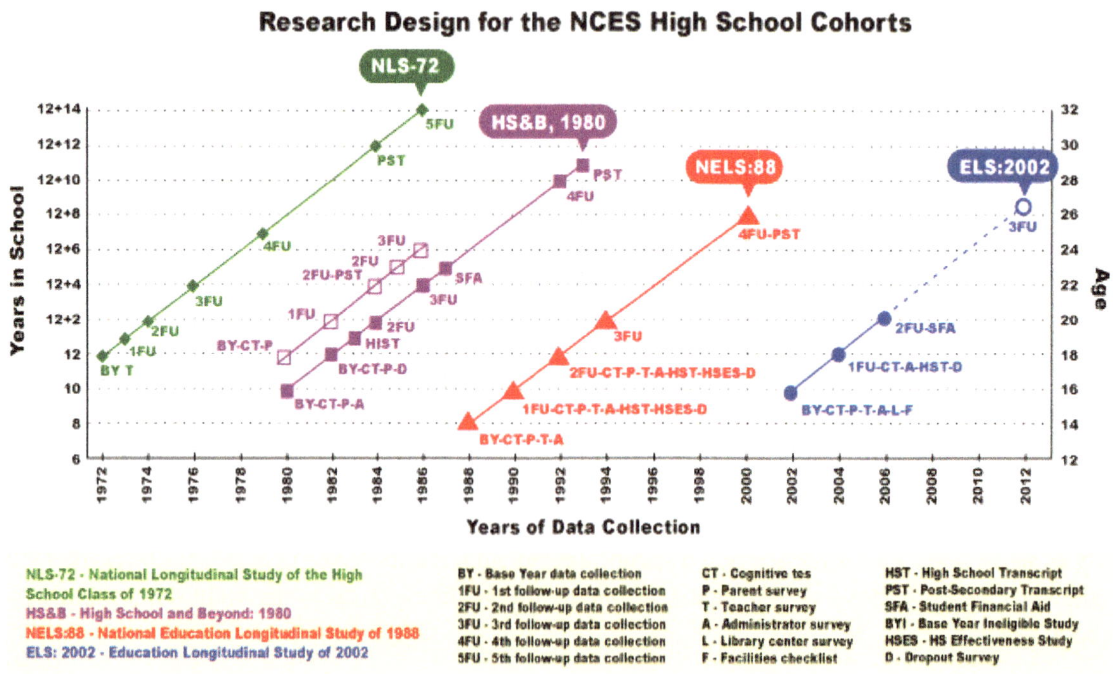

Students' Educational Expectations

The study measured students' educational expectation as a dummy variable to indicate whether students have any expectations of earning a bachelor's degree or higher (BA and/or higher = 1, else = 0).

Negative School Behavior

The study measured school behavior as coding a scale, which includes the following items:

- How many times a student was late for school.
- How many times a student cut or skipped classes.
- How many times a student was absent from school.
- How many times a student got in trouble.
- How many times a student was placed on in-school suspension.
- How many times a student was suspended and put on probation.
- How many times a student transferred for disciplinary reasons.

The possible answers for each item include "never", "1-2 times", "3-6 times", "7-9 times", and "10 or more times." Beginning from "never", the study assigned points between 0 and 5 for each one of these five answer choices. There are seven items in the scale, and a respondent can get a score on this scale between 0 and 35. Lower scores refer to "better" school behavior. The study utilized Cronbach's alpha reliability coefficients to measure the scale. Cronbach's alpha is a measure of internal consistency. In other words, the study measured how closely related a set of items were as a group. It ranges from -1 to 1. A higher value of alpha is often used as evidence that the scale items are reliable. Technically speaking, Cronbach's alpha is not a statistical test; rather it is a coefficient of reliability. In general, a reliability coefficient of .70 or higher is considered "acceptable" in most social science research. Table 1 provides a correlation matrix for the items in this scale and Cronbach's alpha level. The scale is highly reliable given a reliability coefficient of .95.

Time Spent on Homework

Students in ELS were asked how much time they spent on homework each week outside of school. The study quantified time spent on homework as the total number of hours spent on homework each week outside of school.

Family and High School Controls

Socioeconomic status, family structure, and number of siblings were the family background variables. High school controls included high school type and urbanicity.

Socioeconomic status was measured as a composite in ELS based on five equally weighted, standardized components: father's/guardian's education, mother's/guardian's education, family income, father's/guardian's occupation, and mother's/guardian's occupation. This composite measure of SES as provided by NCES, which is a continuous range of variables between 0 and 1.87 was utilized.

Table 1. Correlation matrix of negative school behavior index

		How Many Times Late for School	How Many Times Cut/ Skipped Classes	How Many Times Absent from School	How Many Times got in Trouble	How Many Times Put on In-School Suspension	How Many Times Suspended/Put on Probation	How Many Times Transferred for Disciplinary Reasons
how many times late for school	Pearson's r Sig	1						
how many times cut/ skipped classes	Pearson's r Sig	0.7603 0.00	1					
how many times absent from school	Pearson's r Sig	0.7127 0.00	0.695 0.00	1				
how many times got in trouble	Pearson's r Sig	0.7458 0.00	0.7289 0.00	0.6648 0.00	1			
how many times put on in-school suspension	Pearson's r Sig	0.7935 0.00	0.7887 0.00	0.7314 0.00	0.7549 0.00	1		
how many times suspended/put on probation	Pearson's r Sig	0.7554 0.00	0.7629 0.00	0.6994 0.00	0.7549 0.00	0.8549 0.00	1	
how many times transferred for disciplinary reasons	Pearson's r Sig	0.816 0.00	0.8083 0.00	0.7548 0.00	0.8034 0.00	0.915 0.00	0.8791 0.00	1

Scale reliability coefficient: 0.9545

The study measured family structure as a dichotomous variable indicating students lived in families with two biological or adoptive parents during 2002 when they were eighth graders (Students living in families with two parents =1, not living in families with two parents = 0). The study also measured the number of siblings as the total number of brothers and sisters that students have in the home.

High School Type

High school type is measured as a high school control with dummy variables for public, Catholic and other private schools. The study utilizes public schools as the reference group.

High School Urbanicity

School urbanicity was measured as dummy variables for urban, suburban and rural, with urban as the reference category.

High School Program

The study used the student's self-reported high school program of study. Students were asked if they had to limit themselves to one of the following three choices, which comes nearest to describing their high school program; General, College Preparatory (academic), and Vocational (including technical or business).

Grade Point Average (GPA)

The study measured the student's GPA utilizing the F3TZGPAALL variable in ELS, which indicates the grade-point average (GPA) at all known postsecondary institutions attended as of June 2013. GPA is defined as (earned credit * grade / attempted credit).

Dropouts

To measure how many friends dropped out of high school, the F1S65A variable was used.

College Type

To measure the type of college at which students begin their career, the study included 2-year and 4-year college enrollments.

ANALYTIC STRATEGY

Using ELS:2002 data this chapter examines the college dropout rate across gender and race and estimates those differences in student retention. There are five logistic regression models for identifying the mechanism contributing to gender and race differences in retention. As controls, the models include background variables such as time spent on homework, the number of friends who dropped out of high school, the socio-economic status of parents (SES), the number of siblings, family structure, high school urbanicity (urban, suburban or rural) and finally, high school type (public, Catholic or other private). GPA and educational expectations are separately added into the model as intervening variables.

In the logit model $\ln[p/(1-p)] = a + \beta X + e.$; p is the probability that the event Y occurs, $p(Y=1)$. $p/(1-p)$ is the "odds ratio". $\ln[p/(1-p)]$ is the log odds ratio, or "logit." The slope coefficient (β) is interpreted as the rate of change in the "log odds" as X changes; however, this is not very useful for interpretation. The use of the log odds ratio in this research is an interpretation of the logit coefficient, which is usually more intuitive when used as the "odds ratio." Therefore, the study reports odds ratios and the logit model as follows; $p/(1-p) = \exp(a + \beta \text{Gender/Race} + e)$. In the model, $\exp(\beta)$ is the effect of the independent variable, gender/race, on the "odds ratio" of not having a degree after 7 years they start college.

The logit models to predict college retention are as follows:

Model I: Odds of college dropout $p/(1-p) = \exp(a + \beta_1 \text{gender})$ (the gross gender effect)
Model II: Odds of college dropout $p/(1-p) = \exp(a + \beta_1 \text{race})$ (the gross race effect)
Model III: Odds of college dropout $p/(1-p) = \exp(a + \beta_1 \text{gender} + \beta_2 \text{ race} + \beta_3 \text{ controls})$

Model IV: Odds of college dropout p/(1-p) = exp (a + $\beta_1$gender + β_2 race + β_3 controls + β_4 educational expectation

Model V: Odds of college dropout p/(1-p) = exp (a + $\beta_1$gender + β_2 race + β_3 controls + β_4 educational expectation + β_5 GPA

This chapter reports and compares the odds ratios of each covariate in the model. To facilitate the interpretation of the logistic regression coefficients, the delta-p statistic is used to estimate the change in the probability of college dropout associated with a one unit change in each independent variable. The study tests the results at 0.001, 0.01, and 0.05 alpha levels and use the survey commands in STATA to address the ELS complex survey design to have the correct variance and significance tests.

RESULTS

Table 2 provides descriptive statistics for all dependent and independent variables used in the study. Surveyed students were high school sophomores in 2002, high school seniors in 2004, college sophomores in 2006. In addition, they provided information about whether they received a college degree as of 2013. Those who did not have a college degree seven years after starting college (34%) were considered as "dropouts" in this study. 55% of all students are female and 61% of them are white. The average score of socio-economic status is .52 out of 1.87. The majority of students are from two-parent families (76%), suburban (48%), and public (72%) high schools.

The average number of siblings was 2. Students spent an average of 6.7 hours on homework while not in school each week. The average score on negative school behavior was 10.35 out of 35. The majority of students (68%) identified their high school programs as college preparatory/academic. Compared to students in 2-year colleges (33%), more students were in 4-year colleges (67%) in 2006, the second follow up timeframe. The average GPA was 2.74 from all postsecondary institutions. Students had high parental expectations, in that 94% of parents expected their children to get a bachelor's degree or higher. Similarly, students also had high educational expectations with 93% of them aspiring to obtain a bachelor's degree or higher. The average number of friends who dropped out of high school is 1.46.

Table 3 indicates descriptive characteristics of students who dropped out of college and those who persisted. While 48% of students who dropped out are female, 42% of all students persisted; and similarly, 52% of male students dropped out while 58% of all students persisted. White students constituted the majority (65%) among those who persisted, whereas the majority of dropouts were African American (17%) and Latino (16%). Students that were retained predominantly came from two-parent families and suburban public high schools. The majority of retained students also came from high schools that identified their programs as college preparatory/academic. When looking at students' GPA for all higher education institutions attended, it was expected to find that students who persisted at college have a much higher GPA than those who dropped out of college. Similarly, they also have higher socio-economic status. The number of siblings for each of the two groups of students, those who persisted and those who dropped out, only showed a slight difference.

Compared to students who were retained at college, college dropouts spend less time on homework outside of school and have higher scores on the negative school behavior index. Parental expectations and students' own educational expectations do not seem to differ much among those who persisted and those who dropped out with both groups demonstrating high parental and educational expectations.

Table 2. Descriptive statistics for dependent and independent variables

Variable (n=9,827)	Variable Description	Mean	SD	Min – Max
Dependent variable				
Degree attainment	Indicates whether the student received a known degree from any postsecondary institution as of June 2013 (1 = dropouts, 0 = degree holder)	0.34	.47	0 – 1
Independent variable				
Female*	Gender (1 = Female, 0 = not)	.55	.50	0 – 1
Male	Gender (1 = Male, 0 = not)	.45	.50	0 – 1
White*	Race/ethnicity (1 = White, 0 = not)	.61	.48	0 – 1
African American	Race/ethnicity (1 = African American, 0 = not)	.11	.32	0 – 1
Latino	Race/ethnicity (1 = Latino, 0 = not)	.12	.33	0 – 1
Asian	Race/ethnicity (1 = Asian, 0 = not)	.11	.32	0 – 1
Other	Race/ethnicity (1 = Other, 0 = not)	.05	22	0 – 1
SES	Socio-economic status - composite	.52	.39	0 – 1.87
Two-parent family*	Family composition (1 = two parent families, 0 = not)	.76	.42	0 – 1
Other family	Family composition (1 = other than two parent families, 0 = not)	.24	.42	0 – 1
Siblings	Number of siblings respondent has	2.09	1.28	0 – 6
Urban*	Urbanicity of school locale as indicated in the source data for sampling (1 = urban, 0 = not)	.35	.48	0 – 1
Suburban	Urbanicity of school locale as indicated in the source data for sampling (1 = suburban, 0 = not)	.48	.50	0 – 1
Rural	Urbanicity of school locale as indicated in the source data for sampling (1 = rural, 0 = not)	.17	.37	0 – 1
Public*	High school type (1 = public, 0 = not)	.72	.45	0 – 1
Catholic	High school type (1 = Catholic, 0 = not)	.17	.37	0 – 1
Other private	High school type (1 = other private, 0 = not)	.11	.31	0 – 1
Homework	Time spent on homework out of school	6.70	6.13	0 – 26
Behavior	Negative school behavior index	10.35	2.88	0 – 35
College preparatory	Student self-reported high school program of study (1 = college preparatory/academic, 0 = not)	.68	.47	0 – 1
Vocational/technical	Student self-reported high school program of study (1 = vocational/ technical, 0 = not)	.06	.24	0 – 1
General*	Student self-reported high school program of study (1 = general, 0 = not)	.26	.44	0 – 1
2-year college	College type as highest level of education attempted as the second follow-up (1 = two-year college, 0 = not)	.33	.47	0 – 1
4-year college*	College type as highest level of education attempted as the second follow-up (1 = two-year college, 0 = not)	.67	.47	0 – 1
GPA	GPA at all known institutions attended	2.74	.84	0 – 4
Dropped out friends	How many friends dropped out of high school	1.46	.68	1 – 5
Parental expectation	Higher education expectation of parents (1 = if the parent aspires to get a bachelor's degree or higher, 0 = if not)	.94	.24	0 – 1
Students' educational expectation	Higher education expectation of the student (1= if the student aspires to get a bachelor's degree or higher, 0 = if not)	.92	.27	0 – 1

Note: * indicates a reference group.
Source: ELS 2002.

Table 3. Characteristics of college students who dropped out and who persisted

Variable	Dropped out (n = 3,371) %	Persisted (n= 6,456) %
Female	48.41	42.21
Male	51.59	57.79
White	52.30	64.75
African American	17.03	8.50
Latino	16.20	9.31
Asian	8.96	12.86
Other	5.52	4.58
Two parent family	72.32	78.53
Other family	27.68	21.47
Urban	34.29	35.19
Suburban	47.17	48.98
Rural	18.54	15.83
Public	79.41	68.36
Catholic	12.46	18.79
Other private	8.13	12.86
College preparatory	60.07	72.07
Vocational/technical	8.99	4.86
General	30.94	23.06
2-year college	52.45	77.46
4-year college	47.55	22.54
Dropped out friends	1.62	1.39
GPA	2.07	3.09
SES	.41	.58
Siblings	2.24	2.01
Homework	5.45	7.35
Behavior	10.91	10.06
Parental expectation	.91	.95
Students' educational expectation	1.39	1.35

Source: ELS 2002

While dropped out students have an average of 1.62 friends who dropped out of college, this number is 1.39 for students retained at college.

Table 4 includes logistic regression models to predict the likelihood that a student will drop out of college. The first model (Model I) predicts that being female is advantageous for college retention. Compared to male students, female students are 32% less likely to drop out of college. Model II predicts that the likelihood of African American and Latino students dropping out of college is about two times higher than white students. These first two models seem to indicate that being a male and a racial/ethnic minority in the United States is a disadvantage in terms of postsecondary retention.

Table 4. Logistic regression models to predict college dropout

Variable (n = 9827)	Model I			Model II			Model III			Model IV			Model V		
	β	Sig	SE	β	Sig	SE	B	Sig	SE	β	sig	SE	β	sig	SE
Female	.68	***	.03				.65	***	.03	.64	***	.03	.82	***	.04
African American				2.49	***	.16	1.97	***	.15	1.97	***	.15	.89		.08
Latino				2.15	***	.14	1.48	***	.11	1.48	***	.11	1.16		.10
Asian				.86	*	.06	.83	*	.07	.84	*	.07	.73	***	.07
Other				1.48	***	.15	1.37	**	.14	1.37	**	.14	1.01		.12
SES							.51	***	.04	.52	***	.04	.64	***	.05
Parental expectation							.93		.09	.92		.08	.96		.11
Two-parent family							.97		.05	.96		.05	.96		.64
Siblings							1.07	***	.02	1.06	***	.02	1.07	***	.02
Suburban							.94		.05	.95		.05	1.00		.06
Rural							1.01		.08	1.01		.07	1.07		.09
Catholic							.85	*	.06	.85	*	.06	.78	**	.06
Other private							.87		.07	.87		.07	1.07		.11
Time spent on homework							.98	***	.00	.98	***	.00	.99	*	.00
Negative school behavior							1.06	***	.01	1.06	***	.01	1.03	***	.01
General HS							1.13	*	.06	1.13	*	.06	1.01		.06
Vocational HS							1.29	**	.12	1.28	**	.12	1.14		.12
Dropout friends							1.23	***	.04	1.24	***	.04	1.18	***	.05
2-year college							2.71	***	.14	2.72	***	.14	2.39	***	.14
Educational expectation										.99		.02	.99		.02
GPA													.14	***	.01
Constant	.63	***	.02	.42	***	.02	.25	***	.04	.26	***	.04	50.45	***	11.57
Pseudo R^2	0.0062			0.0247			0.1232			0.1233			0.3347		

Note: Male, white, not-two parent family, urban, public, college preparatory school, 4-year college are the reference groups.
Source: ELS 2002.

Model III predicts the viability of several control variables to test the net effect of gender as well as race. According to the model, male students and racial/ethnic minorities are still more likely to drop out of college compared to their counterparts. Model IV includes students' educational expectations because much of the literature emphasizes the importance of this variable; however, students' educational expectations do not change the results and are not statistically significant in predicting college dropout.

The final model (Model V) then includes students' grade point average (GPA) because that is considered statistically significant to predict college dropout; students who have higher GPA are 86% less likely to drop out of college. The important thing here is that as students' GPA increases, the model predicts that racial ethnic minorities are no longer disadvantaged. It means that net of GPA, African Americans and Latinos are not significantly different from their white peers in terms of the likelihood of dropping out of college. We can argue that inequalities between white students and racial ethnic minorities in retention are largely attributed to academic achievement. According to the final model, there are also several other factors that are important in student retention. These include students' socio-economic background, the type of high school, the number of siblings they have, whether they have negative school behavior, the number of friends who dropped out of high school, and type of college they initially enrolled in.

FUTURE RESEARCH DIRECTIONS

Future research in the statistical analysis area is needed to more fully understand the influence of on-line degree programs and other non-traditional program delivery methods on college persistence rates broken out by gender and race and the other key characteristics listed in Table 3. It is also important to more fully explore retention rates between traditional and nontraditional college students within various institution and program types. Future studies can also contribute to the filed by investigating whether school counselors along with other academic and social support and extracurricular activities would influence college retention across various student groups.

CONCLUSION

Student retention is an important issue in American higher education, and it has a major impact on students' access to employment and earning potential. Furthermore, it also significantly influences the finances of colleges and universities given that many college students do not complete their degree within 6 years (NCES, 2016).

This chapter examined the predictive role of gender and race on students' college retention. As known in the sociology literature, both gender and race are social categories with significant influence on the likelihood of success in higher education, standard of living, health, and lifestyle. The study found that gender and race significantly influence college persistence.

Utilizing a nationally representative dataset, the chapter argued that because male students are more likely to drop out of college, females are more likely to remain in college through graduation; and, this gender effect does not change even after controlling for several other variables. Similar to gender, race is also found to be a significant predictor of student retention. White and Asian students are more likely than African American and Latino students to persist in higher education. A high GPA reduces the likelihood that minority students will drop out, as well as, reducing the gender gap in retention thus eliminating the disadvantage of gender and minority status. Therefore, academic achievement seems to be a crucial indicator in predicting whether a student will persist or drop out. Academic counselors and other higher education staff who interact with students need to provide the necessary support to identify and help those who have low GPA because they can be considered "at-risk" students. If they can increase the "at-risk" students' GPA, then the likelihood of dropping out decreases.

In addition to academic achievement, the chapter described other factors predicting retention, which are socio-economic background, type of high school attended, negative high school behavior, and the number of friends who dropped out of school. Those with a high socio-economic background are much more likely to persist compared to those with low socio-economic backgrounds. Therefore, colleges and universities need to consider and address this important issue, providing financial support to students struggling with poverty and other socio-economic problems.

The type of high school attended is also an important indicator affecting college persistence. Students from Catholic and suburban high schools are at a disadvantage compared to those from public and urban high schools. For example, spending time on homework during high school predicts college retention because those who spent more time are less likely to drop out of college.

Similarly, the chapter discussed that negative school behavior during high school is of great significance and increases the likelihood of dropping out of college. Students who have more friends who dropped out of high school will have a higher chance of dropping out of college compared to those having fewer friends that dropped out of high school, which means that the peer effect is salient for retention.

Over the past few decades there has been a significant focus on the importance of student persistence and retention among postsecondary institutions in the United States. In addition, American colleges and universities are under pressure to improve retention and graduation rates. If these factors indicated by the study can be addressed for students both during high school and then later in the college career, the likelihood for retention increases which is significant for the individual student and for American society as a whole. As society becomes more educated the standard of living for the majority becomes more satisfactory.

REFERENCES

American Institutes for Research. (2010). *College dropouts cost cash-strapped states billions.* Retrieved from http://www.air.org/news/press-release/college-dropouts-cost-cash-strapped-states-billions

Arnett, J. J. (2000). Emerging adulthood: A theory of development from the late teens through the twenties. *The American Psychologist, 55*(5), 469–480.

Astin, A. W. (1984). Student involvement: A developmental theory for higher education. *Journal of College Student Personnel, 25*, 297–308.

Astin, A. W. (1993). *What matters in college? Four critical years revisited.* San Francisco, CA: Jossey Bass.

Baum, S., Ma, J., & Payea, K. (2010). *Education Pays 2010: The Benefits of Higher Education for Individuals and Society.* College Board Trends in Higher Education Series.

Bennet, C., & Okinaka, A. (1990). Factors Related to Persistence Among Asian, Black, Hispanic, and White Undergraduates at a Predominantly White University: Comparison Between First and Fourth Year Cohorts. *The Urban Review, 22*(1), 33–59. doi:10.1007/BF01110631

Berger, B. J. (2001). Understanding the Organizational Nature of Student Persistence: Empirically-Based Recommendations for Practice. *Journal of College Retention, 3*(1), 3–21. doi:10.2190/3K6A-2REC-GJU5-8280

Berlin, L.G., Furstenberg, F.F., & Waters, M.C. (2010). Introducing the Issue. *The Future of Children*, *20*(1), 3–17. doi:10.1353/foc.0.0042 PMID:20064752

Buchmann, C., & DiPrete, T. A. (2006). The growing female advantage in college completion: The role of family background and academic achievement. *American Sociological Review*, *71*(4), 515–541. doi:10.1177/000312240607100401

Child Trends Data Bank. (2015). Parental Expectations for Their Children's Academic Attainment. Retrieved from https://www.childtrends.org/wp-content/ uploads/2015/10/115_Parental_Expectations.pdf

Grabmeier, J. (2006). Better grades and greater incentives help explain why women outpace men in college degrees. Retrieved from http://researchnews.osu.edu/archive/ womcolge.htm

Hu, S. (2011). Reconsidering the relationship between student engagement and persistence in college. *Innovative Higher Education*, *36*(2), 97–106. doi:10.1007/s10755-010-9158-4

McLanahan, S., Haskins, R., Paxson, C., & Sawhill, I. (2010). Transition to Adulthood. *The Future of Children*, *20*(1).

National Center for Educational Statistics. (2016). *The Condition of Education 2016* (NCES 2016-144).

Nora, A. (1987). Determinants of Retention Among Chicano College Students: A Structural Model. *Research in Higher Education*, *26*(1), 31–59. doi:10.1007/BF00991932

Obama, B. (2010). *Remarks by the President on Higher Education and the Economy at the University of Texas at Austin*. Retrieved from https://obamawhitehouse.archives.gov/the-press-office/2010/08/09/ remarks-president-higher-education-and-economy-university-texas-austin

Pascarella, E. T., & Terenzini, P. T. (2005). *How college affects students: A third decade of research* (Vol. 2). San Francisco, CA: Jossey-Bass Higher & Adult Education.

Sciarra, T. D., Seirup, J. H., & Sposato, E. (2016). High school predictors of college persistence: the significance of engagement and teacher interaction. *The Professional Counselor*, *6*(2), 189–202. doi:10.15241/ds.6.2.189

Settersten, R. A., & Ray, B. (2010). What's Going on With Young People Today? The Long and Twisting Path to Adulthood. *The Future of Children*, *20*(1), 19–37.

Thomas, L. (2002). Student retention in higher education: The role of institutional habitus. *Journal of Education Policy*, *17*(4), 423–442. doi:10.1080/02680930210140257

Tinto, V. (1975). Dropout from higher education: A theoretical synthesis of recent research. *Review of Educational Research*, *45*(1), 89–125. doi:10.3102/00346543045001089

Tinto, V. (1994). *Leaving college: Rethinking the causes and cures of student attrition* (2nd ed.). Chicago, IL: University of Chicago Press. doi:10.7208/chicago/9780226922461.001.0001

Tinto, V. (2004). *Student retention and graduation: Facing the truth, living with the consequences*. Washington, DC: The Pell Institute for the Study of Opportunity in Higher Education.

U.S. Department of Labor. (2015a). Labor Force Statistics from the Current Population Survey. Retrieved from http://www.bls.gov/cps/demographics.htm

U.S. Department of Labor. (2015b). *Earning and Unemployment Rates by Educational Attainment, 2015.* Retrieved from https://www.bls.gov/emp/ep_chart_001.htm

KEY TERMS AND DEFINITIONS

Complex Sampling Design: Probability samples that do not meet the assumptions of simple random samples (SRS) are referred to as complex sampling designs (CSD).

Cronbach's Alpha: Cronbach's alpha is a measure of internal consistency and was utilized in the study to measure how closely related a set of items were as a group.

Cronbach's Alpha Reliability Coefficients: Cronbach's alpha is not a statistical test; rather it is a coefficient of reliability. It ranges from -1 to 1. A higher value of alpha is often considered to be good evidence that the scale items are reliable, and a reliability coefficient of .70 or higher is considered "acceptable" within most social science research studies.

Delta-p Statistic: The delta-p statistic is used to estimate the change in the probability of college dropout associated with a one unit change in each independent variable.

Design Weight: The number of people in the target population represented by that sample member when selected for the study.

The Education Longitudinal Study of 2002 (ELS, 2002): ELS 2002 follows a nationally representative cohort of US based students from 2002, when they were high school sophomores, through their postsecondary education. The researchers collected information from the same students in 2004 when they were high school seniors and 2006 when the majority of the students were in college. One additional follow-up was conducted in 2012, six years after the second follow-up or eight years after high school graduation. The ELS includes 16,197 10th grade students in 752 schools, which represent 3.4 million students as of 2002 in the US.

Logit Model: In the logit model $\ln[p/(1-p)] = a + \beta X + e.$; p is the probability that the event Y occurs, $p(Y = 1)$. $p/(1-p)$ is the "odds ratio". $\ln[p/(1-p)]$ is the log odds ratio, or "logit." The slope coefficient (β) is interpreted as the rate of change in the "log odds" as X changes. The use of the log odds ratio in this research is an interpretation of the logit coefficient, which is usually more intuitive when used as the "odds ratio."

Panel Weight: Adjusts for sample member nonresponses to maintain representativeness of the analysis sample.

Simple Random Sampling (SRS): A subset of a statistical population in which each member of the subset has an equal probability of being selected.

Socioeconomic Status: Socioeconomic status was measured as a composite in ELS based on five equally weighted, standardized components: father's/guardian's education, mother's/guardian's education, family income, father's/guardian's occupation, and mother's/guardian's occupation.

Chapter 4
The Role of Values, Beliefs, and Culture in Student Retention and Success

Julie M Little
Taylor University, USA

Scott Gaier
Taylor University, USA

Danielle Spoutz
Taylor University, USA

ABSTRACT

Culture is comprised of a shared set of values and beliefs, and is known to contribute to organizational success. But how do these factors contribute to individual success within the framework of academia? The purpose of this chapter is to better understand the intersection of values, beliefs, and culture within the area of student retention and success in higher education. As both universities and colleges struggle to identify aspects to increase graduation completion rates in various environments, it is essential to examine the most basic factors that often contribute significantly to this area. This chapter defines individual values and beliefs, the development of shared values, beliefs and culture, the impact of each, and the role of each within the larger topic of student retention and success.

INTRODUCTION

We understand the goal of most colleges and universities to be the successful academic completion of the degree a student may be pursuing. But this simple concept has many facets. In order for a student to successfully complete his or her academic interest, they must persist through the challenges a college experience may offer. For some, this may mean challenges with early adulthood, responsibility, determining one's own self, growing in community with others, or understanding questions and self-doubt that may occur. For others, the challenges may come in a more mature form including balancing work and family, with their academics as well as financial and health issues.

DOI: 10.4018/978-1-5225-2998-9.ch004

Student retention is often discussed as a comparison between these two groups, first-time degree-seeking students and "returning students". Returning students are identified as having spent at least one semester out of school for any number of reasons including personal issues and those related to the institution or college experience. Student retention and success factors are most often tracked using the initial group, first-time degree-seeking students, as it is most desirable for institutions to see them start and finish their degree program within the traditional four-year timeframe.

What compels first-time students to successfully complete their programs? The study of retention factors within this group in recent years is increasing. Various elements have been examined. As Dumbrigue, Moxley & Najor-Durack explain: "retention is a complex personal, social, and academic enterprise. It requires an institutional-community-personal partnership. This partnership links students to proactive support systems" before disruption occurs (2001, p. 12). There are many contributing aspects to student success, which occur at personal and organizational levels, and are deeply rooted in the very essence of the student and the student experience. This chapter seeks to delve into the essential elements of retention, including the individual components of values and beliefs, and the broader institutional concept of culture.

FACTORS OF RETENTION: VALUES, BELIEFS, AND CULTURE

As previously discussed, underlying foundational elements must be present in order for students to persist within their degree programs. These elements include personally held values and beliefs, as well as the development of shared culture. The individual factors of values and beliefs contribute to both the connection and the identification of students with the institution. Meanwhile, the university's culture impacts students on a continuous basis.

To best understand exactly what culture is and how it might contribute to both student retention and success, we must first consider the various definitions of the term. Culture has been defined by many, often from varying perspectives, from psychology to sociology to anthropology. Western society has attempted to define culture from the ethic or minority viewpoint in order to better understand the multicultural world in which we now exist. But as Mantovani (2000) revealed, understanding culture is a "universal problem" (p. 1). For our purposes within this chapter, culture will be considered as the set of shared assumptions, which produce understanding and behavior, resulting in the very identity or essence of the group, including an organization.

The realization that culture is the compilation of several contributing factors including an individual's values and beliefs; and then demonstrated within the framework of norms, artifacts and behavior within a group, provides a deeper understanding of the structure that exists (Hogan & Coote, 2013; Homburg & Pflesser, 2000; Schein, 2010). Culture is known to have a profound impact upon and within a group, whether it be a small independent group or a large complex organization such as an academic institution. Group success has been directly tied to the existing culture. Notably, the concept of "individual-fit" within the culture has become increasingly important and therefore, studied, with respect to retention.

As the study of culture becomes more prominent within several environments including our general society, organizational systems, and within the educational realm, there is greater awareness to the complexity of its structure. As Schein (2010) theorized, culture is a multi-dimensional entity with many layers. It is now understood that culture begins with individual values and beliefs, often those of the leader of the group or the organization, and then manifests itself within the developed norms and behaviors of those within the group. Finally, it is expressed in the ceremonies, rituals, and artifacts the group conducts and displays. Consequently, culture cultivates and directly impacts the institution's ability to promote retention and student success.

VALUES

Values are an elusive concept that can explain many aspects of an individual or community's core elements. Values can be described as a set of ethical standards, meaning, significant resonating concepts, or evaluation structures. In all these descriptions, value is seen as a way of measuring how one lives. Communities and individuals exhibit their values through where they spend their time. Especially in higher education, students exhibit their values through what they do with their free time.

Values are also a component of culture. Kuh and Whitt define values as "widely held beliefs or sentiments about the importance of certain goals, activities, relationships, and feelings" (1988, p. 39). These scholars view values as the second level of culture, contributing to the bond of a community. Many higher education institutions share values such as academic integrity, open access or selective admission criteria and diversity. These values are tightly connected to the beliefs of the university as well, through laying the foundations and structures that allow beliefs to take hold. Between the values and beliefs of an institution, students are provided a context in which they can make sense of their experience at the university. These values also shape the way the community interacts with one another, providing a foundation for social interaction.

When students discover their values, they are able to further develop their identities. Identity development is an essential aspect of a college education. Universities challenge their students to develop their own values and perspectives on difficult topics (Magolda, 2004). Through the development of their values, students achieve growth and self-authorship. Baxter Magolda articulates the importance of values in an individual's self-authorship, as the ability to define one's beliefs, identity, and social relations (Evans et. al, 2010). The process of discovering self-authorship needs to be accomplished in a supportive community, one that allows members to question their values and beliefs in order to refine their identities.

Espoused values are those that are adopted by an individual, but not completely into the individual's mindset as their own. In higher education, these values are seen as trends in the overarching industry (Kuh & Whitt, 1988). Espoused values are typically perceived in the form of wishes rather than concrete values.

Development of Individual to Shared Values

Each university has a unique culture to that campus; and that culture maintains a certain set of values. Similarly, every individual has a culture they personally identify with and appreciate. When a student arrives at a campus, they are introducing their culture to the university, and vice versa. "Acculturation is the dual process of cultural and psychological change that takes places as a result of contact between two or more cultural groups and their individual members" (Berry, 2005, p. 698). Through this exchange, the students shape their campus as much as the campus shapes the students (Kuh, 1990). Acculturation differs from assimilation in that acculturation is an ongoing process (Berry, 2005). The introduction of university values to students socializes them to the culture of the institution (Kuh & Whitt, 1988). These values give way to understanding the recognized behavioral norms, which are established by the university's beliefs.

One of the most important influences on a student is their peer group (Astin, 1993a; Elliot & Healy, 2001). When students enter the college setting, they assimilate to their peer groups in order to fit into the group structure and culture. These peer groups are then assimilated through finding other group members who are most like them. Finally, these peer groups are then adopted into the broader university community through acculturation of the university. Through acculturation, students will be introduced to new values and beliefs that challenge and shape them. This change and shaping process continues throughout the student's college career, as the students' values are changed due to the introduction of new and different values.

An example of how university values influence student values is found in Astin's study on multiculturalism (1993b). In this study, Astin focused on diversity and multiculturalism in higher education. He suggests, if institutions emphasis diversity through both academics and non-academics, students' satisfaction with their college experience increases (Astin, 1993b). The increase in student satisfaction indicates that the student has the adopted the university's values as their own.

Impact of Values

Through defining and assessing the development of values, we understand the importance and impact values have on students. The greatest effect that values have are on the individual's self-concept, or how the student relates with the university, and how the university's values influence students. These are all important considerations for college retention.

Students' values are important to retention for multiple reasons. According to Astin, the college experience is likely to change a student's values and beliefs towards certain subjects (1993a). The university is where a young adult's values are refined, challenged, and then solidified. If a student does not feel like they are developing at a university, they will look elsewhere to receive the growth required. In addition, if a student is not satisfied with their experience, specifically with how the university challenges them or the university's values, they will look for another institution to find that satisfaction (Elliot & Healy, 2001). Satisfaction is then a key element to the student experience. But as Elliot and Healy (2001) explain, "…student satisfaction is generally accepted as a short-term attitude resulting from an evaluation of a student's educational experience. Student satisfaction results when actual performance meets or exceeds the student's expectations…" (p. 2). The difficulty lies in understanding the student's expectations and how the university can then exceed the individual student's expectations. While satisfaction is a short-term measure, it contributes to the student's long-term, overall perception of the university.

Because of the short-term nature of satisfaction, "Retention activities… tend to focus more on how best to keep current students satisfied and coming back year-after-year…" (Elliott & Healy, 2001, p. 9). In a higher education setting, both academics and non-academics are valuable to students. Universities need to emphasis how they challenge students in both realms. The overall educational experience is important to student retention, especially with elements of values displayed in both academic and extracurricular areas.

Lotkowski, Robbins, and Noeth found a difference in the relation of academic and non-academic factors to performance and retention (2004). They found that academic factors have a stronger relationship with retention; rather than values, beliefs, and motivation which tended to relate more directly to performance. Because retention is closely tied with satisfaction, which relates with performance (Elliott & Healy, 2001), academics, values, beliefs, and motivation are all necessary elements to study when institutions assess student persistence. Persistence is best defined as "...voluntary continuation of a goal-directed action in spite of obstacles, difficulties, or discouragement..."; and is closely tied to happiness, or as described for our purposes, student satisfaction (Peterson & Seligman, 2004, p. 229). Research also demonstrates support in the thesis that both academic and non-academic factors play a significant role in a student's identification.

Academics play a large role in student identity development. Through their course work, students are preparing for careers and vocations. Students who identify with their majors are more likely to persist in their education (Perez, Cromley, & Kaplan, 2014). Their persistence is due to the enjoyment of the course content and the perception of adequate preparation for a future career. Those who do not identify with their majors are less likely to persist in that major, and possibly the university, if they do not reach self-authorship. Research supports a correlation between students' academic identification and academic success (Walker, Greene, & Mansell, 2006). Undergraduates who value academic achievement are more likely to stay in college due to the motivation to succeed (Walker, Greene, & Mansell, 2006). However, "Students who master course content but fail to develop adequate academic self-confidence, academic goals, institutional commitment, achievement motivation, and social support and involvement may still be at risk of dropping out..." (Lotkowski, Robbins, & Noeth, 2004, p.10; Gloria, Kurpius, Hamilton, & Willson, 1999; Zaleski, 1988). These students have not successfully linked their values to course work, making the purpose behind attending college and pursuing a degree harder to understand.

As previously mentioned, non-academics also have an influence on student retention. With the integration of non-academics, students are able to receive whole person education, which aids in the satisfaction of the student. Astin also speaks to the role of non-academics in retention, "Retention in college (degree attainment) is enhanced by a number of factors reported in earlier studies: the residential experience, and student involvement with peers and with faculty" (1993a, p.242). The influence of community members has a large impact on a student's identity development and their satisfaction (Lounsbury et al., 2009). In Elliott and Healy's study, the two most influential themes of student satisfaction were student-centeredness, how much students felt welcomed and valued, campus climate, and a sense of belonging to the campus community (2001). Through these themes, students are able to explore and challenge their personal values in refinement for self-authorship.

A student's identification with an institution's values is not the only way to retain students. Universities with a supportive campus culture will find a greater number of students are invested in the institution. With the supportive culture, students are able to question values openly, allowing others to speak into their process of self-authorship. Student's and student groups' values can differ from their institution and many times do (Kuh & Whitt, 1988). This difference allows both parties to reconsider and reevaluate their own values, which may allow for new values to take hold or for current values to grow stronger.

An additional method to encourage retention occurs with university selectivity through the admissions process. Selectivity is very important when assessing the incoming students and faculty members (Kuh & Whitt, 1988). This selectivity allows for the continuation of values that every community member adheres to within the institution. Once there is an establishment of core values, universities can cultivate a culture-rich campus that its students can identify with, thereby, increasing the retention rate of the institution.

BELIEFS

Beliefs are a set of accepted truths where one can place their confidence and trust (Beliefs, 2017). Developing over time and based on an individual's experience, beliefs tend to shape one's worldview or outlook. Based on what one holds to be true, perspectives can differ. When a group of people agree on the same truths, a community is formed. Many private religious universities have a statement of faith as their central declaration of what they hold to be true. This declaration is what the community is founded upon, and the core of who they are as a collective group. While beliefs may be most prevalent in religious institutions, secular institutions also have core beliefs, often in the form of mission statements, that guide both their students and the larger university community.

Beliefs are also a key component of culture. Kuh and Whitt describe beliefs as the third level of culture, behind the level two values and the level one artifacts (1988). "The core of culture is comprised of assumptions and beliefs shared–to some degree–by members of the institution that guide decision making and shape major events and activities..." (Kuh & Whitt, 1988, p. 42). The community beliefs are intended to guide the group's behaviors and values. However, not every member of the community is going to agree with the shared beliefs. While these individuals can still be a part of the community, they will not experience a sense of belonging because of the unconsciously deep nature of beliefs.

The community culture provides a framework for understanding the core beliefs. The artifacts of a culture often display the community's beliefs. Values are seen to be the bridge between artifacts and beliefs (Kuh & Whitt, 1988), giving context to why a community does what it does. Through the use of values to acculturate individuals to the community, learned behaviors are established. These learned behaviors are another element of culture, and they extend the community's beliefs through the actions of the individual members.

Development of Individual to Shared Beliefs

As beliefs are the accepted truths individuals use to shape their views, it is not unusual that over the course of the college experience students' beliefs may change. During their time as a student, young adults are often challenged in their beliefs (both in and out of the classroom); and they will either react by strengthening or adopting different beliefs. Through Astin's study, he found "...students' views on social issues generally tend to change in the direction of the dominant view of the group..." (Astin, 1993a, p. 164). The beliefs of students' surroundings, specifically those of an institution or peer group, heavily impact the students' personal beliefs. The peer group acts as a membership group in which students conform to the established set of beliefs or norms; the institution, or peer group at large, acts as a reference group to which the students will compare themselves (Antonio, 2004). The resulting effect of these groups is a significant impact on a student's personal development throughout their college career.

The college experience itself also tends to build and attract students with liberal beliefs. Astin revealed, "An environment that places a high value on teaching, learning, discovery, artistic expression, independence of thought, critical thinking and freedom of speech and expression naturally tends to attract such people..." (1993b, p. 155). The values and beliefs found within higher education tend to produce similar beliefs in its' graduates. Young adults will be introduced to various perspectives and beliefs, depending on the institution, during the course of their studies. As Astin explains, students will then typically find themselves graduating college with a more liberal perspective on the world, and often, with an altered set of core beliefs.

Impact of Beliefs

As one would imagine, student's beliefs play a large role in their perspective on life, the world, and themselves. As DeWitz, Woolsey and Walsh communicate, "...a college student can feel a fit with the institution, a fit with peers, feel supported by faculty and by students, but without a purpose or goals, the student may lack a clear sense of personal meaning..." (2009, p. 21). Without a purpose or set of established goals, students will be unsure of what to study and where their own strengths fit within the society-at-large. Identifying these strengths is critical to the purpose of higher education: educating and preparing students to graduate as contributing members of society. The discovery of purpose can redefine a student's perspective, give way to deeper personal beliefs, and provide motivation for a student to persist in their education.

Due to the deep connection beliefs have to identity and purpose, students may not completely agree with the beliefs of the university they attend. This concept of partial identification does not seem to have a significant influence on the student's retention, but could eventually contribute to their satisfaction of the university experience. Once a student is registered in college, the major factors that contribute to the overall retention include "...level of academic self-confidence, ...and level of academic and social integration into the institution..." (Lotkowski, Robbins, & Noeth, 2004, p. 4). The level of a student's belonging and self-confidence are the most important factors when considering the impact of beliefs on student retention. These key concepts are found both in academics and non-academics, thereby demonstrating that both the academic and extracurricular realm positively contribute towards retention to graduation.

Students who feel they 'matter' at an institution tend to have a higher sense of belonging. Through expressing their beliefs and still feeling accepted by a community, students will feel that they belong at their institution. They will feel a sense of significance through having other students listen to them and critically think through their perspective. This acceptance increases a student's trust level with their peers and with their institution. Student satisfaction is considerably influenced by the concept of trust (Elliot & Healy, 2000), and can be built upon through exceeding a student's expectations in this core retention concept.

Many scholars have discovered the power of the belief in one's own academic ability in order to succeed in college (Walker, Greene, & Mansell, 2006). Students who lack academic self-confidence tend to lose motivation to persist in their college career (Peterson & Seligman, 2004). On the other hand, those who know themselves well, have set academic and social goals coming into their university experience, and believe in themselves will experience an easier transition into college (DeWitz, Woolsey, & Walsh, 2009). These students understand their purpose in attending college and find their motivation to succeed from within that determination.

Scholars have also found a strong relationship between GPA, academic self-confidence, and achievement motivation (Lounsbury, Fisher, Levy & Welsh, 2009; Lotkowski, Robbins, & Noeth, 2004). The belief in academic performance will put forth a self-confidence that motivates students to achieve a strong GPA within the major field of study. Sometimes these students use comparison as that motivation, or they truly want to work hard and do well for other personal reasons. This belief in academic self-confidence has been identified as self-efficacy by many scholars, and has been continually found as one of the significant sources for predicting retention.

Self-Efficacy Belief

As previously discussed, research has proven that there is one belief that can predict both persistence and retention, and that is the belief of self-efficacy. "Self-efficacy is defined as individuals' confidence in their ability to successfully complete a task…" (DeWitz, Woolsey, & Walsh, 2009, p. 19). This specific belief can be found in both academics and non-academics within the institution. Research indicates confidence in the ability to succeed is an important indicator of retention and student success. The self-efficacy belief has also been linked to the strength of a student's perception on their purpose of life (DeWitz, Woolsey, & Walsh, 2009). Individuals with a higher level of self-efficacy are "…more likely to seek challenges, persist in the face of those challenges, and adopt effective strategies to mediate those challenges…" (Walker, Greene, & Mansell, 2006, p. 4). These students develop a resiliency in all aspects of their lives from their self-efficacy belief (Bandura, 1977).

According to Eccles and Wigfield, to be successful, there are two types of expectations for the individual (2002). The first is outcome expectations, which signifies understanding that behaviors lead to the intended outcomes (Eisenberger, 1992). The second is efficacy expectations, which is the belief that one can accomplish the behaviors required to achieve their goals. Scholars have suggested these efficacy expectations to be connected to persistence. "For example, high personal academic expectations predict subsequent performance, course enrollment, and occupational aspirations choice…" (Eccles & Wigfield, 2002. p. 111). Similarly, performance and persistence in a range of academics relates to the strength in student's self-efficacy beliefs (DeWitz, Woolsey, & Walsh, 2009). Students' self-efficacy concepts are vital components in their college success. Self-efficacy beliefs reach into every aspect of the college experience. If a student's beliefs are not strong within this area, then it could be a central clue or indicator for the university to determine the student's overall persistence.

CULTURE

Culture is an established part of everyday life. It is found within ethnic groups, organizations, and even families. Culture can generally be defined as the behaviors and beliefs characteristic of a particular social, ethnic, or age group. Further clarified, culture is the

…pattern of shared basic assumptions learned by a group as it solved its problems of external adaptation and internal integration, which has worked well enough to be considered valid and, therefore, to be taught to new members as the correct way to perceive, think and feel in relation to those problems. (Schein, 2010, p. 18)

As we think about culture within higher education, culture has been defined as:

A widely shared, generally enduring set of beliefs and values represented by the language people use; institutional norms and established patterns of behavior by students, faculty, administrators, and others; and certain aspects of the physical environment that manifest institutional values and priorities. Taken together, these properties represent accepted patterns of thinking and behaving that tend to become shared among faculty, staff, and students at the institution. (Kuh, Kinzie, Schuh, & Whitt, 2010, p. 341)

Although, it manifests in both behavior and belief, culture is an abstraction that defines, and powerfully guides the group. Schein asserts "...perhaps the most intriguing aspect of culture as a concept is that it points us to phenomena that are below the surface, that are powerful in their impact but invisible and to a considerable degree unconscious..." (2010, p. 14). These below the surface, invisible, and unconscious phenomena shape and guide the very essence of the organization—manifestation of identity, giving rise to protect who we are and what we do, including a resistance to change. Consequently, the culture of an organization is a powerful force for shaping and guiding the organization and individuals who comprise the organization.

A good tool for identifying and assessing culture is to research culture. One design for researching culture is the ethnographic design. According to Creswell, "Ethnographic designs are qualitative research procedures for describing, analyzing, and interpreting a culture-sharing group's shared patterns of behavior, beliefs, and language that develop over time (2012, p. 462). Through the ethnographic design a researcher—or in the case of institutions, a leader—seeks to answer two questions: (1) What must people in this setting know and do in order to make this system work? And (2) How do new comers acquire cultural being? (Creswell, 2013). The two questions should take into consideration the overall retention culture. For example, what must people in this setting [the institution] know and do in order to make this system [of retention] work? Properly understanding the culture should reveal the cultural assumptions regarding retention and student success. Once leadership has identified the shared assumptions for retention, then it is possible to evaluate the overall health and necessary direction of retention projects, including next steps for developing a retention culture through influencing the shared assumptions. In other words, according to Schein "if elements of a culture become dysfunctional, it is the unique function of leadership to perceive the functional and dysfunctional elements of the existing culture and to manage culture evolution and change in such a way that the group can survive in a changing environment" (2010, p. 22).

Development of Shared Culture

Institutional culture characterized by a "...sustained and principled commitment to student success..." is more likely to generate student success (Yorke & Longden, 2004, p. 142). Consequently, if institutions are to develop a retention and student success culture, then leaders need to engage specific activities and create shared experiences for the individuals within the institutions. Since culture is built upon a pattern of basic assumptions, then influencing a culture requires the influencing of these shared assumptions. Although, Schein warns, "...culture change, in the sense of changing basic assumptions, is difficult, time-consuming, and highly anxiety-provoking..." (2010, p. 33). Regardless, if the institution seeks a culture of retention and student success, then retention and student success should be taught, modeled, and rewarded.

Teaching Retention and Student Success

The basic assumptions for retention and student success need to be taught throughout the institution. "All members of the college community need to have a stake in the success of policies and practices to reduce student departure..." (Braxton, Hirschy, & McClendon, 2004, p. 68). According to Kuh et at. "...when they [presidents] and other senior level administrators make student success an institutional priority, the focus of attention of others at the institution, as well as resources, tend to follow..." (2010, p. 335). The first step is to evaluate the current assumptions regarding retention and student success. Determine what is being done—what is beneficial and what needs to go or be changed. The next step is to determine implementation of strategies for retention and student success. What needs to be done? When does it need to be done? And who needs to be involved? Communication, networking, and teamwork is of the upmost importance to foster student success and retention (Gabriel, 2008). Communication will reinforce the culture—is there trust, openness, etc.? Good communication reinforces the values of the organization.

There are a variety of communication strategies that are effective for teaching the institution about retention and student success. These can come from a centralized office for retention, the president, department chairs, faculty, and so forth. The goal is for retention and student success to be so well-defined that people within the organization can teach one other regardless of the individual or the department. This process reinforces culture and the patterns of basic assumptions the comprise culture. Messaging is vitally important because of the inherent confusion associated with retention and the amount of data available for retention. Additional strategies for teaching about retention include campus-wide initiatives, workshops, and departmental meetings.

Modeling Retention and Student Success

According to Scott (2003), one of the central features of organizations is the social structure, "the patterned or regularized aspects of the relationships existing among participants in an organization" (p. 18). Scott draws upon the work of Davis to help understand the social structure. Davis notes the potential of a "...double reality - on the one hand a normative system embodying what ought to be, and on the other a factual order embodying what is..." (1949, as cited by Scott, 2003). In other words, the social structure contains two realities, one governed by what ought to be and the other by what is the actual pattern of relationships among participants. Furthermore, Scott establishes the cultural-cognitive as another component of organizations. Consequently, the social system contains three components: normative structure, cultural-cognitive structure, and behavioral structure. A simplified way of understanding these components are as norms, beliefs, and behaviors. They are interconnected, and as Scott points out, these "...structures are always in a state of dynamic tension - each existing and changing somewhat independently of the other while at the same time exerting continuing influence on the others..." (2003, p. 20).

Consequently, as institutions seek to generate a culture of retention through the changing of shared assumptions, they should seek to establish clarity and congruence in the norms, beliefs, and behaviors of the organization as it relates to retention and student support objectives. As mentioned in the previous section, it's important that these stated elements are accurate, accessible or in other words, easily understood and clear. Hence, without clear normative structure it is difficult for the institution to operate with quality and efficiency; and the institution will be marked with confusion and chaos. On the other hand, behavioral structure is marked by the collective basic understanding of what the organization is all about–both how to get things done and how to navigate within the organization. Often the behavior is not consistent with, or can even be contrary to the normative structure: what actually is happening is not congruent with what ought to be. Organizations may intend things one way, but operate much differently.

Creating a culture of retention requires congruence and consistency between the normative, cultural-cognitive, and behavioral structures. This happens through communication and modeling. Leaders should be able to communicate with clarity the cultural assumptions, and model these as well. What is modeled contributes to what actually happens. Leaders who care about retention and student success engage in activities and interactions, which promote these throughout the culture. "Through their actions and words, senor administrators and faculty model preferred ways [emphasis added] of interaction, making decisions, and responding to challenges. Equally important, every day, individuals make small gestures that create and sustain a caring community for students…" (Kuh et al., 2010, p. 157).

Rewarding Retention and Student Success

The things institutions value most, they reward. If an institution wants to generate a culture of retention, then retention needs to be rewarded (Tinto, 2006). Reward can include a variety of strategies. One strategy can be awareness. Communicating throughout the organization what individuals or departments are doing to increase retention helps to not only encourage the institution wide retention initiative and culture, it also serves as a way to edify and encourage the individuals who are contributing to the work. This can also serve to increase morale. Lack of mention or even praise can negatively impact retention by communicating that retention is not an important or praiseworthy mission. Addition strategies can include standard pay and promotion procedures, tenure requirements, and service awards.

BUILDING A CULTURE OF RENTENTION: RIGHT DATA AND RIGHT THINKING

One of the challenges with generating a culture of retention and student success is the complexity and multiplicity of approaches to student care and retention. Furthermore, there are external pressures from governments, accreditation organizations, and so forth. There are also internal demands such as dependence upon retention revenue to meet budget requirements. Needless to say, retention and student success possess a myriad of elements and pressurized demands. As pressure mounts and complexity grows, the ability to think clearly becomes difficult. What is needed to maintain the health of an organization is the pursuit of right data characterized by thinking rightly about data—as McAfee and Brynjolfsson note "…big data's power does not erase the need for vision or human insight…" (2012, p. 65) while Yorke and Longden warn that retention data must be used "wisely" (2004, p. 74).

Right Data

The information age has created a wealth of data—data mining, big data, predictive analytics, and so forth—data that should be utilized to form strategies and decision making (Daniel, 2015; Kuh et al., 2010). The wealth of data available provides angles that we were never before able to access, and insights that were formerly impossible. Unfortunately, the amount of data also creates a significant potential for bias and noise—also known as veracity (Daniel, 2015). Having access to an abundance of data can also generate the false expectation that more data will deliver better results. Furthermore, researchers perceive there remains a willingness to rely on experience and intuition instead of data (McAfee & Brynjolfsson, 2012). A healthy retention and student success culture uses data in order to make informed decisions (Kuh et al., 2010). Right data includes identifying and collecting data that is thorough, accurate (i.e., valid), and current.

Thorough data collection is an emphasis on the breadth and depth of data necessary to understand retention. The primary threat to thorough data is incomplete data. Accurate data collection focuses upon identifying and collecting data that is factual. "If data is to be used to make accurate decisions in time it becomes necessary that it should be available in accurate, complete and timely manner" (Katal, Wazid,

& Goudar, 2013, p. 406). The primary threat to accurate data is false data. Current data collection emphasizes data that is the latest or most recently collected. The primary threat to current data collection is obsolete data.

Data collection is the foundation and backbone for data analysis. Cleansing and organizing data is important for getting right data (McAfee & Brynjolfsson, 2012). Additionally, there are several methods for analyzing data. These range from qualitative designs emphasizing thick, rich descriptions; to quantitative descriptive designs geared for describing the data; to quantitative inferential designs focusing on making inferences from the data (Creswell, 2012, 2013). Without quality data, all of these designs will lead to inferior results; and in turn, to poor decision making, an unhealthy culture, and a lack of focus on retention.

Right Thinking

Data is merely a tool. As mentioned previously, the amount of data provides excellent potential for problem-solving. It also provides opportunity for noise, confusion, and false expectations that problems can be resolved with more and more data. Data is only information. What we do with this information, and more importantly how we think about the information, will ultimately determine the answers to our problems (McAfee & Brynjolfsson, 2012). Incorrect thinking will lead to wrong solutions regardless of the quality of data. Consequently, institutions must strive to think rightly about data. Thinking rightly includes the aspects of truth-seeking, integrity, humility and a willingness to fail.

Truth-Seeking

Because of the wide variety of data available and the breadth and depth of data, it becomes necessary to be truth-seekers of the data. Truth-seeking is a mindset or disposition associated with critical thinking and is defined as "being eager to seek the best knowledge in a given context, courageous about asking questions, and honest and objective about pursuing inquiry even if the findings do not support one's self interests or one's preconceived opinions" (Facione, Sánchez, Facione, & Gainen, 1995, p. 8). Truth-seeking, pertaining to thinking rightly about data, involves identifying all possible data, weighing the data, and trusting the data. Identifying data requires correct, appropriate, and quality data (i.e., research) designs. Individuals responsible for institutional research should have training in research design including qualitative, quantitative, and mixed methods. The quality of data will be only as good as the quality of the design. Poor designs threaten data. Truth-seekers rely on quality data designs and appropriate data analysis to inform how they think about data. They strive for thoroughness, ask good questions, and are good listeners. They ultimately seek to weigh the data and understand exactly what the data has to say. The primary threat to truth-seeking can come from multiple areas including one's own ego, a willingness to be satisfied with something short of the entire truth, willingness to be deceived by others and our self, and an overall lack of understanding of data design, analysis, and interpretation.

Integrity

One of the primary threats to right thinking is the deception to self and others of one's own thoughts. Hence, the best lies are 99 percent true. Thinking rightly about data requires integrity. Integrity can be defined in various ways. For the purpose of this chapter, the concept of self-awareness and honesty are combined to produce the concept of integrity. From a basic dictionary rendering, integrity is often synonymous with honesty. Whereas, self-awareness is a component of emotional intelligence (Goleman, 2004). Taken together, those who possess self-awareness "are honest—with themselves and with others" (Goleman, 2004, p. 3). As it pertains to right thinking about data, self-awareness is knowing what we are thinking. Awareness directly combats mindlessness and ignorance. Often, there are times that we are not aware of what we are thinking (i.e., mindlessness) or succumb to the temptation of ignoring what we truly are thinking (i.e., ignorance). Once we develop awareness, much like pursuing truth, we pursue honesty. Honesty is being truthful with one's self about our knowledge of our thoughts. In essence, we own what we are thinking. This intertwined process of self-awareness and honesty manifesting in integrity helps us to think rightly

Willingness to Fail

Retention and student success have so many varying inputs, possibilities, and solutions. All of this variability can create a lack of clarity about what needs to be done or what direction an institution should head in order to increase success. Furthermore, retention has turned into a high-stakes, pressurized subject. Many institutions often rely upon high retention rates in order to help fund the institution or to meet budget. Taken together, the complexity of retention and the pressurized situations, it can be extremely challenging to think rightly about the data. The 'noise' of the data, the drumming of talking heads, and a variety of options and solutions can create confusion in the best leaders. The unknown can often be perceived as a fearful and dangerous place where mistakes may consume the leader. However, in an environment of uncertainty, a leader must take responsibility and move the organization forward. Consequently, what is necessary is a leader's ability to wade into the unknown, and perhaps even fail. As long as the leader stays true to this process leaders continue to learn more and more about data, and how to think rightly about retention and student success.

Leaders who are unwilling to learn from mistakes or even embrace failure will struggle to think rightly about data. Burger and Starboard (2012) assert that leaders need to be able to know how to fail in order to improve their leadership and their responsibilities. One strategy for helping leaders to utilize failure more effectively is to fail forward. Failing forward is an emphasis on learning from mistakes and failures along the way so that learning takes place - mistakes and failures will most certainly happen as a part of the natural learning process. Leaders realize that the correct or best answer lies ahead and cannot be identified nor implemented without the tutoring that comes from mistakes and failures. One of the best concepts or strategies that is associated with the importance of failure as a part of the learning process is Schwartz's concept of productive stupidity - an emphasis on choosing to be ignorant, wrestling with the unknown, yet learning along the way (2008). Schwartz concludes "The more comfortable we become with being stupid, the deeper we will wade into the unknown and the more likely we are to make big discoveries…" (2008, p. 1771).

Without a mindset that embraces and learns from mistakes or a willingness to fail, it becomes very difficult to think rightly about data. Finally, keep in mind, that one of the main challenges or threats to thinking rightly about data is either our insecurities or our ego. Both point to our unwillingness to be wrong and prevents further learning.

Humility

Humility is the ability to place the institution and students' needs as more significant than one's own interests. Jim Collins (2001) in his book *Good to Great: Why Some Companies Make the Leap and Others Don't* detailed characteristics which differentiated or distinguished great leaders from good leaders. He notes that the very best leaders were marked by fierce resolve and humility; leaders who were very driven and embodied a deep willingness to put the company's needs before their own. If we strive to have our institutions' and our students' best interests in mind, then the likelihood to think rightly about data or interpreting data increases. Otherwise, our own pride or arrogance gets in the way of how we think about and interpret data. Lack of humility goes hand-in-hand with an unwillingness to be wrong. It is very difficult, if not impossible, to think rightly about data if we insist on being right or having things our own way. When the goal is turned into what is best for the institution and for our students, not who gets credit, then we typically are unrestrained and can do great things as we think about data.

Summary of Recommendations on Culture

If you want to create a culture of retention and student success, then you need to identify the current shared assumptions in the culture regarding retention and student success. Then determine if these shared assumptions are consistent with the shared assumptions necessary for a healthy or (your) ideal retention and student success culture. Identify what can remain and what needs to change in the process. Remember influencing and changing the shared assumptions within a culture is hard work, takes time, and can create anxiety. Once leaders identify the shared assumptions for the retention and student success culture, these shared assumptions should be taught, modeled, and rewarded. Furthermore, any retention culture should be built upon right data and characterized by people who think rightly about the data.

The primary focus of this section was to equip practitioners with the necessary content, tools, and information for generating a culture of retention and student success. For thorough understanding of culture and leadership, readers are strongly encouraged to review Edgar Schein's book *Organizational Culture and Leadership* (2010). Schein, the Society of Sloan Fellows Professor of Management Emeritus at the MIT Sloan School of Management, is one of the foremost thinkers and experts on organizational culture. Furthermore, readers are encouraged to understand how to research culture for themselves to aid them in their efforts within their institution. John Creswell's books - *Qualitative Inquiry & Research Design: Choosing Among the Five Approaches* (2013) and *Educational Research: Planning, Conducting, and Evaluating Quantitative and Qualitative Research* (2012) - contain excellent material for understanding how to research culture via ethnographic research.

A FINAL THOUGHT: EMPHASIS ON DIVERSITY

One final aspect related to this chapter's discussion uncovered by Astin was a trend in the university's emphasis on diversity and student satisfaction within their culture (1993a; 1993b). He shared in one study,

...emphasizing diversity either as a matter of institutional policy or in faculty research and teaching, as well as providing students with curricular and extracurricular opportunities to confront racial and multicultural issues, are all associated with widespread beneficial effects on a student's cognitive and af-

fective development. In particular, such policies and experiences are associated with greater self-reported gains in cognitive and affective development (especially increased cultural awareness), with increased satisfaction in most areas of the college experience, and with increased commitment to promoting racial understanding. (Astin, 1993b, p. 154)

This discovery touches each area that has been examined through this chapter: culture, values, and beliefs. Through both academic and non-academic realms, students need to be introduced to perspectives that are different from their own. This introduction is essential to their development as a contributing member of society, to one's self-authorship, and to a student's detection of their purpose. "All of us are cultural beings, with culture influencing the development of our beliefs, perspectives, and behaviors" (Rodriguez, 1995, p. 33). We now live in a post-modern, multicultural, pluralistic society. In order to prepare students well, they need to engage in diversity issues to understand where they 'fit' into the larger conversation (or culture). The university's emphasis on diversity supports what students need through their satisfaction level, and this will be positively reflected in their overall retention.

FUTURE RESEARCH DIRECTIONS

A longitudinal study on both academic and non-academic impact on retention in respect to culture, values, and beliefs development is required. Ideally, this study would be comprised of at least three to five institutions of similar size and organization, and track student responses for their full term of education. Methods utilized would include a survey with short answer options where appropriate. A pilot study at one institution could be conducted to further develop and refine the survey instrument before the larger, longitudinal study launch. Results would aid researchers in identifying specific areas of both the academic and non-academic areas impact on culture, values, and belief development within individual college students, while providing further understanding of those areas in relation to retention.

Additional study in the area of university leadership's influence in regard to culture, values, and beliefs development within institutions with improved retention. The development of case studies centered around leadership practices, not strategies, within these institutions would enhance the understanding of leader impact within the contributing areas (culture, values, and beliefs) and their overall effect on student retention. Surveys, personal interviews, and focus groups would provide adequate triangulated data for researchers to begin to determine leader affect.

CONCLUSION

The continued examination of the factors of student retention and success within the academic environment leads us back to a foundational core of elements that are often the very basis of student satisfaction. As this chapter discussed, both individual and organizational components contribute to this initiative. At the individual level, values set the stage for a student's standards, derived meaning, and evaluation method. Development of shared values occurs as a student begins the process of challenging their own personal values with those of the institutions' as well as their newly found peer groups. Beliefs, themselves, provide a viewpoint in which individuals perceive their circumstances. As values are often the underpinning for beliefs, both work together to provide a student with the necessary social environment and context. These elements begin to further the student's development of, and provide them access to, their acculturation of the institution. Finally, culture results in the culmination of both the concepts of individual to shared values and beliefs. It also provides the framework for appreciation of core beliefs, which leads to the use of artifacts, linked together by individual and shared values.

Culture, a concept that is found in various facets from ethnic groups to organizations, provides the means for administrators to address student retention and success issues. As Aguirre reinforces this objective, "...culture thus socializes persons to a common worldview, perceptions of the environment, and value of orientation..." (2000, p. 34). Culture, itself, gives the institution opportunity to impact student retention by providing an environment that not only appreciates the concepts of student retention and success, but also firmly grounds its retention practices so that they are taught, modeled, and rewarded. The institution must also practice the ideas of "right data" and "right thinking" as collecting and analyzing data can often be an overwhelming process at best. The manner in which data is collected must be thorough, accurate, and current. Likewise, the method of analysis of that same data must involve the aspects of truth-seeking, integrity, a willingness to fail, and humility. As institutions practice these concepts in efforts to continue to develop a culture of student retention and success, they avoid the pitfalls of implementation concerning wrong solutions and wasted efforts.

REFERENCES

Aguirre, A. Jr. (2000). *Women and Minority Faculty in the Academic Workplace: Recruitment, Retention, and Academic Culture. ASHE-ERIC Higher Education Report, 27(6).*

Antonio, A. L. (2004). The influence of friendship groups on intellectual self-confidence and educational aspirations in college. *The Journal of Higher Education*, *75*(4), 446–471. doi:10.1353/jhe.2004.0019

Astin, A. W. (1993). *What matters in college?: Four critical years revisited* (Vol. 1). San Francisco: Jossey-Bass.

Astin, A. W. (1993). Diversity and multiculturalism on the campus: How are students affected? *Change: The Magazine of Higher Learning*, *25*(2), 44–49. doi:10.1080/00091383.1993.9940617

Bandura, A. (1997). *Self-efficacy: The exercise of control*. New York: Freeman.

Berry, J. W. (2005). Acculturation: Living successfully in two cultures. *International Journal of Intercultural Relations*, *29*(6), 697–712. doi:10.1016/j.ijintrel.2005.07.013

Braxton, J. M., Hirschy, A. S., & McClendon, S. A. (2004). *Understanding and reducing college student departure* (A. J. Kezar, Ed.). (Vol. 30). San Francisco, CA: Wiley Periodicals, Inc.

Burger, E. B., & Starbird, M. (2012). *The 5 elements of effective thinking*. Princeton, NJ: Princeton University Press.

Collins, J. (2001). *Good to great: Why some companies make the leap...and others don't*. New York: HarperCollins Publishers Inc.

Creswell, J. W. (2012). *Educational research: Planning, conducting, and evaluating quantitative and qualitative research* (4th ed.). Boston, MA: Pearson Education, Inc.

Creswell, J. W. (2013). *Qualitative inquiry and research design: Choosing among five approaches* (3rd ed.). Thousand Oaks, California: Sage Publications, Inc.

Daniel, B. (2015). Big Data and analytics in higher education: Opportunities and challenges: The Value of Big Data in Higher Education. *British Journal of Educational Technology, 46*(5), 904–920. doi:10.1111/bjet.12230

Davis, K. (1949). *Human Society*. New York, NY: Macmillan.

DeWitz, S. J., Woolsey, M. L., & Walsh, W. B. (2009). College student retention: An exploration of the relationship between self-efficacy beliefs and purpose in life among college students. *Journal of College Student Development, 50*(1), 19–34. doi:10.1353/csd.0.0049

Dumbrigue, C., Moxley, D., & Najor-Durack, A. (2001). *Keeping students in higher education: Successful practices and strategies for retention*. Routledge.

Eccles, J. S., & Wigfield, A. (2002). Motivational beliefs, values, and goals. *Annual Review of Psychology, 53*(1), 109–132. doi:10.1146/annurev.psych.53.100901.135153 PMID:11752481

Eisenberger, R. (1992). Learned industriousness. *Psychological Review, 99*(2), 248–267. doi:10.1037/0033-295X.99.2.248 PMID:1594725

Elliott, K. M., & Healy, M. A. (2001). Key factors influencing student satisfaction related to recruitment and retention. *Journal of Marketing for Higher Education, 10*(4), 1–11. doi:10.1300/J050v10n04_01

Evans, N. J., Forney, D. S., Guido, F. M., Patton, L. D., & Renn, K. A. (2010). *Student development in college: theory, research, and practice*. San Francisco: Jossey-Bass.

Facione, P. A., Sánchez, C. A., Facione, N. C., & Gainen, J. (1995). The disposition toward critical thinking. *The Journal of General Education*.

Gabriel, K. F. (2008). *Teaching unprepared students: Strategies for promoting success and retention in higher education*. Sterling, VA: Stylus Publishing, LLC.

Gloria, A. M., Kurpius, S. E. R., Hamilton, K. D., & Willson, M. S. (1999). African American students' persistence at a predominantly white university: Influence of social support, university comfort, and self-beliefs. *Journal of College Student Development, 40*, 257–268.

Goleman, D. (2004). What makes a leader? *Harvard Business Review, 82*(1), 1–11. PMID:10187249

Hogan, S. J., & Coote, L. V. (2013). Organizational culture, innovation, and performance: A test of Schein's model. *Journal of Business Research*. doi:10.1016/j.jbusres.2013.09.007

Homburg, C., & Pflesser, C. (2000). A multiple-layer model of market-oriented organizational culture: Measurement issues and performance outcomes. *JMR, Journal of Marketing Research, 37*(4), 449–462. doi:10.1509/jmkr.37.4.449.18786

Katal, A., Wazid, M., & Goudar, R. H. (2013). Big data: Issues, challenges, tools and good practices. *Presented at the International Conference on Contemporary Computing* (IC3), Noida, India. Retrieved from https://doi.org/http://ieeexplore.ieee.org/document/6612229/

Kuh, G. (1990, November). Influence of collegiate environments on student learning. *Paper presented at the Association for the Study of Higher Education Annual Meeting*, Portland, Oregon.

Kuh, G. D., Kinzie, J., Schuh, J. H., & Whitt, E. J. (2010). *Student success in college: Creating conditions that matter*. San Francisco, CA: Jossey-Bass.

Kuh, G. D., & Whitt, E. J. (1988). ASHE-ERIC Higher Education Report: Vol. 1. *The Invisible Tapestry. Culture in American Colleges and Universities*. Washington, D.C.: Association for the Study of Higher Education.

Lotkowski, V. A., Robbins, S. B., & Noeth, R. J. (2004). *The Role of Academic and Non-Academic Factors in Improving College Retention. ACT Policy Report*. American College Testing ACT Inc. doi:10.1037/e420492008-001

Lounsbury, J. W., Fisher, L. A., Levy, J. J., & Welsh, D. P. (2009). Investigation of character strengths in relation to the academic success of college students. *Individual Differences Research*, *7*(1), 52–69.

Magolda, B. (2004). Self-authorship as the common goal. In *Learning partnerships: Theory and models of practice to educate for self-authorship*.

Mantovani, G. (2000). *Exploring borders: Understanding culture and psychology*. London: Routledge.

McAfee, A., & Brynjolfsson, E. (2012). Big data: The management revolution. *Harvard Business Review*, *90*(10), 59–68. PMID:23074865

Merriam-Webster.com. (2017). Beliefs. Retrieved April 18, 2017, from https://www.merriam-webster.com/dictionary/beliefs

Perez, T., Cromley, J. G., & Kaplan, A. (2014). The role of identity development, values, and costs in college STEM retention. *Journal of Educational Psychology*, *106*(1), 315–329. doi:10.1037/a0034027

Peterson, C., & Seligman, M. E. P. (2004). *Character strengths and virtues: A handbook and classification*. Washington, DC: American Psychological Association.

Rodriguez, B. M. (1995). Creating inclusive and multicultural communities: Working through assumptions of culture, power, diversity, and equity. *Multicultural education: Strategies for implementation in colleges and universities, 4*, 33-40.

Schein, E. H. (2010). *Organization culture and leadership* (4th ed.). San Francisco, CA: Jossey-Bass.

Schwartz, M. A. (2008). The importance of stupidity in scientific research. *Journal of Cell Science*, *121*(11), 1771. doi:10.1242/jcs.033340 PMID:18492790

Scott, W. R. (2003). *Organizations: Rational, natural, and open systems* (5th ed.). Upper Saddle River, NJ: Pearson Education, Inc.

Tinto, V. (2006). Research and practice of student retention: What next? *Journal of College Student Retention*, *8*(1), 1–19. doi:10.2190/4YNU-4TMB-22DJ-AN4W

Walker, C. O., Greene, B. A., & Mansell, R. A. (2006). Identification with academics, intrinsic/extrinsic motivation, and self-efficacy as predictors of cognitive engagement. *Learning and Individual Differences*, *16*(1), 1–12. doi:10.1016/j.lindif.2005.06.004

Yorke, M., & Longden, B. (2004). *Retention and student success in higher education*. Berkshire, England: Open University Press.

Zaleski, Z. (1988). Close relationships and acting for self-set goals. *European Journal of Social Psychology*, *18*(2), 191–194. doi:10.1002/ejsp.2420180210

KEY TERMS AND DEFINITIONS

Acculturation: The on-going process of cultural and psychological change due to the interaction of both different groups and/or individuals.

Beliefs: A set of accepted truths where one can place their confidence and trust.

Culture: Set of shared assumptions which produce understanding and behavior, resulting in the very identity or essence of the group, including an organization.

Graduation Completion: Often related to the statistics and metrics used to calculate those that finish degree pursuits at any given institution.

Student Retention and Success: The ability of the institution to maintain student enrollment and have students continue through to graduation.

Student Satisfaction: The capability of student goals to be fulfilled.

Values: A way of measuring how one lives through a set of ethical standards, meanings, significant resonating concepts, or an evaluation structure.

Section 2

Supporting and Encouraging Student Success

Chapter 5
Factors That Influence the Retention of Freshmen Students at Historically Black Colleges and Universities: A Theoretical Assessment

Samuel L. Hinton
Independent Researcher, USA

ABSTRACT

The research study was guided by such theories as social interactionist, student involvement framework, and student attrition framework that were to form the frame of references or foundation for the study. The social interactionist framework used students' backgrounds characteristics; shared values and a friendship to explain behavior while the student attrition framework used intentions attitudes and believe. In integrating the two theories, the researcher was able to predict or explain the study outcomes and link them to the existing body of knowledge. This research presented the theoretical framework of the study on factors that affected retention among freshmen students at Historically Black Colleges and Universities. The research problem in this chapter was factors that affected retention among freshmen students at HBCUs. This research problem anchored the entire study and formed the basis of constructing the study's theoretical framework. The key variables in this study were the freshmen students, theories, retention rates, and the factors affecting retention.

FACTORS THAT INFLUENCE THE RETENTION OF FRESHMEN STUDENTS AT HBCUS: THEORETICAL EXPLANATION

This chapter outlines the issues and challenges that influence the retention of freshmen students at Historically Black Colleges and Universities (HBCU's). A brief review of the literature yields numerous theoretical frameworks that have been studied and applied to determining the various factors that

DOI: 10.4018/978-1-5225-2998-9.ch005

affect the rate at which HBCUs are capable of retaining Black students. This chapter explores various studies, approaches and strategies that have been utilized to substantiate solutions and that underscore the importance of retention work specific to the HBCU setting. Exploring and understanding the work of a range of scholars and experts strengthens the perspective and recommendations outlined within the paper. This chapter methodically outlines these theories, their importance, how they apply to the topic of retention, how they apply to first-year students, and then explores the application of these theories and approaches to students at HBCUs.

LITERATURE REVIEW

The first year of a student's college career is critical for long-term success and instrumental for future positive academic experiences. Students incorrectly perceive their admittance to college as the single most difficult element of the college-going process, but first-year students are often confronted with negotiating new responsibilities as well as developing a new identity as a college student. In the first year, students make the transition to college; hence, they must find their place within the greater campus community. Notably, there has to be social, academic and emotional adjustments, which depend on the students' ability to manage their own limited resources, their course work as well as changing family relationships. Broadly speaking, freshman success entails developing academic and intellectual competencies, negotiating relationships, developing own identity, establishing a healthy lifestyle, pursuing a career choice, and developing a personal sense of civic duty.

The first-year in college is a challenging period for students, and it is particularly difficult for non-traditional students who may have come from underprivileged backgrounds. One of the important obstacles for these groups of students is a social, intellectual, and cultural disequilibrium within the college environment. Disequilibrium challenges students' earlier assumptions as well as beliefs as they meet new people, ideas, and experiences. Secondly, the tension between challenge and support is equally important for traditionally under-represented groups. Institutions have the chance to express commitment to their students by continually communicating the high level of concern that exists for the well-being and development of its students. Ability of a student to adapt to college will depend more on the student's perception of an institution's level of commitment to the welfare of its students (Braxton, Hirschy & McClendon, 2004). Nicpon et al. (2007) noted that students with high-perceived level of social support are less likely to feel lonely, but commit to higher levels of educational goals. However, another study by Conley & Hamlin (2009) revealed that a student of color who is from first-generation background encounter compounded obstacles to completing their college degree because most of college campuses expect such students to adapt successfully on their own to the new, often puzzling and culturally alienating environment.

Among freshmen Mexican American and Latino students, a successful first year is noteworthy. It is because of demographic trends as well as persistent educational gap between Mexican Americans and other Latinos, and Latinos with the White majority. Even though Latinos make up the largest as well as the fastest growing minority group in the US, their level of educational attainment remains lowest among other ethnic groups (Liu, 2011). Therefore, rising Latino enrollment presents a unique opportunity for HBCUs to attract these students to their institutions. HBCUs have, for a long time, been known for offering an education that not only uniquely meets black students' aspirations, desires and goals but is also culturally relevant to them (Esmieu & Martinez, 2014). Previous studies have, however, failed to present the HBCU context as a viable environment for first-year students. HBCUs remain some of the most academically supportive environments for students because of the increased faculty attention as well as encouragement. Besides, campus environment is often culturally supportive and welcoming to diverse students.

HBCUs are the only institutions that were created in the United States for the purpose of educating Black citizens. Until the mid-1960s, HBCUs were, with very few concessions, the only higher education choice for most African Americans. However, the push to have the HBCUs integrated with those that have been admitting the whites all along as the Civil Right groups were agitating led to a decrease in the enrollment to HBCUs. In addition, there was a shift of the initial focus to have every middle-class black obtain education (Gasman, 2013). Currently, the enrollment threshold for HBCUs is nearly 11 percent of Black students within the United States (Gasman, 2013). This is irrespective of the fact that they account less than three percent of the higher learning institutions within the country. Some of these institutions are strong and flourishing while others are struggling.

HBCUs are degree-granting institutions that were established before 1964 with the main objective of educating Black Americans. Currently, there are 100 HBCUs in 19 states, 51 of which are public while 49 are private nonprofit institutions. From 1976 to 2014, the enrollment at HBCUs increased by 32 percent from 223,000 to 294,000 (National Center for Education Statistics, 2016). Even though HBCUs were initially founded to educate Black students, they have traditionally enrolled other students apart from Black Americans. This diversity has increased with time. By 2014, there were 21 percent of non-Black students enrolled in HBCUs in the US, compared to 15 percent in 1976. Among Black students that are enrolled in institutions that grant degree, the percentage of those enrolled at HBCUs dropped over time, from 18 percent that was recorded in 1976 to 8 percent in the year 2014 (National Center for Education Statistics, n.d).

Dwyer (2006) lists some of the positive outcomes of joining HBCUs as higher retention rates than in predominantly white institutions, better psychosocial development, high level of satisfaction with the activities that are offered in campus, greater academic growth as well as maturity. Moreover, students relate better with staff and can look forward to a higher degree. In another research, Flemming (2004) compared black students learning in HBCUs with their peers in predominantly white institutions (PWIs) and recorded that HBCUs are more effective than PWIs in promoting high aspirations and confidence, and a higher level of intellectual engagement among blacks in HBCUs than those in PWIs. Moreover, research by Gilbert et al. (2006) involving 154 African-American undergraduates attending an HBCU recorded that as opposed to African-African students at predominantly white institutions where racial identity was found to be indicative of psychological stress, those that study in institutions where they were not in the minority did suffer in that same way. Harper's (2012) research highlighted the results of the pressures that students experience because of "loneliness" or the psychological burden of having to navigate a racially politicized space that is occupied few peers, role models or even guardians from one's same ethnic or racial group. In the study, students in PWIs only were found to be overwhelmed by the pressure of having to be the spokesperson for the minorities, particularly blacks, and their loneliness negatively affected their achievement and sense of belonging. The problems identified with the PWIs are likely to be the reason behind why more non-Blacks are joining HBCUs.

Today HBCUs across the nation have at least 20% non-Black student population. Consequently, there have been some worries that the changing composition in these colleges and universities endangers the aspects of the institutions that make them unique (Gasman, 2013). For instance, in 1950, Blacks made up almost all HBCU enrollment, but by 1980 they represented 80% of total enrollment. Likewise, Latino's enrollment at HBCUs has increased in the last 30 years, especially in the regions of the country where the Latino population is expanding rapidly. In 2011, Asian American enrollment in HBCUs had increased by 60% to 4,311 from 2001. White student enrollment in HBCUs ranged between 10-13% in the last 20 years (National Center for Education Statistics, 2011). Female undergraduate students still outstrip male undergrads at HBCUs. The lack of balance in student enrollment is associated with significant gender imbalance on individual HBCU campuses.

HBCUs are playing a significant role in educating their students to become global citizens. To achieve this objective, the institutions provide study abroad programs as well as more foreign language minors and majors. The study abroad programs diversify students' learning, broaden their perspectives, and enhance cultural understanding (Twombly, Salisbury, Tumanut & Klute, 2012). Twombly and colleagues further report that lower income and students from Black families are privileged to go abroad for their studies at a comparatively lower cost than their white colleagues as well as those whose families belong to upper middle class. However, the researchers pose a complex question on how students and parents at HBCU's can be helped to identify new funding opportunities to support study abroad opportunities for their students.

Another important issue is the need to learn new languages to enable HBCU graduates to pursue employment in a labor market that requires global perspective and global competencies such as multiple language skills. Notably, foreign languages that are offered at HBCUs include Spanish, French, Chinese, Japanese, and Arabic, which are all important to international commerce. However, merely offering the foreign languages is not enough. There is the need for HBCUs to have the languages-related programs enhanced through giving the students and opportunity to study diversity of languages (Campbell-Whatley, 2003; Valle-Riestra, Shealey & Cramer, 2011).

THEORETICAL FRAMEWORK

It is important to understand the backgrounds of the key theories on first year experience approaches. For instance, the reader might want information why certain students may have not been successful and their unique experiences at their HBCUs. The theories are only as useful as the evidence that is provided. It is crucial that available data supports the offered approaches to increasing student retention, and subsequently, increased enrollment of students in HBCUs. -

Table 1.

Ethnic/Racial Group	Percentage Enrollment in HBCUs
Black/African American	76
White	13
Race/Ethnicity Unknown	5
Hispanic/Latino	3
Asian, Native Hawaiian, American Indian	1
Two or more races	1
Non-resident Alien	
Source: (Gasman, 2013)	

The study will focus on three of these frameworks for effectiveness. These will include: (a) the interactionist framework, (b) the student involvement framework, as well as the (c) student attrition framework. The interactionist theory focuses on background characteristics, shared group values, friendship, as well as how they influence the retention of students in various institutions of learning (Siedman, 2009).

THE INTERACTIONIST FRAMEWORK

Carter & Fuller (2015) explain the theory of symbiotic interactionism, as a framework and perspective in sociology that explains how society is created and maintained through repeated interactions among individuals. The emergence of the theory was because of the need to respond to the principle perspectives that dominated sociology in the when. The main positivist approaches that were previously applied tended to examine society from the top-down standpoint, focusing on the effect of macro-level institutions as well as social structures, and how the structures impose and constrain individuals. However, symbolic interactionism was created to understand the operation of society from the bottom-up, shifting the focus to micro-level processes that arise during face-to-face encounters so as to explain the operation of society. The basic tenets of symbolic interactionism, when summarized succinctly, encompass individuals' actions that are based on the meanings that objects have for them (Carter & Fuller, 2015). That is, interaction takes place within a given social and cultural background; meanings emerge from interactions with other individuals and with society; and meanings are continually created via interpretation during any interaction process. Symbolic interactionism is essential for understanding construction of gender and sexuality. It is also valuable in promoting self and identity (Cinoğlu & Arıkan, 2012).

Even though minority students have continued entering college at a higher rate than in the previous years, they also continue to leave at a higher rate than non-minorities. Attrition causes a severe loss of resources by society, by students, as well as colleges, which have to spend more on programs and services that help in retaining and graduating students (Seidman, 2005). Whenever a student leaves school prematurely, any debt that is incurred must be repaid regardless of the student not graduating, and the college will lose future funding for tuition, fees as well as auxiliary services generated over time (Seidman, 2005). Moreover, the surrounding community that supports the college, movie theaters, and restaurants among other service providers also suffers an economic loss associated with students.

According to Chickering and Reisser (1993), the ability of Historically Black Colleges and Universities to retain students will be dependent on the students' background characteristics, commitments, social integration, goals, as well as academic integration. The theory suggests that Black students enroll at HBCUs while possessing certain characteristics based on their family background. The characteristics include such aspects as the parents' level of education and their socio-economic status, the gender and age attributes of the individual student, the student's educational experiences prior to attending college, such as grades attained at high school, rank, as well as test performance (Ortiz & Santos, 2009).

The hypothesis of the theory is that student's goals, as well as level of commitment in the learning process are influenced directly by their academic and social experiences (Allport, 1971). Such experiences have a direct effect on the level to which students in Historically Black Colleges and Universities can be integrated into their colleges or universities' social as well as academic systems. This is because the likelihood of students persisting up to the time they complete their course is dependent on the level to which they are integrated academically and socially (Kasbeek, 2013).

Notably, the Tinto (1975, 1987, 1993) retention model hypothesize that individual pre-entry college characteristics such as family background, ability, skill, as well as prior schooling constitute individual goals and commitments which interact over time with institutional experiences, both formal and informal social and academic systems of an institution. Therefore, extent to which an individual becomes socially and academically integrated into the academic social and academic system of an institution will determine if that individual will be integrated into informal or formal social and academic systems of an institution and this also determines leaving behavior.

Positive experiences and interventions will strengthen persistence through heightened individual intentions as well as commitments while negative experiences are likely to weaken intentions and commitments. For instance, intentions can include a student wanting to earn a degree in a given field while commitment is the student's aspiration of completing the degree as well as willingness to spend time and energy in acquiring that degree. Therefore, the greater the level of student integration into academic and social systems of the institution, the greater is his or her subsequent commitment to the institution and positive retention rate (Seidman, 2005).

Tinto (1993) expanded the theory to bring in the fact that students who want to succeed in persisting until the completion of their course must seek to have them dissociated from the past relationships. The following will enable students to go through the transition into the new lifestyle, both socially and intellectually. This is also affected by the intention of each individual student. The intention has an impact on the institutional commitment and the individual's goals. A student must, therefore, be able to integrate himself/herself both normatively, into the intellectual environment, and socially, into the social system of the institution (Sutherland, 2002; Burns, 2006).

THE STUDENT ATTRITION FRAMEWORK

The theory of attrition as contributed by Bean (1985) is based on the turnover of an individual learning institution, as well as the attitude-behavior interaction models. The attrition model considers the effects of external factors on the likelihood that non-traditional students will persist. The factors, many of which are beyond the control of an institution, often affect students by putting pressure on their resources, time, and sense of wellbeing. Bean's model emphasizes that student interactions and the level of their integration combine with subjective evaluations of the institution, educational processes, as well as experience to directly influence satisfaction and indirectly influence intentions to persist. At the same, external factors that institutions cannot control including opportunity to transfer, financial constraints, and family commitments, directly influence intentions to leave or drop out. Importantly, external attitudinal, and interaction factors collectively influence departure or persistence.

The factors that affect retention include institutional environmental factors, student demographic characteristics, psychosocial and study skills factors, academic preparation and success factors, integration and fit, student finances, as well as environmental pull factors. Bean & Metzner (1985) indicate that the major difference between the attrition process of traditional and non-traditional students is that the external environment more than social integration variables affects non-traditional students compared with traditional student. Raab & Adam (2005) in their report indicate that issues regarding financial aid, especially with respect to the first-generation students in college have been associated with the dilemma that continued to overshadow academic needs as a major impediment retention and success of freshmen.

Bean (1985) argued that just as an adult can decide to quit a work place, a student could decide to quit college. Bean's theories emphasize the role played by the students' intentions, attitudes, and beliefs (Bronstein, 2007). According to Bronstein, student retention in any institution will depend upon student's background characteristics, perception concerning the quality of the institution, grades, as well as the practical worth of the education being pursued (Bronstein, 2007). This model put more emphasis on the impact that the behavioral intentions have on students' retention. Accordingly, Historically Black Colleges and Universities students' intention to remain in their various learning institutions are greatly dependent on their attitudes that are, in turn, determined by the experiences that an individual student has with his/her institution (Williams, Howell & Hricko, 2006).

Bean (1985) also suggested such external factors as finances, responsibilities at the family level, number of hours an individual student is required to be at work, and the ease with which students can transfer and join another learning institution are important elements in student's decisions to continue or leave school. All these factors affect the probability of a student dropping out of a college or a university based on their direct impact on the students' intention to leave. This means that apart from the HBCU students' level of satisfaction with their colleges, background, and attitude, various environmental and organizational factors do affect their drop out intention (Schuh, Kones & Harper, 2010).

To gauge the effectiveness of retention strategies and approaches it is important to be able to compare how other student groups and other institutions are doing with retention. Additionally, there may be a need for more information regarding the strategies that other institutions are using to address similar retention issues and challenges. Likewise, it is important to know if there are Historically Black Colleges and Universities that are doing well with retention, the approaches they are using, and whether these successful approaches might be instructive and applicable at other institutions.

However, other psychologically based approaches that can be applicable to the topic include (a) the copying behavioral theory, (b) the self-efficacy theory, and (c) the attribution theory. The copying behavioral theory concerns a student's ability to carry out self-evaluation in order to be able to cope with a new learning environment (Bandura, 1986). The level of adjustment that a student exhibits will determine whether he/she is able to cope with life in the HBCUs. On the other hand, the self-efficacy theory is based on the perception of an individual student of the Historically Black Colleges and Universities on his/her ability to perform various tasks as required of them while in school (Snowmont, McCown, & Biehler, 2012). It is also partly dependent on a student's perceived ability to handle certain situations that he/she may have experienced or have seen others go through. Finally, the attribution theory has its focus on the locus of control's concept (Leary & Hoyle, 2009). The concept refers to a student's ability to process his/her experience looked at from both external and the internal perspectives. The external perspectives focus on such factors as the luck and fate of Black students while the internal perspectives my focus on such personal attributes as students' skills and attitudes (Deci & Ryan, 1985). Additional frameworks relevant in the study include those based on the cultural and economic factors. This means that some students can leave school based on their inability to pay for their studying. The following fact has been common among the Black students who join most UK colleges and universities (Gallien & Peterson, 2006). As stated by Darryl (2010), finances do affect Historically Black Colleges and Universities students' experience both cognitively and non-cognitively. Whenever, the retention of students in these institutions are approached from a cultural framework perspective, the dropout rate of students is not analyzed at as a psychological phenomenon, but rather a socio-cultural phenomenon (Palmer & Wood, 2012). The argument here is that students of Historically Black Colleges and Universities voluntarily decide to leave an institution based on various cultural forces; such as the kind of experiences they have had with their families or neighborhoods, the racial and ethnic groups with which they identify themselves, their social class and religious affiliations, as well as institutional culture of their college or university where they study (Darryl, 2010; Palmer & Wood, 2012).

INDIVIDUAL FACTORS THAT AFFECT RETENTION AMONG BLACK STUDENTS

Numerous factors influence the rate at which Black students are retained in their various institutions of higher learning. The major factors include: the level of involvement, background characteristics, place of residence, and the first-year college grades of students among others.

Student's Level of Involvement

Austin's (1999) student involvement theory refers to the amount of energy that an individual student has devoted to both social and academic life while in college. Those students who are able to assign much of their time to studies while also spending the same amount of time in their learning institutions are likely to stay in their institutions until graduation (Andrea & Gosling, 2005). This also applies to students who take part in various student organizations, as well as those who frequently interact with members of the faculty (Lindsay & Justiz, 2001). On the other hand, the uninvolved students who neglect their studies and avoid participating in extracurricular activities have higher chances of dropping out of the institution before the completion of their course. That relates to students who spend the limited amount of time they have in college, as well as those who limit their contact with other students and faculty members (Upcraft & Gardener, 1990).

It is obvious that students who are involved in their HBCUs are directly and positively impacted with higher retention rates. Darryl (2010) reported that the levels to which students are involved in their colleges and campuses activities affect their integration into the schools' academic and social systems. This, in turn, promotes persistence. Yeakey and Henderson (2000) also restated Austin's argument that students who are more involved academically are more likely to be satisfied with most of the college life aspects (Bataille & Brown, 2006). However, being intensely involved in studies within the colleges and universities can subject HBCUs' students to isolation from their peers and reduce the influence of their peers to remain in their colleges. Studies showed that Black students who are actively involved in various social activities are easily integrated into their college's social environment and, thus, have higher chances of persisting (Fleming, 2012; Skipper, 2003).

Background Characteristics of Students

Darryl (2010) emphasized that students who join institutions of higher learning possess certain individual characteristics that impact retention and student success. Some students are not academically prepared, while others are either prepared or underprepared. Equally, a student is either a male or female in gender and has parents who possess different education levels (Gasman & Baez, 2008). Additionally, students are of different age groups and they earn varying amounts of income. All these characteristics have certain levels of impact on the probability of students being retained in HBCUs. However, the level of parents' education, grades attained while at high school, socio-economic status, and gender is the strongest retentions influences among Black students.

There various factors that can affect African American retention positively and negatively including having other African Americans as roommates, low-esteem, lack of parental support, low social expectations, and poor financial backgrounds (Seidman, 2005). Among American Indian students, the factors that affect retention either positively or negatively include academic preparations, aspirations as well as the level of interaction with staff and faculty. Proactive services such as academic preparation and engagement with staff and faculty help to retain the students. Among Hispanic students, lack of academic preparedness, lack of community and being away from the members of family, lack of information regarding financial aid as well as inadequate knowledge on how to apply for the aid significantly affect retention rate among Hispanic students (Seidman, 2005).

According to Darryl (2010), though the influence of gender on retention has been controversial, it is proven that gender greatly influences Black students' willingness to persist. This, in turn, directly influences their willingness to stay in their institutions to the completion of their courses. Whereas men are less likely to persist, the opposite is true with women. However, this has remained a controversy with the seminal scholar Tinto. Tinto's research stated that women's patterns of withdrawal are greatly affected by their social issues and not academic ones, as is the case with men. Thus, unlike male students who often leave school only upon being forced to do so, female students do leave voluntarily in most cases. The Black male students are mostly influenced by the academic experiences that they had had prior to joining college, as well as their personal attributes (Tinto, 1993).

The level of education of parents is another factor that has a large influence on retention of Black students. Darryl (2010) noted that the level of parents' education of Black students greatly impacts the students' retention. The researcher further says that compared to the HBCUs students whose parents did not attend college; students whose parents attended college are more inspired to attain degrees. Therefore, they are most likely to remain in school until the completion of their courses. Moreover, those students whose parents failed to attend college have a very low likelihood of being involved in the activities of the institution. This also affects their ability to persist in the school environment. These students experience a lot of difficulties throughout college since they lack the necessary preparation. They earn grades that are lower compared to their privileged counterparts. With the humiliations that come with their failures, most of them will opt to drop out a college.

Another factor that affects the probability of HBCU students being retained is the level of preparation that a student obtained prior to attending college. Academic preparedness has widely been defined in relation to the performance of students prior to their joining of colleges and their performance while there. There is a close relationship between students' performance while in high school and their performance during their first year at HBCUs. Those students who fail to perform well in high school have a high likelihood of failing their first-year exams at HBCUs and are less likely to persist. This factor is considered substantial and consistent in predicting the undergraduate persistence among Black students. Studies on the relationship between academic preparedness and retention among Black students reveal that retention is closely related to students' performance while in high school (Brown & Bartee, 2009).

Place of Residence

Darryl (2010) observed that students who live in a school environment are likely to remain in their colleges and universities until the completion of their course. Landefeld (2009) also stated that students who live within the residence of their colleges or universities are seen to gain a lot when it comes to art, interpersonal self-esteem, as well as liberalism. Moreover, HBCU students staying on campus have more time for interaction with their colleagues and faculty members. They may take part in the government of students, extracurricular activities, as well as social fraternities. Equally, they usually receive more satisfaction from the friendships, as well as the relationships they have with other students. This means that a student who lives within the environs of the campus is likely to be much more involved than students living off campus. Thus, the probability of being retained to graduation is boosted for resident students (Benjamin, 1997).

First Year College Grades

The rate at which students drop out is highest right after completion of the first year of study. Therefore, Darryl (2010) noted that the most efficient measure that can be used to improve the rate of graduation among Black students is to come up with various intervention programs that target this critical timeframe. Such programs should be aimed at assisting students in adapting to the social life of their institutions' right from their first year of study (Pryor, 2009). The average grade of students completing their first year of study in HBCUs has, thus, been a measurement tool to predict the expected retention rate of students.

However, it is suggested that the interpretation of grades should be limited to their depiction of the performance of each student in relation to others at a time (Green Jr., 2016). Thus, it should never be used as a determinant of what students have learnt. Therefore, there is the need for the replacement of the grading system currently used in use at most colleges. Assessments need to properly reflect the level of growth and development among students (Jordan, 2012). This can be achieved through conducting repeated pre and post-testing, as this approach is more efficient in measuring program effectiveness. Nevertheless, schools should be cautious about not downplaying performance since the grades that students attain are important in understanding learning, motivation and other critical retention factors. Accordingly, students with poor grades can be dismissed, while those with good grades get an opportunity to further their career in the universities. There is, thus, the need for integration of the two approaches (Darryl, 2010).

Financial Barriers

Based on the magnitude of the effect of students' economic stability on their rate of retention, it is necessary to fully explore this important retention indicator. Although it is difficult for Black students to finance their education, those aspiring to join HBCUs have had to find adequate financial sources. The stability of the funding source that a student is relying on will affect the rate of retention, as well as their ability to persist to the completion of their courses (Jones, 2001). Educational funding is subjected to strict federal and state regulations. The same is true of various HBCU institutions that also have their own regulations for the various kinds of financial aid that is available to their students. Thus, a good percentage of Black students depend on their families as their main financial source for paying for their education. This, in turn, contributes to the already poor families' heavier responsibilities. Thus, most students from low and middle-income families drop out of their institutions at the end of or even before the completion of their first year of study (Berg, 2010).

The challenge to finance higher education that Black students face has been exacerbated over time due to the increasing number of students who are seeking access to college and university education. Chau (2012) stated that the proportion of college students receiving financial aid is constantly increasing. Chau (2012) further explains that based on this challenge, at least one out of every five students drop out of college immediately after or during their first year of study. Overall, it is indicated that 30% of the first-year students who register either in HBCUs or PWIs failed to report back to their colleges or campuses for their second year of study. This retention failure accounts for over $6.2 billion in state appropriations for college and university education. About $1.5 billion is provided in the form of grants. Loss of this significant state revenue for post-secondary education has a measurable impact and effect on future college funding (U.S. Department of Education, 2012).

According to Pan and Bai (2009), the challenges that come with securing financial aid influence the rate of students' persistence in colleges and universities. Scholars noted that the finance-related choices directly and indirectly influence students' retention and persistence decisions. Financially, persistence factors affecting the retention of Black students in HBCUs include monetary measures used for financial aid, housing costs, tuition costs, and other related living costs (Hale, 2006).

Another challenge related to financial aid has been the new regulations for a student's eligibility to receive financial aid. This has become more important since the beginning of July 2012, when the US Department of Education strengthened its requirements around satisfactory academic standards and satisfactory progress. Besides the initial requirements, the Department for Education provided that if a student wants to apply for the government's financial aid, he/she will have to complete an academic plan. The plan made added hurdles and complexity for most HBCUs students, the majority of which had not performed well in high school (U.S. Department of Education, 2012). For instance, due to the under-preparedness of most Black students, it is a challenging task for them to acquire an acceptable copy of an evaluation on their intended courses and this is a requirement stated in the S.A.P. Academic Plan.

In addition, the former President of the USA, President Barack Obama signed into law the *Consolidated Appropriation Act* in 2012. This Act limits the Lifetime Pell grant and this limitation had a disproportionate impact on HBCU students. Since 2012, the provisions of the Act have also limited the total number of students able to access the Federal Pell Grant program since they can only do so for twelve semesters rather than the 18 semesters of eligibility under the prior rule (U.S. Department of Education, 2012).

CONCLUSION

HBCUs were initially developed to cater for the educational needs of students from economically challenged backgrounds, particularly Blacks. However, it is evident that over the years, HBCUs have continued to attract even students from white communities. For instance, the enrollment of whites in these learning institutions now stand at more than 13 percent, an indication of how various groups now perceive them. Students adapt more and faster to their institutions of learning based on the types of environments that such the institutions provide. For instance, HBCUs have been reported to provide financial support to students, promote learning environments that prevent emotional and psychological distress. It is highly possible that moving forward, the number of students attending HBCUs will continue to grow as they become the preferred centers for learning for both white and non-white students.

REFERENCES

Allport, G. (1971). *Personality: A psychological interpretation*. London: Constable.

Austin, A. W. (1999). Student involvement: A development theory for higher education. *Journal of College Student Development*, *40*(5), 518–529.

Bandura, A. (1986). *Social foundations of thought and action: A social cognitive theory*. Englewood Cliffs, NJ: Prentice Hall.

Bataille, G., & Brown, B. (2006). *Faculty career paths: Multiple routes to academic success and satisfaction*. Wesport, CT: Praeger Publishers.

Bean, J. P. (1985). Interaction effects based on class level in an explanatory model of college student dropout syndrome. *American Educational Research Journal*, *22*(1), 35–64. doi:10.3102/00028312022001035

Bean, J. P., & Metzner, B. S. (1985). A conceptual model of nontraditional undergraduate student attrition. *Review of Educational Research*, *55*(4), 485–540. doi:10.3102/00346543055004485

Benjamin, L. (1997). *Black women in academy: Promises and perils*. Gainesville, Fl: University Press of Florida.

Berg, G. (2010). *Low-income students and perpetuation of inequality: Higher education in America*. Burlington, VT: Ashgate Pub.

Braxton, J. M., Hirschy, A. S., & McClendon, S. A. (2004). Toward understanding and reducing college student departure. ASHE-ERIC Higher Education Report, 30(3).

Bronstein, S. (2007). Supplemental instruction: supporting persistence in barrier courses [Thesis Ed.D.]. University of Massachusetts.

Brown, C., & Bartee, R. (2009). *The broken cisterns of African American education: academic performance and achievement in the post-Brown era*. Charlotte, NC: Information Age Pub.

Burns, P. (2006). *Success in college: From C's in high school to A's in college*. Lanham, MD: Rowman & Littlefield Education.

Campbell-Whatley, G. D. (2003). Recruiting and retaining of culturally and linguistically diverse groups in special education: Defining the problem. *Teacher Education and Special Education*, *26*(4), 255–263. doi:10.1177/088840640302600402

Carter, M., & Fuller, C. (2015). Symbolic interactionism. *Sociopedia*. doi:10.1177/205684601561

Chau, J. (2012). More students are enrolled in college and on financial aid, annual report shows. The Chronicle of Higher Education. Retrieved March 3 2012, from http://chronicle.com/article/More-Students-Are_Enrolled-in/131320/

Chickering, A., & Reisser, L. (1993). *Education and identity. San*. Francisco: Jossey-Bass Publishers.

Cinoğlu, H., & Arıkan, Y. (2012). Self, identity and identity formation: From the perspectives of three major theories. *Journal of Human Sciences*, *9*(2), 1115–1131.

Conley, P. A., & Hamlin, M. L. (2009). Justice-learning: Exploring the efficacy with low-income, first-generation college students. *Michigan Journal of Community Service Learning*, *16*(1), 47–58.

Darryl, T. R. (2010). The influence of student involvement with campus life on the retention of African American students enrolled at Public Historically Black University. Retrieved from http://repository.lib.ncsu.edu/ir/bitstream/1840.16/5067/1/etd.pdf

Deci, E., & Ryan, R. (1985). *Intrinsic motivation and self-determination in human behavior*. New York: Plenum Publishing. doi:10.1007/978-1-4899-2271-7

Dwyer, B. (2006). Framing the effect of multiculturalism on diversity outcomes among students at historically black colleges and universities. *Educational Foundations*, *20*(1-2), 37–59.

Esmieu, P., & Martinez, A. (2014). *Expanding the HBCU legacy: Enrolling and supporting the rising Latino population*. US: Palgrave Macmillan. doi:10.1057/9781137480415.0012

Fleming, J. (2012). *Enhancing minority student and retention and academic performance: What we can learn from program evaluations*. San Francisco, CA: Jossey-Bass.

Flemming, J. (2004). The significance of historically black colleges for high achievers: correlates of standardized test scores in African American students. In M. Brown & K. Freeman (Eds.), *Black colleges: new perspectives on policy and practice* (pp. 29–53). Westport, CT: Praeger.

Gallien, L., & Peterson, M. (2006). *Instructing and mentoring the African American college student: Strategies for success in higher education*. Boston: Pearson/Allyn and Bacon.

Gasman, M. (n. d.). The changing face of Historically Black Colleges and Universities. University of Pennsylvania. Retrieved from http://www.gse.upenn.edu/pdf/cmsi/Changing_Face_HBCUs.pdf

Gasman, M., & Baez, B. (2008). *Understanding minority-serving institutions*. Albany: Sate University of New York Press.

Gilbert, S., So, D., Russell, T. M., & Wessel, T. (2006). Racial identity and psychological symptoms among African Americans attending a historically black university. *Journal of College Counseling*, *9*(2), 111–122. doi:10.1002/j.2161-1882.2006.tb00098.x

Green, D. F., Jr. (2016). Expanding the dialogue on writing assessment at HBCUs: Foundational assessment concepts and legacies of Historically Black Colleges and Universities. Retrieved from http://www.ncte.org/library/NCTEFiles/Resources/Journals/CE/0792nov2016/CE0792Expandingpword.pdf

Hale, F. (2006). *How black colleges empower black students: Lessons for higher education*. Sterling, VA: Stylus Publishing.

Harper, S. R. (2012). *Black male student success in higher education: A report from the National Black Male College Achievement Study*. Philadelphia: University of Pennsylvania. Retrieved from http://works.bepress.com/cgi/viewcontent.cgi?article=1049&context=sharper

Jordan, Z. (2012). In A. Inoue, M. Poe, & P. Lang (Eds.), *Students' Right, African American English, and Writing Assessment: Considering the HBCU. In Race and Writing Assessment* (pp. 97–110).

Kasbeek, D. (2013). *Reframing retention strategy for institutional improvement*. San Francisco, CA: Jossey-Bass Publishing.

Landefeld, T. (2009). *Mentoring and diversity: Tips for students and professionals for developing and maintaining a diverse community*. Dordrecht, New York: Springer.

Leary, M., & Hoyle, R. (2009). *Handbook of individual differences in social behavior*. New York: Guilford Press.

Lindsay, B., & Justiz, M. (2001). *The quest for equity in higher education: Toward new paradigms in an evolving affirmative action era*. Albany: State University of New York Press.

Liu, M. C. (July, 2011). Investing in higher education for Latinos: trends in Latino college access and success. Denver, CO: *National Conference of State Legislatures*.

National Center for Education Statistics. (2011). HBCU enrollment rates component. Retrieved from nces.ed.gov/ipeds/tablefiles/tableDocs/IPEDS201112Tablesdoc.xlsx

National Center for Education Statistics. (2016). Status and trends in the education and ethnic groups 2016. Retrieved from https://nces.ed.gov/pubs2016/2016007.pdf

National Center for Education Statistics. (n. d.). Historically Black Colleges and Universities. Retrieved from https://nces.ed.gov/fastfacts/display.asp?id=667

Nicpon, M. F., Huser, L., Blanks, E. H., Sollenberger, S., Befort, C., & Robinson-Kurpius, S. E. R. (2007). The relationship of loneliness and social support with college freshmen's academic performance and persistence. *Journal of College Student Retention*, *8*(3), 345–358. doi:10.2190/A465-356M-7652-783R

Ortiz, A., & Santos, S. (2009). *Ethnicity in college: Advancing theory and improving diversity practices on campus*. Sterling, VA: Stylus Publishing.

Ozuna, T. G. (2012). Examining the first-year experience and perceptions of sense of belonging among Mexican American students enrolled in a Texas HBCU. Unpublished doctoral dissertation, University of Texas, Austin, Texas.

Palmer, R., & Wood, L. (2012). *Black men in college: Implications for HBCUs and beyond*. New York: Routledge.

Pan, W., & Bai, H. (2009). A multilevel approach to assessing the interaction effects on college student retention. *Journal of College Student Retention*, *11*(2), 287–301. doi:10.2190/CS.11.2.g

Raab, L., & Adam, A. J. (2005). The university college model: A learning-centered approach to retention and remediation. *New Directions for Institutional Research*, (125): 86–106.

Schuh, J. H., Kones, S. R., & Harper, S. R. (2010). *Student services: A handbook for the profession*. San Francisco, CA: Jossey-Bass Publishing.

Seidman, A. (2005). *Minority students' retention: Resources for practitioners*. New Directions for Institutional Research. Wiley Periodicals Inc.

Seidman, A. (2009). *College student Retention: Formula for student success*. Lanham, MD: Rowman & Littlefield Publishers.

Skipper, T. (2003). *Involvement in campus activities and the retention of first year college students*. Columbia, SC: National Resource Center.

Snowmont, J., McCown, R., & Biehler, R. (2012). *Psychology applied to teaching*. Belmont, CA: Wadsworth.

Sutherland, T. (2002). *Active learning toolkit*. Sarasota, FL: American Accounting Association.

Tinto, V. (1975). Dropouts from Higher Education: A Theoretical Synthesis of the Recent Literature. *Review of Educational Research*, *45*(1), 89–125. doi:10.3102/00346543045001089

Tinto, V. (1987). *Leaving College: Rethinking the Causes and Cures of Student Attrition*. Chicago: The University of Chicago Press.

Tinto, V. (1993). Leaving college: Rethinking causes and cures of student attrition. Chicago, IL: University of Chicago Press.

Twombly, S. B., Salisbury, M. H., Tumanut, S. D., & Klute, P. (2012). *Study Abroad in a New Global Century: Renewing the Promise, Refining the Purpose*. San Francisco, CA: Jossey-Bass.

Upcraft, L., & Gardner, J. (1990). *The freshmen year experience: Helping students survive and succeed in college*. San Francisco, CA: Jossey - Bass Publishers.

U.S. Department of Education. (2012). Federal student financial aid handbook. Retrieved from www.ifap.ed.gove/fsahandbook/attachments/1112FSAHbkVol1Ch1.pdf

Valle-Riestra, D. M., Shealey, M. W., & Cramer, E. D. (2011). Recruiting and retaining culturally diverse special educators. *Multidisciplinary Journal of Teaching and Learning*, *1*(2), 68–87.

Williams, D., Howell, S., & Hricko, M. (2006). *Online assessment, measurement, and evaluation: Emerging practices*. Hershey, PA: Information Science Publishing. doi:10.4018/978-1-59140-747-8

Yeakey, C., & Henderson, R. (2000). *Surmounting all odds: Education opportunity, and society in the new millennium*. Greenwich, CT: Information Age Publishing.

KEY TERMS AND DEFINITIONS

Attrition Theory: The Theory of Attrition as contributed by Bean (1985) is based on the turnover of an individual learning institution. Bean's (1985) theory of attrition also incorporates the attitude-behavior interaction model. The attitude-behavior attrition model considers the effects of external factors on the likelihood that non-traditional students will persist.

Consolidated Appropriation Act of 2012: The Consolidated Appropriation Act of 2012 limits the Lifetime Pell grant award amount. Provisions of the Act also limit the total number of students that can access the Federal Pell Grant program. Pursuant to the Act, students can only have access to Pell grant awards for twelve semesters rather than the 18 semesters of eligibility under the prior rule (U.S. Department of Education, 2012).

Coping Behavioral Theory: The coping behavioral theory concerns a student's ability to carry out self-evaluation in order to cope with a new learning environment (Bandura, 1986).

Disequilibrium: Disequilibrium challenges students' earlier assumptions as well as their values and beliefs as they meet new people and try to establish new relationships, are exposed to new ideas, and have new experiences.

Historically Black College and Universities (HBCU): HBCUs are degree granting institutions that were established before 1964 with the main objective of educating Black Americans. Currently, there are 100 HBCUs in 19 states, 51 of which are public while 49 are private nonprofit institutions.

Interactionist Theory: The interactionist theory focuses on background characteristics, shared group values, friendships, and how these factors influence the retention of students in various institutions of learning (Siedman, 2009).

Predominantly White Institutions (PWIs): Post-secondary institutions where white student enrollment is at or above 50%.

Self-Efficacy Theory: Self-efficacy theory is based on the perception of each individual student of his or her ability to perform the various tasks required to remain enrolled in school (Snowmont, McCown, & Biehler, 2012).

Symbiotic Interactionism: Carter and Fuller (2015) explain the theory of symbiotic interactionism, as a framework and perspective in sociology that explains how society is created and maintained through repeated interactions among individuals. The emergence of the theory was because of the need to respond to the principle perspectives that had previously dominated sociology.

Tinto's Retention Model: Tinto's Retention Model hypothesizes that individual pre-entry college characteristics such as family background, ability, skill, and prior schooling interact with each student's institutional experiences, both formal and informal, and the social and academic systems of the institution. The extent to which an individual becomes integrated into the academic, social and academic systems of an institution determines if that individual will be retained (Tinto,1975, 1987 & 1993).

Chapter 6

Looking Beyond the Academic Institution for Retention and Student Engagement Models:

A Study of Employee Engagement in a Rapidly Growing Worldwide Marketing Company

Alexandra McDermott Wilcox
University of Southern California, USA

Ruth Claire Black
Namseoul University, South Korea

ABSTRACT

Student engagement, retention, and success are not issues limited to traditional academic settings. Corporations and non-traditional business entities allocate significant resources to identify and develop viable solutions that positively impact employee engagement, retention and success. While most corporate training and employee development programs are driven by corporate mandates to improve efficiency and cost containment, the most dynamic retention and student success elements within these programs deserve further study and exploration. Indeed, the possible transfer of these effective retention and student success elements from the corporate training and development space to other learning, training and student development settings, can yield significant benefits to any institution or organization that seeks to improve its initiatives focused on training and education.

DOI: 10.4018/978-1-5225-2998-9.ch006

INTRODUCTION

Student engagement, retention, and success are not issues limited to traditional academic settings. Corporations and non-traditional business entities also allocate significant resources to identify and develop viable solutions that positively impact employee engagement, retention and success. While most corporate training and employee development programs are driven by corporate mandates to improve efficiency and cost containment, the most dynamic retention and student success elements within these programs deserve further study and exploration. Indeed, the possible transfer of these effective retention, engagement and success elements from the corporate training and development space to other learning, training and student development settings can yield significant benefits to any institution or organization that seeks to improve its initiatives focused on training and education. Moreover, while the application of corporate training approaches and strategies might appear to be most applicable to workforce development, these strategies can serve as an excellent template to inform the full range of traditional institutional academic programs, including degree completion and college preparation bridge programs.

A comprehensive corporate training and retention study of an international multi-level marketing distributorship program took place in 2015 and 2016 as part of a dissertation study. This paper re-examines the findings from this study in terms of how the strategies apply to retention and student success in academic programs and institutions, with a special focus on institutions that serve non-traditional and underserved student communities. Looking at models and approaches to retention that exist outside of academia can help to identify best practices, innovative approaches and viable alternatives to existing approaches to retention and student success.

REVIEW OF THE LITERATURE

The significance of employee engagement strategies and training programs in corporations should not be overlooked when searching for best practices and innovative approaches to student engagement and retention challenges. As Gallup concluded in its 2013 engagement study, lack of employee engagement has far-reaching consequences in our current global business landscape. In 2013, Gallup, pursuant to its ongoing study of employee engagement, published "The State of the Global Workplace: Employee Engagement Insights for Business Leaders Worldwide," which revealed that only 13% of workers worldwide are engaged. Gallup defines engagement as a worker who is "emotionally invested" and focused on creating daily value for the organization (Gallup, 2013). Since the 1990s, Gallup has administered its Q12 survey to over 25 million workers, in 195 countries, in 70 languages. In its Q12 survey, Gallup measures four stages of engagement, and asks workers to consider the following: (1) what do I get from this role; (2) how do I contribute; (3) do I belong; and (4) how can I make improvements, innovations, or contributions. In its latest study, Gallup administered the Q12 survey to an additional 230,000 workers in 142 countries. Gallup then grouped workers' responses into three categories: (1) engaged; (2) not engaged; and (3) actively disengaged. After analyzing the results, Gallup made several key findings about engagement and the cost of disengagement, including (1) disengagement drains world economies, (2) engagement varies widely from region to region, (3) engagement leads to the perception of job growth, (4) poor management practices hinder organizational performance and engagement levels, (5) engagement leads to feelings of positivity and living better, and (6) education is often associated with higher engagement levels.

Gallup concluded, through its findings, that engagement is critical to an organization's productivity, profitability, and customer satisfaction. Further, that these three elements are ultimately the key to a company or organization's longevity and financial success. This conclusion and finding on engagement is equally applicable to the full range of academic institutions and organizations. Further, this finding is in line with prevailing academic theories on self-efficacy, motivation and organizational development (Eccles & Wigfiled, 2002; Bandura & Locke, 2003; Maddux & Gosslin, 2015). The level of a student's engagement is critical to course and program retention and to self-reported levels of student success as well as a student's motivation and confidence to tackle and overcome academic obstacles (Bandura & Locke, 2003; Maddux & Gosslin, 2015).

After analyzing the data, Gallup found that organizations ranking in the top 25% for engagement, report higher productivity, more profitability, better customer ratings, and less turnover and absenteeism. Gallup also discovered that organizations in its database that have an average of 9.3 engaged employees for every actively disengaged employee yielded 147% higher earnings per share (EPS) as compared to their competition. By contrast, organizations with an average of 2.6 engaged employees for every actively disengaged employee experienced 2% lower EPS compared with their competition.

Given the strong parallel between the academic literature on task non-cognitive factors, value, motivation, confidence and self-efficacy, and the Gallup employee engagement findings, it is logical to suggest that academic institutions and programs should explore business-based strategies for retention and employee success and look for best practices that can be applied to an academic course, program or institution, in such a way as to meaningfully impact retention and student success.

Figure 1. Earnings per share comparison emphasizing impact of engagement

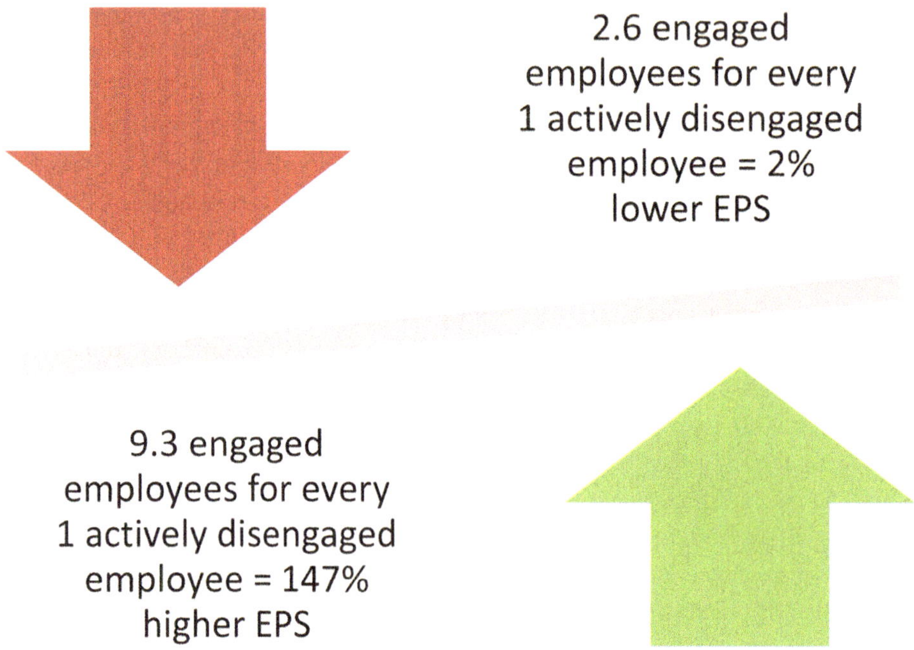

THE COST OF FAILING TO ENGAGE AND RETAIN STUDENTS AND EMPLOYEES

While organizations experience the fiscal impact of engagement, Gallup contends that engagement is actually a much bigger issue, and has worldwide economic implications. In its latest study, Gallup reduced disengagement to a monetary value and made a multi-country comparison. Gallup estimates that active disengagement in the United States, for example, costs approximately $450 billion to $550 billion per year. Active disengagement in Germany costs approximately €112 billion to €138 billion per year (US$151 billion to $186 billion). In the United Kingdom, active disengagement costs approximately £52 billion and £70 billion (US$83 billion and $112 billion) per year. After making this multi-country comparison, Gallup urged business leaders to do what it takes to engage their workers.

Like traditional academic institutions, corporations are constantly tasked with how to create the optimal environment to achieve maximum engagement. If an individual is engaged, he or she is less likely to leave. One way to create maximum "stickiness" with employees, is to provide employees with diverse and engaging training opportunities. However, the challenges of rolling out corporate trainings are significant and include: content and design considerations, adult learner challenges, the mode of delivery and type of training, course evaluation, cultural and language barriers, and a robust implementation and evaluation plan. While traditional academic and business environments may differ in many important respects, one thing is clear: the cost of failing to engage, retain, and help students or employees succeed is unimaginably high. Notably, the study on which this paper is based, explored many cross-over applications within academic institutions that were focused on improving student retention. The 2015-2016 dissertation study, on which this chapter is based, focused on the critical need to identify new ways to engage independent distributors within a $1B international multi-level marketing company, because the cost of failure was not only high for the organization, but for the new entrepreneurs as well.

MULTI-LEVEL MARKETING AS AN ENTREPRENEURIAL BUSINESS OPPORTUNITY

The global economic crash of 2008 prompted many people to explore new entrepreneurial opportunities, including many who avoided traditional businesses and, instead, pursued opportunities and pathways to financial and business success within multi-level marketing organizations. However, the majority of these individuals were not successful in their endeavors within multi-level marketing organizations. A unique facet of multilevel marketing organizations is that they often turn over 100% of their distributors annually (Berry, 1997; Yarnell & Yarnell, 1998). As a result, retention of successful distributors is a significant issue for all multi-level marketing organizations and it was the central focus of the 2015-16 study. Another important issue is how to onboard new distributors and how to quickly facilitate the success of new distributors so that they will remain with the company and dedicate themselves to aggressively growing their own distribution networks. Another key issue within the study was how to support independent minded entrepreneurs and help them to quickly overcome the business challenges they would face as they developed their organizations.

These issues were further magnified because the distributors were working primarily in developing countries and most lacked English language fluency, even though the company produced training only in English. Therefore, the company needed to address a wide variety of organizational contexts across extremely varied cultural and international settings, while consistently and effectively encouraging new and developing distributors to remain engaged. These unique corporate structure factors when combined with the diverse socio-economic, cultural and language backgrounds of the study subjects created an opportunity to explore retention and success in an unusual and unique context. This paper seeks to communicate the findings from the study to the broad academic audience of institutions, administrators, faculty members and staff that are charged with improving retention and student success within their program, organization or university.

INDEPENDENT DISTRIBUTOR ENGAGEMENT STUDY AT A HIGHLY SUCCESSFUL AND PROFITABLE MULTI-LEVEL MARKETING ORGANIZATION

A 2015-2016 interview based study of a large independent distributor of a $1B multi-level marketing company examined worker engagement, training, and self-reported success measures of high-level distributors. Both the independent distributor and the multi-level marketing company were experiencing unprecedented growth. The distributors were on pace to add over two hundred thousand distributors within a year and expected to grapple with significantly higher distributor growth over the next several years. The study focused on one of the company's largest distributors. This large distributor was a vital component of the company's extraordinary growth and success and it was considered a model for other distributors within the multi-level marketing company.

Through a range of distributor interviews the study identified gaps in knowledge and skills, motivation, and organizational resources throughout the distributor's organization. However, despite these gaps, the distributor continued to significantly outperform the market and to continue to grow exponentially. This unprecedented rate of sustained growth and network expansion, despite the identified knowledge and skills gaps, was worthy of study, assessment and further exploration. However, one key element of the study was to understand what qualities, values, beliefs and/or assets of the downline distributors enabled this strong and consistent growth pattern in spite of the identified issues, challenges and training gaps. Further, could these qualities, values, beliefs or assets inform retention and success in other environments and organizations?

This study interviewed 12 seasoned global distributors in an international and culturally diverse multi-level marketing organization. The purpose of the study was to identify the knowledge and skills, motivational, and organizational needs necessary for seasoned distributors to train new distributors based upon best practices principles in a standardized manner. The study showed that experienced and up-line distributors needed to have a systematic method for providing entrepreneurial and leadership trainings to new and less experienced distributors. A review of the literature on engagement confirmed the negative impact of disengagement on the economy and on society generally and the organization specifically. Therefore, it was vital that distributors lead their own global teams by promoting engagement and by understanding the value in developing entrepreneurial and leadership training programs in order to prioritize education and training for their global teams. The proposed solution was a lead-the-leader training program specifically designed to address the training and engagement needs of independent entrepreneurs working within the multi-level marketing organization and globalized market. The final aspect of the study delivered a detailed plan for implementation and evaluation of the proposed lead-the-leader training program.

DEVELOPING IMPACTFUL AND ENGAGING TRAINING IS DIFFICULT ACROSS MULTI-NATIONAL SCALE

The study interview asked distributors about their approach to leadership development on their downline teams and found that leadership development philosophies and activities differed dramatically across the organization. While it was important for experienced distributors to be able to convey entrepreneurial and leadership trainings to their downline distributors to compete globally, the interviews confirmed that there was no systematic method for providing leadership trainings to distributors to ensure they were able to lead their own global teams. Most of them relied on one-on-one trainings to develop key leaders, which was shown to be problematic once a team reached thousands of members across multiple countries, and then further complicated by cultural and language barriers. Moreover, the inability of the distributor to effectively and efficiently develop its leaders had a negative impact on their profitability and growth.

A parallel to the academic institutional setting could be inferred because faculty, academic advisors, and retention coaches are charged with developing engaged relationships in a one-on-one model. The research suggests that individually cultivated relationships are a successful approach to retention, especially if the relationship is developed through coaching (Black, 2016). However, effective retention efforts on a large scale are difficult to identify; most institutional efforts are extremely costly, and eventually demonstrate little to no impact (Kelly & Schneider, 2012). Kelly and Schneider (2012) found that, across a diverse set of institutions, targeted retention interventions occasionally proved successful, but the effects were generally small or so minor as to question whether the investment of time and financial resources to the project was worthwhile. This poor track record of large scale institutional retention projects is another reason that academia should look to business environments and other non-traditional leadership training settings for retention solutions, best practices, and innovative approaches to promoting retention and supporting student success.

ADULT LEARNER ISSUES

Perhaps the most direct parallel from the multi-level marketing study to academic retention issues is in the prevalence of adult learner issues in both contexts. In the context of the multi-level marketing organization, the distributors who join the enterprise are adults. The average age of the interviewees was forty. A critical component of the study was to understand the adult learner, when considering the implementation of distributor training programs.

In describing the approaches to take in designing curriculum for adult learners in either a business or an academic setting, a discussion of andragogy is important. According to Knowles, Holton and Swanson (2011) andragogy encompasses the methods and approaches taken in designing curriculum for adult students as they illustrated several aspects of adult learners that make them different from traditional learners. First, adult learners do not excel in a rigid, uncompromising and authoritative environment. Second, in many adult classes, the teacher and the student have equal authority, and at times, it is difficult to discern who benefits from the class more – the adult learner or the teacher. Finally, adult learners are typically intrinsically motivated and they can only meet their goals, needs, and interests through furthering their education (Knowles et al., 2011). Indeed, adult learners need to know why it is important to learn something (Fidishun, 2000).

As a result, in designing effective adult online curriculum, the point of inquiry for teachers, instructors, faculty members, or corporate trainers needs to begin with student needs and interests. Moreover, according to Stilborne and Williams (1996), adult learners learn only what they need to learn, they learn by comparing past experiences with new ones, and they need immediate feedback regarding their progress. Further, the adult learner wants to engage in practical learning experiences because adults do not learn in the same manner as younger peers. Since adults learn in different ways, the online curriculum designer and instructor must consider various learning strategies before designing online curriculum and instruction for adults.

While there are many learning strategies that could be considered, four particular strategies are important in this context, they include contextual, experiential, self-directed and transformational learning. Contextual learning engages the adult learner in tasks featuring real-world scenarios with which they can relate using their prior experience, culture, and profession (Bransford, Brown, & Cocking, 1999). Since adults like to relate new experiences with prior experiences, contextual learning can provide adults with a rich and fulfilling learning experience.

Experiential learning, on the other hand, is the result of making meaningful connections between what the adult learner already knows and new concepts that are introduced. Experiential learning involves action, reflection, abstraction and application (Itin, 1999). To have a successful experiential learning experience, the instructor must make clear the new knowledge, help the adult learner connect the new knowledge to prior knowledge, and assist the adult learner in an analysis of the learning experience.

Self-directed learning has also proven to be a successful strategy in engaging adult, online students. However, to be successful, self-directed learning requires that the adult learner be intrinsically motivated and goal-oriented (Merriam, 2001). Self-directed learning is often the best learning strategy to employ in learning environments that require a high level of independence from adult learners.

Finally, transformational learning uses critical reflection as its foundation. Transformational learning can occur gradually or suddenly and changes the way learners see the world and themselves (Baumgartner, 2001). Adult learners, through critical analysis, reflect on unique scenarios that force them to engage in their own learning experiences by using their own goals and reference points.

While it is important to consider various learning strategies when designing online curriculum for the corporate training setting, it is also critical for designers and instructors to be cognizant of whether the adult students know how to navigate within the online platforms. Because many adults do not yet have the required technological skills, the adult learners need to first master the skills required to engage in online learning in order to be successful (Fidishun, 2000).

The distributor study highlighted that it is important, when designing corporate training modules, to provide collaborative learning spaces, online access to resources in multiple formats, and multiple ways to express key ideas to engage differentiated learners (Stilborne & Williams, 1996). Because these elements of corporate training along with adult learning issues and concepts apply directly to academic institutions, further study of the efficacy of corporate training programs for adult learners could inform academic institutions and programs on a wide variety of retention and student success issues.

ASYNCHRONOUS VERSUS SYNCHRONOURS ONLINE LEARNING ENVIRONMENTS

The significant preference of the interviewed distributors in the study was to be able to access training with peer and senior distributors in a group led, video-based environment. However, the distributors also indicated the desire for flexible access to the trainings as needed, similar to a library, for resources to address the various challenges and issues in their downline personnel. This stated preference for both asynchronous and synchronous training that can be used in a variety of ways across a range of issues requires a careful assessment of the benefits of both asynchronous and synchronous learning.

Through research, scholars determined that courses offered in certain asynchronous formats provided students with a sense of community and an ability to reflect on material in order to form a proper response, along with increased collaboration among classmates (Knapcyzk, Frey & Wall-Marencik, 2005). Asynchronous online learning environments are flexible and self-paced as learners can access the prerecorded coursework with a variety of devices and learning management systems. Notably, asynchronous learning is unique as it does not require a set time or date for accessing the materials.

As courses began to transition online, many educators tried to use synchronous formats to mimic traditional, on-campus courses. In synchronous classes, the instructor and students are logged in at the same time and communicate with each other directly (Shi & Morrow, 2006).

Online formats include webinars, live lectures, video conferencing and chat rooms. Synchronous courses are highly interactive and involve real-time sharing of information. The challenge, however, is that synchronous courses require all participants to be online at the same time.

Several studies found that the most successful synchronous courses have a high level of participant interaction, which allows students to understand and retain the information thereby increasing their course achievements (Offir, Lev, & Bezalel, 2008). To achieve high interactivity throughout the course, the instructor, must therefore be well-versed in the technical aspects of course delivery (Reushle & Loch, 2008).

Even though asynchronous and synchronous online learning environments are widely used, many scholars debate the effectiveness of each delivery and format as well as the impact it has on student learning. In her study, Skyler (2009) used both asynchronous and synchronous elements in an online course and subsequently measured student learning, student satisfaction, and student confidence regarding use of technology. The results of the study found that both methods of online delivery were effective; however, almost three-fourths of the students surveyed stated that they would rather take a course with synchronous lectures versus a course with asynchronous text-based learning elements. Finally, Skylar's (2009) study demonstrated that students enrolled in an online course perceived an increase in their technological skills as a result of the course.

Even though asynchronous and synchronous online delivery formats are both effective, like the stated preference of the distributors, many students prefer synchronous elements within a training or course module. Thus, when developing online curriculum in either the business training or the academic arena it is important to consider incorporating synchronous elements into the program to maximize engagement and satisfaction among both academic student and professional training course participants. Successful program delivery and engaged learning in both areas likewise requires that there are significant and easily accessible resources to assist with the technological aspects of the training or class delivery.

ON-DEMAND TRAINING

The clear preference of the distributors was easily accessible and just-in-time (JIT) training across a wide range of business, social and cultural issues, which allows a student to access digital learning opportunities when the student needs them (Davenport, 2005). JIT learning usually consists of modules that are delivered in a variety of ways, including web-based, face-to-face, or hybrid formats and can also be delivered with low-tech tools or high-tech systems (Paradise, 2008).

According to many scholars, IBM led the market on the implementation of JIT learning. According to employees at IBM, JIT education and training offers employees a way to integrate learning with work in order to enhance employee performance (Davenport, 2005). In addition, on-demand learning opportunities can be personalized and relate to a worker's role, goals, and activity.

JIT learners are typically adults who value learning, prefer to control how it is accomplished, and want to exercise power over when it is scheduled and when they can use it (Hartley, 2000). Successful on-demand learners share the following characteristics: they take initiative, desire information, are flexible, and are oriented around goals, careers, and results. With respect to leadership curriculum, in particular, even though some scholars contend that managerial skills and leadership training are not as effective in an on-demand format, many practitioners believe it is a good substitute when traditional training methods are unavailable or cannot meet the needs of the organization in a timely manner (Comacchio & Scapolan, 2004).

On-demand trainings offer many benefits to participants. On-demand training caters to independent, adult learners who can choose when to engage and access the training. For many of the distributors that participated in the study this type of training is ideal. Moreover, even though scholars criticize the use of on-demand training for leadership skills, this method of training delivery was both the most feasible and the most requested. Thus, despite the criticisms within the academic literature of this type of training, feasibility and student preference are the most important elements and should be given the most weight when designing adult learning programs. Therefore, online degree programs should carefully consider the benefits of incorporating on-demand material as part of course and program curriculum.

ON-THE-JOB TRAINING AND DEVELOPMENT

The fast-paced environment of the multi-national, multi-level marketing organization required that new distributors be able to quickly and easily learn how to be effective and have the ability to access on-the-job training. In traditional on-the-job training, trainees experience work as a primary observer. As a result, the training is unstructured and follows the flow of the work being done by the trainer (Sisson, 2002). Generally, an on-the-job trainer is a skilled worker who may even be an expert in his or her job; however, the trainer is not a trained instructor. Sisson (2002) contends that on-the-job training is not an effective method since it is not standardized, relies heavily on the trainer to convey the information in a show-and-tell format, and as a result, produces inconsistent results.

Despite criticisms of using on-the-job training, Sisson (2002) shares four reasons it is still the dominant training method: (1) it is a hands-on training; (2) includes real-world practice; (3) is a simple approach; and (4) requires no corporate formalities and relies heavily on informal interactions among workers. Sisson (2002) also advocates using a hands-on training approach instead of a traditional on-the-job training approach. To be effective, hands-on training requires six steps: (1) prepare for the training; (2) open the training session; (3) present the training subject; (4) practice the training skills; (5) evaluate trainee performance; and (6) review the training subject.

According to Sisson (2002), in preparing for the training, the trainer needs to have the materials ready to distribute, set up the training area, and simply get ready. In opening the session with the trainee, the trainer should determine the trainee's skill level, explain the objective of the training, and share reasons why the training is important. In presenting the subject, the trainer should start a two-way demonstration by going step-by-step. The trainer should conclude this step by demonstrating twice and setting the trainee up to practice the skill.

In order to practice the skill, the trainer should set up something realistic. After the trainee attempts the skill, the trainer must evaluate his or her performance. In evaluating the performance, the trainer should observe critically, offer words of encouragement, provide coaching, stress the importance of quality over expeditiousness, and ask the trainee a lot of questions. Finally, in reviewing the training subject, the trainer should review the trainee's performance, set up additional practice opportunities, offer guidance, follow-up with the trainee frequently, and encourage the trainee to ask the trainer questions as the trainee masters the skill.

According to Sisson (2002) the HOT POPPER training method is best accomplished individually or in small groups. To that end, the biggest limitation to the hands-on training approach is the inability of a trainer to address large groups since many participants will not be able to view the live demonstrations or participate fully. Given this limitation of scale, employers and academic institutions should consider incorporating elements of the hands-on training approach outlined by Sisson (2002) within other training and teaching approaches.

Even though Sisson (2002) establishes the benefit of standardizing on-the-job training for adult learners, in various non-traditional academic settings, like vocational and workforce development programs, these methods are rarely used (Kuipers, Meijers, & Gundy, 2011). According to Kuipers, et al (2011), an effective vocational and career-oriented learning environment differs in five significant ways from a traditional learning environment. First, in an effective vocational and career-oriented learning environment, the primary focus is not on information transfer, but instead the focus is on gaining actual work experience. Second, in an effective vocational learning environment, there must be a "dialogue" between teacher and student, instead of a traditional "monologue" teaching approach. Third, for students to effectively learn a craft, it is critical to focus on providing students with consistent guidance instead of decision-making opportunities. Fourth, to be most effective, in a vocational learning environment, an understanding and appreciation of student emotion is equally as important as cognitive tasks. Finally, an effective and productive vocational and career-oriented learning environment provides students with the skills to make choices and engage in mutual participation instead of defining a standard-learning direction.

Indeed, the dynamic success of the multi-level marketing distributors in the study and the rate of growth that the organization has been able to achieve and maintain are so impressive that a reexamination of on-the-job style training utilization within certain types of academic settings should be undertaken to determine if it can have a meaningful and sustained impact on retention and student success.

DESIGNING SUCCESSFUL TRAINING PROGRAMS AND FOCUS ON LEADERSHIP DEVELOPMENT

Many scholars agree that leadership curriculum is difficult to design. Allen, Miguel and Martin (2014) contend there are four main reasons preventing the design and implementation of effective leadership curriculum: (1) there is a lack of empirical and theoretical research on designing effective leadership curriculum; (2) there is a lack of transferable leadership skills from student to manager and leader; (3) there is no well-developed leadership development theory; and (4) faculty do not use existing research or evidence-based teaching in delivering their leadership courses. In examining the challenges that exist in leadership development curriculum, Allen et al. (2014) identify and propose a model for developing curriculum called the "Know, See, Plan, Do Model (KSPD)." Even though the authors contend that the KSPD model was designed for use in higher education settings, they acknowledge that their model can also be used in organizational environments (Allen et al., 2014).

Allen et al. (2014) explain their model through each of its core components - know, see, plan, and do. With respect to "Know", they explain that at the beginning of every leadership development curriculum, students should be well steeped in leadership theory, concepts and literature. According to the authors, students may learn about women in leadership, power and leadership, authenticity, transformational leadership, and ethics, among many other key concepts and theories. Through this foundation, students are then able to ascend to the next rung of the leadership development curriculum ladder (Allen et al., 2014).

In the "See" phase, Allen et al. (2014) expect students to integrate new leadership information into existing mental schema. In this phase, students begin to practice the leadership concepts they obtained in the "Know" phase. The authors suggest that in practicing these concepts, the students may elect to keep a daily diary of leadership events that occur, students may watch videos to examine what types of leadership styles are being expressed, or in an online setting, students might blog about specific leadership concepts they observed. The "See" phase takes time for students to practice what they learned.

Since the goal of leadership development curriculum is to move students from novices to experts, they must synthesize the information they have learned and understand how to properly analyze a problem before they can solve it. According to Allen et al. (2014) since leaders are often confronted by unique problems, they must be adept at problem-solving skills. Leaders must be able to define the problem, generate solutions, and develop the best plan to solve it. In the "Plan" phase, students are tasked with developing skillful interventions. The "Plan" component requires students to be able to forecast potential outcomes and evaluate options available for taking action. The authors suggest that students can practice this skill by examining a past failure and serving as class consultants to each other. This type of practice allows students to critically reflect on and analyze the past failure and generate alternative solutions.

In the "Do" phase of the development of leadership curriculum, Allen et al. (2014) emphasize the importance of applying knowledge from different perspectives and then taking the appropriate action. In this phase, the authors highlight the importance of experiential learning, which enables students to acquire knowledge in four phases: Experience, Reflect, Think and Act. The goal, in the "Do" phase, is to provide students with an opportunity to implement a plan. Allen et al. (2014) suggest that to test this component, instructors should provide students with the opportunity to complete a larger project and give as examples breaking a Guinness world record, assisting a non-profit, or planning an event.

The Know, See, Plan, Do model proposed by Allen et al. (2014) provides a sound basis for the development of engaging leadership curricula and can be applied to a wide range of subject matter areas and professional majors. This structure intentionally creates an engagement focus within the curriculum. The model also emphasizes action and practical application, both of which support engagement and are key to building a student's confidence in both a work and an academic setting. The study of the multi-level marketing distributor illustrates the value of creating curriculum around an engagement focus.

THE COURSE MUST BE CLEAR AND COMPELLING

Another critical finding of the distributor study was that it was critical for trainers to identify key internal and external stakeholder needs. Once the trainers acquired an understanding of stakeholder needs, the training could then be tailored to meet those needs. If stakeholder needs are properly identified and addressed, training participants will complete the training with a renewed motivation to transfer the skills they learned to their global businesses.

As the seasoned distributors were extremely busy building their businesses and their downline organizations they emphatically and repeatedly stated that they had little time to engage in anything other than the operation of their businesses. Given the extremely busy schedules and business growth priorities of the interviewed distributors, it was quickly apparent that any training had to have clear and relevant course objectives that would quickly resonate with the distributors and help them to understand the value of the training and encourage them to make the training a priority. The decision to attend or complete a training program was directly linked to how applicable the distributor felt that the training would be to their most immediate issues and challenges. Only issues or challenges of high priority to the distributors would cause them to prioritize the training over their work of recruiting and building their downline networks.

To address this strong perspective about prioritizing training in an ultra-busy and dynamic work environment, the distributor study recommended that the company hire an expert consultant to help develop clear and relevant training course objectives that could be quickly scanned by the distributor. In creating these objectives, the consultant needed to define the roles of key individuals in the training program, define the characteristics of a successful trainer, and identify the importance and significance of the training. Careful creation of course objectives would also allow the trainers and the company to measure the effectiveness of the training to determine whether to modify any key areas or to add or subtract training topics.

This finding from the study is consistent with a prominent reason that students give for stopping-out or electing to discontinue their academic programs. If the student did not believe that their program of study was interesting, engaging, or sufficiently related to their desired career path they became less interested and less motivated to complete the course or program (Black, 2016). Improving retention rates, especially among working adult student communities, is often as simple as carefully matching the student to the program that most directly meets their immediate needs and near-term career goals, while ensuring that the student understands exactly how each course will help them to meet these goals.

Both the study finding and the literature suggest that institutions should work hard to ensure that students select an academic or a professional development program that will have a direct and positive career impact. The student in the academic environment and the distributor in the business environment must understand exactly how each course within the program will further their career goals. The course or training program must have a near term impact that the student can measure. This simple and consistent alignment of courses with the specific career goals or the immediate challenges of the distributors will lead to increased program retention, as well as, increased satisfaction with the program (Black, 2016).

Students will also subsequently report higher career satisfaction metrics and have stronger alumni relationships with their institution (Black, 2016). Thus, for training to be successful within the multi-level marketing distributorship and for coursework to be engaging and relevant for students, it is important for the trainers and instructors to share with the adult training participants or the adult students what the training program or the course attempts to address and the impact that completing the course will have on their academic careers or their global businesses' success (Knowles et al., 2011).

LANGUAGE BARRIERS

The multi-level marketing distributor study examined a highly successful organization operating in a cross-border, cross-cultural, highly virtual, global environment. Thus, the consideration of language was highly appropriate and important in the context of developing a new training and education initiative for the organization. Many scholars note that globalization is requiring organizations to examine language more closely. According to Charles (2007) "only language can enable individuals and companies to communicate" (p. 261). Many scholars explored the impact on organizations that adopted a common corporate language (Louhiala-Salminen, Charles, & Kankaanranta, 2005, 2012).

Scholars Louhiala-Salminen et al. (2012) recognize the use of language as a corporate asset and claim that the "true cost of the language barrier cannot be measured in terms of translating and interpreting but in damaged relationships" (p. 264). Since language is a key to building cultural knowledge, enhancing trust, and building rapport, both the multi-level marketing company and its successful distributors must develop strategic plans regarding use of language in various settings, both internal and external to the organization. The results from some of the distributor interviews showed that language can unintentionally cause power imbalances and inequality among the distributors due to the diverse international organization with proficient and less proficient language distributors working within the same network or even within the entire company. Limited language skills can inhibit knowledge sharing that is critical to successful operations and the achievement of organizational goals as noted in the distributor interviews. If issues of language and cultural barriers persist, distributors begin to disengage from the organization, and they will eventually quit.

Like the distributors in the study, many students experience feelings of isolation due to limited language skills. Globalization is not only changing business boarders, but academic boarders as well. Universities and colleges throughout the world participate in a fluid exchange of students among their global campuses. In much of the academic literature, there is a tendency toward over-generalization with respect to the needs and experiences of international students (Carroll, 2015). However, Jones (2017) contends that by over-generalizing the international student experience, academic institutions, faculty and staff, start seeing "deficits" in international students, including, for example, less developed critical thinking skills and communication skills. This "deficit-focused" thinking ignores diversity among students, both domestic and international, and is not a solid indicator of student needs at an academic institution (Jones, 2017). Instead, Jones (2017) suggests that the organizational structures need to change at the institutional level. Indeed, to bridge the gaps among diverse student populations, both in language and culture, there should be interdisciplinary connections among offices dealing with access, equality and diversity issues, in order to reimagine the student experience (Jones, 2017). To solve issues of language and culture, businesses and academic institutions must look toward integrated service delivery and training opportunities that celebrate the heterogenous nature of student and worker populations.

CULTURAL BARRIERS

As most successful seasoned distributors within the study led teams of 40,000-50,000 distributors in 40 to 50 countries, it was important to consider the multicultural dimensions of a new education and training initiative. Within the study, cultural barriers were potentially systemic and they were a top concern of high-level distributors. The strong recommendation from the study was that when designing online curriculum for global distributors, the company must consider and anticipate potential cultural roadblocks that might prevent successful implementation.

To avoid dire unintended consequences, online curriculum designers and instructors must identify ways to engage students and deliver culturally sensitive and appropriate learning experiences (Gaudelli, 2006). Studies show that students who engage in an environment that optimizes their learning preferences tend to score higher on learning assessments, as well as, learning attitude scores versus students who learn in an unsuitable environment (Shih, Lui, & Sanchez, 2013). Moreover, if developing online curriculum is a goal for an organization, the curriculum designers must keep in mind the orientation that adult students' desire "life-centeredness" (Alalshaikh, 2015).

The consideration of cultural issues was particularly important within the multi-level marketing study because the company operates in many countries, with many cultures and languages and it was evident early in the study that a one-size-fits-all solution to cultural issues and barriers whether real or perceived would not work. Thus, the recommendation of the study was to develop several pilot programs and to make modifications to the pilot programs as data became available and as new countries with new cultural issues joined the company.

Like many international businesses, academic institutions are tasked with identifying the reasons why international student engagement is lower, in comparison to local students. However, according to Straker (2016), academic institutions, faculty and staff are focusing their efforts in the wrong place. Instead of focusing on the differences between international students and local students in terms of language and culture, Straker (2016) suggests academic institutions focus instead on participation. Moreover, academic institutions must also recognize that the international student experience is not generalizable. Students, even if they come from the same country, have different needs (Jones, 2017). As a result, the institutional focus should instead be on the transition to higher education and the provision of supports and services for all students. By operating laterally, instead of vertically, academic institutions are in a better position to offer comprehensive support to a socially, culturally, and internationally diverse student population (Jones, 2017).

FACTORS AFFECTING WOMEN AND MINORITY LEARNERS

The international multi-level marketing organization that was the subject of the dissertation study was unique, especially because its founder is a minority woman. Notably, the majority of distributors that provided interviews for the study were women, many of whom operated their distributorship networks in emerging and developing economies. A key element of the study was to identify the important factors and opportunities in the company culture to develop its women entrepreneurs and to understand the factors that could impede their performance and retention.

According to scholars, many women entrepreneurs often have the ability to transform many of the world's economies as they explore untapped market niches; and, as a result, emerge as key players in a difficult world economy (Apergis & Pekka-Economou, 2010). However, there are barriers to the success of women entrepreneurs, including lack of education, finances, training, environment, and mentorship (Terjesen & Elam, 2012). For rural women, these challenges are compounded including among other factors, lack of education, experience, professional skills in modern business contexts, quality control systems, and most importantly, the desire to accept entrepreneurial risk (Lassithiotaki, 2011).

Studies show that training can be a pivotal factor in motivating women to become entrepreneurs but may only meet the needs of some women (Bauer, 2011; Bhardwaj, 2013). In his article, Bhardwaj (2013) highlights the results of several studies that demonstrate business education has a positive effect on women entrepreneurs' income potential. Moreover, training builds their confidence, and, with flexible training programs and interest based skills training, women are more apt to engage in entrepreneurial endeavors (Bhardwaj, 2013). According to the study of the multi-level marketing company, for many of the women distributors, joining this company was their first foray into the business world. It was critical for the company to acknowledge potential issues facing its women entrepreneurs and to provide education and training to help them combat gender-oriented business challenges. Appropriate training can pave the way for women to become successful business leaders in a range of business, entrepreneurial and academic settings.

Like the participants in the 2015-2016 study, women who return as adult learners to traditional academic institutions are seeking empowerment, increased self-confidence, and greater employment opportunities (Shea & Stone, 2014). However, research on female returning students reveals the complex issues they face when returning to traditional academic settings, particularly women who are older or women who come from low socio-economic backgrounds. Accordingly, it is critical for academic institutions to provide adequate support and guidance for returning adult women learners, who tend to lack an understanding of traditional academic procedures, including: knowledge of enrollment processes, time management skills, and financial aid (Shea & Stone, 2014). Notably, students do not experience transitions, like these, in a "time-bound" phase (Shea & Stone, 2014). Thus, it is critical for academic institutions with an eye towards retention and engagement to provide the necessary supports for returning women students to succeed. These supports may include assistance with issues of academic rigor, writing, adaptation to a new social landscape, forming an identity as a student, time management skills, and preparing for exams, among many other complex issues.

One effective individualized retention and student success support for returning adult students, most of whom are female and struggling to balance life and work challenges while returning to school to complete their education, is the use of professional coaching. Coaching supports students in overcoming obstacles and persisting through challenges and works to keep students accountable to themselves and to support them in reaching goals and overcoming challenges and obstacles to success (Black, 2016). Emerging research suggests that counseling and academic advising support offered through professional coaching can significantly improve students' performance and persistence (Black, 2016; U.S. Department of Education, 2014). No one strategy fits everyone, and proactive coaches can carefully customize academic advising and counseling strategies to fit the needs of each student and the challenges and issues they are facing (Bettinger & Baker, 2013). Improving the graduation rates and the individual experiences of female non-first-time students will have a profoundly positive impact on the individual, her workplace, family and community, and coaching plays a key role in helping students persist to completion of a baccalaureate degree (Black, 2016). Further, improving the graduation rates of non-first-time students, the lowest graduating student group, will have the largest positive impact on institution-wide graduation rates and on improving the number of baccalaureate degree holders nationwide (Lumina Foundation, 2015).

FUTURE RESEARCH DIRECTIONS

Comprehensive retention and student success studies within highly specialized and boutique corporate and business environments are rare. The same is true of retention and worker success studies within highly stigmatized and marginalized business and industry models such as multi-level marketing, door to door sales and telemarketing companies. With few exceptions, these types of companies and organizations have been deemed unworthy of mainstream academic inquiry and assessment. Yet, these highly specialized, stigmatized and marginalized businesses and organizations offer unique environments within which to study retention, success, motivation to learn, resilience and the perspective of the learner towards training and education. The broad application of the findings from this study of an international multi-level marketing company suggest that academic researches should actively seek out unique corporate and organizational structures that are designed to support diverse socio-economic, cross cultural and multilingual workers and then study them through multiple and distinct lens of inquiry.

As this study demonstrates, the correspondingly innovative approaches to retention of workers or learners within these types of companies and organizations are instructive across a range of learning and educational settings. More studies of the approaches, strategies and best practices to improve retention and self-reported worker or learner success within these types extremely non-traditional or fringe business and learning environments are necessary to help academic institutions and learning organizations develop and assess new, innovative and impactful approaches to the age-old question of why people leave.

CONCLUSION

Looking closely at training and development within an international multi-level marketing organization can inform retention and student success issues and challenges in degree completion programs, workforce development programs, online programs and other academic degree and extended education environments. Multi-level marketing organizations have dramatically changed the lives of millions of people and provided significant professional opportunities where there were few traditional business and industry jobs. If success is possible in the multi-level marketing company, then surely it must be possible within a well-supported, well-funded, educational institution that is targeting underserved populations.

The 2015-2016 study of an international multi-level marketing company sought to address the age-old question of why people leave. The study explored this question within the highly stigmatized multi-level marketing industry where people who had been marginalized from traditional business and industrial job opportunities found a way to thrive and succeed. A significant element of the study also looked at how training and education impact entrepreneurial engagement, retention and success. The research and findings related to success of non-traditional or marginalized workers and how to train, support and retain them within the multi-level marketing organization inform retention and student success in non-traditional academic programs and institutions.

Many of the key considerations related to training program development within the highly innovative and non-traditional multi-level marketing organization are equally important considerations for similarly situated academic institutions. The 2015-2016 study of a multi-level marketing distributor can provide key insights into retaining and engaging students within academic programs focused on workforce development or retraining and vocational degree programs. While traditional academic and business environments may differ in many important respects, the cost of failing to engage, retain and help students and employees is simply too significant to ignore.

REFERENCES

Alalshaikh, S. (2015). Cultural Impacts on distance learning, online learning styles, and design. *Quarterly Review of Distance Education*, *16*(3), 67–75.

Allen, S. J., Miguel, R. F., & Martin, B. A. (2014). Know, see, plan, do: A model for curriculum design in leadership development. *SAM Advanced Management Journal*, *79*(2), 26.

Apergis, N., & Pekka-Economou, V. (2010). Incentives and female entrepreneurial activity: Evidence from panel firm level data. *International Advances in Economic Research*, *16*(4), 371–387. doi:10.1007/s11294-010-9277-9

Bandura, A., & Locke, E. A. (2003). Negative self-efficacy and goal effects revisited. *The Journal of Applied Psychology*, *88*(1), 87–99. doi:10.1037/0021-9010.88.1.87 PMID:12675397

Bauer, K. (2011). Training women for success: An evaluation of entrepreneurship training programs in Vermont, USA. *Journal of Entrepreneurship Education*, *14*, 1.

Baumgartner, L. M. (2001). An update on transformational learning. *New Directions for Adult and Continuing Education*, *2001*(89), 15–24. doi:10.1002/ace.4

Berry, R. M. (1997). *Direct Selling: From door-to-door to network marketing*. Butterworth Heinemann.

Bettinger, E. P., & Baker, R. B. (2013). The effects of student coaching: An evaluation of a randomized experiment in student advising. Retrieved from http://www.insidetrack.com/wp-content/uploads/2014/09/aera-educational-evaluation-policy-analysis.pdf

Bhardwaj, B. (2014). Impact of education and training on performance of women entrepreneurs: A study in emerging market context. *Journal of Entrepreneurship in Emerging Economies*, *6*(1), 38–52. doi:10.1108/JEEE-05-2013-0014

Black, R. (2016). Coaching to Increase Retention in Online Degree Completion Programs. Retrieved from http://digitallibrary.usc.edu/cdm/compoundobject/collection/p15799 coll40/ id/298507/rec/2

Carroll, J. (2015). *Tools for Teaching in an Educationally Mobile World*. London: Routledge.

Charles, M. (2007). Language matters in global communication: Article based on ORA lecture, October 2006. *Journal of Business Communication*, *44*(3), 260-282.

Comacchio, A., & Scapolan, A. (2004). The adoption process of corporate e-learning it Italy. *Education + Training*, *46*(6/7), 315–325. doi:10.1108/00400910410555222

Davenport, R. (2005). A new shade of BIG BLUE IBM reinvents its workforce. *T + d*, *59*(5), 34.

Eccles, J. S., & Wigfield, A. (2002). Motivational beliefs, values, and goals. *Annual Review of Psychology*, *53*(1), 109–132. doi:10.1146/annurev.psych.53.100901.135153 PMID:11752481

Fidishun, D. (2000, April). Andragogy and technology: Integrating adult learning theory as we teach with technology. In *Proceedings of the 2000 Mid-South Instructional Technology Conference*.

Gallup (2013). State of the global workplace. Retrieved from http://www.gallup.com/services/176735/state-global-workplace.aspx

Gaudelli, W. (2006). Convergence of technology and diversity: Experiences of two beginning teachers in web-based distance learning for Global/Multicultural education. *Teacher Education Quarterly*, *33*(1), 97–116.

Hartley, D. E., & Cone, J. (2000). *On-demand learning: Training in the new millennium*. Amherst, Mass: HRD Press.

Jones, E. (2017). Problematising and reimagining the notion of international student experience. *Studies in Higher Education*, *42*(5), 933–943. doi:10.1080/03075079.2017.1293880

Kelly, A. P., & Schneider, M. (2012). *Getting to graduation: The completion agenda in higher education*. Baltimore, MD: John Hopkins University Press.

Knapczyk, D. R., Frey, T. J., & Wall-Marencik, W. (2005). An evaluation of web conferencing in online teacher preparation. Teacher Education and Special Education. *The Journal of the Teacher Education Division of the Council for Exceptional Children*, *28*(2), 114–124. doi:10.1177/088840640502800205

Knowles, M. L., Holton, E. F., & Swanson, R. A. (2011). *The adult learner* (7th ed.). Houston, TX: Gulf.

Kuijpers, M., Meijers, F., & Gundy, C. (2011). The relationship between learning environment and career competencies of students in vocational education. *Journal of Vocational Behavior*, *78*(1), 21–30. doi:10.1016/j.jvb.2010.05.005

Lassithiotaki, A. (2011). rural women and entrepreneurship: A case study in Heraklion Crete prefecture, Greece. *Journal of Developmental Entrepreneurship*, *16*(2), 269–284. doi:10.1142/S1084946711001835

Louhiala-Salminen, L., Charles, M., & Kankaanranta, A. (2005). English as a lingua franca in Nordic corporate mergers: Two case companies. *English for Specific Purposes*, *24*(4), 401–421. doi:10.1016/j.esp.2005.02.003

Louhiala-Salminen, L., & Kankaanranta, A. (2012). *Language as an issue in international internal communication: English or local language? if English, what English?* Public.

Lumina Foundation. (2015). A stronger nation through higher education. Retrieved from http://www.luminafoundation.org/publications

Maddux, J., & Gosselin, J. (2015). *Self-Efficacy*. Oxford Bibliographies in Psychology; doi:10.1093/obo/9780199828340-0088

Merriam, S. B. (2001). Andragogy and self-directed learning: Pillars of adult learning theory. *New Directions for Adult and Continuing Education*, (89), 3-14. doi:10.1002/ace.3

National Research Council. (1999). How people learn: Brain, mind, experience, and school. Washington, D.C: National Academy Press.

O' Shea, S., & Stone, C. (2014). The hero's journey: Stories of women returning to education. *The International Journal of the First Year in Higher Education, 5*(1), 79. doi:http://dx.doi.org.libproxy1.usc.edu/10.5204/intjfyhe.v5i1.186

Offir, B., Lev, Y., & Bezalel, R. (2008). Surface and deep learning processes in distance education: Synchronous versus asynchronous systems. *Computers & Education*, *51*(3), 1172–1183. doi:10.1016/j.compedu.2007.10.009

Paradise, A. (2008). *Informal learning: Overlooked or overhyped?* Alexandria: American Society for Training & Development, Inc.

Reushle, S., & Loch, B. I. (2008). Conducting a trial of web conferencing software: Why, how, and perceptions from the coalface. *TOJDE/ Turkish Online Journal of Distance Education*, *9*(3), 19–28.

Rueda, R. (2011). *The 3 dimensions of improving student performance*. New York: Teachers College Press.

Shi, S., & Morrow, B. V. (2006). E-Conferencing for Instruction: What Works? *EDUCAUSE Quarterly*, *29*(4), 42–49.

Sisson, G. R. (2002). *Hands-on training: A simple and effective method for on-the-job training*. San Francisco: Berrett-Koehler Publishers.

Skylar, A. A. (2009). A comparison of asynchronous online text-based lectures and synchronous interactive web conferencing lectures. *Issues in Teacher Education*, *18*(2), 69.

Stilborne, L., & Williams, L. (1996). Meeting the needs of adult learners in developing courses for the Internet. In *Proceedings of INET '96*.

Straker, J. (2016). International student participation in higher education: changing the focus from international students to participation. *Journal of Studies in International Education*, *20*(4), 299–318. doi:10.1177/1028315316628992

Terjesen, S., & Elam, A. (2012). Women entrepreneurship: A force for growth. *International Trade Forum*, (2), 16.

United States Department of Education. (2014). Pathways to success, integrating learning with life and work to increase national college completion. Retrieved from http://www2.ed.gov/about/bdscomm/list/acsfa/ptsreport2.pdf

Yarnell, M., & Yarnell, R. R. (2010). *Your first year in network marketing: Overcome your fears, experience success, and achieve your dreams!* Three Rivers Press.

KEY TERMS AND DEFINITIONS

Andragogy: The chapter uses the definition of andragogy developed by Knowles, Holton and Swanson (2011). Andragogy encompasses the methods and approaches taken in designing curriculum for adult students.

Coaching: The chapter uses the definition of coaching from the International Coach Federation website. The International Coach Federation defines professional coaching as "partnering with clients in a thought-provoking and creative process that inspires them to maximize their personal and professional potential" (International Coach Federation, 2015). The International Coach Federation website lists the core competencies of professional coaching as goal setting and managing incremental progress towards goals, as well as, developing individual accountability measures regarding progress towards one or more collaboratively established goals or skill proficiencies.

Contextual Learning: The chapter uses the definition of contextual learning developed by Bransford, Brown and Cocking (1999). They define contextual learning as engaging the adult learner in tasks featuring real-world scenarios with which they can relate using their prior experience, culture, and profession as context (Bransford, Brown, & Cocking, 1999).

Engagement: The chapter utilizes the definition of engagement that appeared in the 2013 Gallup Report: The State of the Global Workplace: Employee Engagement Insights for Business Leaders Worldwide. The report defines engagement as a worker who is "emotionally invested" and focused on creating daily value for the organization (Gallup, 2013).

Experiential Learning: The chapter utilizes the definition of experiential learning put forth by Itin (1999). Experiential learning involves action, reflection, abstraction and application (Itin, 1999).

HOT POPPER Training Method: The HOT POPPER training method is a form on on-the-job training discussed by Sisson (2002).

Just-in-Time (JIT) Training: The chapter utilizes the definition of just-in-time (JIT) training from Paradise (2008). JIT learning allows a student to access digital learning opportunities when the student needs them. JIT learning usually consists of modules that are delivered in a variety of ways, including web-based, face-to-face, or hybrid formats and can also be delivered with low-tech tools or high-tech systems (Paradise, 2008).

Know See Plan Do (KSPD) Model: Allen Miguel and Martin's (2014) four-part model for designing curriculum. Allen et al. (2014) explain their model through each of its core components – know, see, plan and do.

Multi-Level Marketing: The chapter utilizes the Federal Trade Commission (FTC) definition of multi-level marketing that appears on its website. "In multilevel or network marketing, individuals sell products to the public — often by word of mouth and direct sales. Typically, distributors earn commissions, not only for their own sales, but also for sales made by the people they recruit."

On-the-Job Training: The chapter utilizes the descriptive definition of on-the-job training from Sisson (2002). Sisson explains on-the-job training as, trainees experiencing work as a primary observer. Usually, the training is unstructured and follows the flow of the work being done by the trainer. Generally, an on-the-job trainer is a skilled worker who may even be an expert in his or her job; however, the trainer is not a trained instructor.

Self-Efficacy: The chapter utilizes Albert Bandura's definition of self-efficacy from his seminal 1977 work, Self-Efficacy: Toward a Unifying Theory of Behavioral Change. Bandura's definition can be summarized as an individual's confidence in their ability to achieve desired goals and intended outcomes.

Transformational Learning: The chapter utilizes the definition of transformational learning put forth by Baumgartner (2001). Transformational learning changes the way that learners see the world and themselves.

Chapter 7
Anonymous Online Learning in Korea and Integrating Anonymous Learning Elements

Alan Cromlish
Namseoul University, South Korea

ABSTRACT

This paper explores anonymous online learning as a tool to overcome specific teaching and learning issues within Korean post-secondary institutions. The chapter utilizes a survey of a small group of ESL students at a single Korean university to better understand student preferences and opinions about non-traditional learning options and opportunities in Korea. While many students in Korea have not been exposed to online learning, the students surveyed expressed interest in learning online and they were especially interested in collaborative learning opportunities. As more online classes and online learning opportunities start to become available in South Korea, this study explores anonymous online learning as an effective tool to overcome some significant and distinct teaching and learning challenges at Korean post-secondary institutions. The anonymous online learning suggestions and approaches in the paper can be implemented within fully online courses and blended classes but they can also be used as stand-alone online components of traditional face to face and ESL courses.

INTRODUCTION

Technology is no longer a sidebar in higher education. Today it is a critical component of teaching and learning in any program and across all program delivery modalities. Students utilize tablets, smart phones, and laptops throughout class and they move seamlessly between their devices. Wi-Fi enabled classrooms are now norm and poor or intermittent Wi-Fi is a sure way to draw the ire of students and faculty alike. These technology and smart device proliferations are forcing educators to rethink teaching techniques and to look for new and interesting ways to further integrate technology supported learning into their courses and programs.

DOI: 10.4018/978-1-5225-2998-9.ch007

Online forums and online submissions have changed how students interact with each other and how students interact with the instructor. These technologies have driven changes in student to instructor and student to student interactions can be used to an educator's advantage. These advantages are more evident when the changes are combined with accessibility to the unlimited information on internet to better facilitate student learning and to promote and enhance authentic learning experiences. Although Korea is viewed as a highly-evolved, technology driven economy and country it is both interesting and unfortunate that Korea has not taken advantage of the vast array of technology tools available in online learning. Korea is well behind the United States and other developed countries such as Australia and New Zealand in supporting and encouraging the proliferation of technology across the curriculum.

REVIEW OF THE LITERATURE

South Korea has a rich and deep history. The country has strong early influences from China dating back to the Gojoseon Kingdom as late as 300 BC (Daehwan, 2003). During the Goryeo Dynasty, from 918 to 1392 AD, Buddhism was the dominant religion, but as the dynasty fell, Confucianism started to become more prominent and blended with previously dominant Buddhist beliefs eventually creating Neo-Confucianism (Daehwan, 2003; Koh, 2003). The succession of the Joseon dynasty, from 1392 to 1910 AD, saw the rise of King Sejong who favored Confucianism and helped place its roots in Korea through his focus on education and scholarships (Koh, 2003; Weidman & Park, 2000). The Gukjagam, the highest educational institution built during the Goryeo Dynasty, was enhanced, relocated to modern day Seoul, and renamed the Sungkyunkwan (Lee & Yi, 2002; Weidman & Park, 2000). The Sungkyunkwan became the most important center for training Confucian scholars who were also political and cultural leaders. King Sejong also introduced Hangul, Korea's current language. With the introduction of Hangul, literacy in Korea rose and so did Confucian beliefs which ultimately shaped the language to include hierarchical language and placed importance on Confucian based relationships (Daehwan, 2003; Koh, 2003; Weidman & Park, 2000; Zhang, Lin Nonaka, & Beom, 2005).

The Confucian influence led to teacher centered classrooms that provide little room for students to participate in discussion with classmates or the teacher because of the relationships that are emphasized by Confucian teachings (Bang & Kim, 2016; Daehwan, 2003; Zhang, Lin Nonaka, & Beom, 2005). There is little praise given along with frequent scolding (Bang & Kim, 2016). This criticism of the students in conjunction with the hierarchical relationships built into the classroom through the language and social norms has caused students to abhor school (Bang & Kim, 2016; Joen & Kim, 2012; Yun, Kim, & Kim, 2009).

Korea's traditional mindset and feelings of needing to hold onto past ideas directly conflicts with their advanced information communication technology infrastructure (Geert Hofstede, 2017). In the past, Korea lacked online courses, but with the creation of K-MOOC, Korea is rapidly developing online courses to better utilize the advancements made through their ICT infrastructure (K-MOOC, 2017). This new emerging method of educational instruction in South Korea provides new opportunities for educators to provide greater motivation and participation from their students compared to traditional classes (Warschauer, 1995, Zheng & Warschauer, 2015).

SOUTH KOREA: ULTRA CONNECTED AND TECHNOLOGY FOCUSED BUT TRAILING IN ONLINE LEARNING

South Koreans are some of the most connected people in the world and the country enjoys some of the fastest internet systems and connection speeds in the world. Korean citizens are so tied to the internet that the country requires citizens to register their identification number, full name and email addresses with the government. Yet, online learning and online courses and programs remain uncommon and underutilized in Korea. Korean online class adoption significantly lags that of the United States, Canada, U.K., or Australia. However, according to the ICEF Monitor (2012), recent trend data indicates that Korean universities are beginning to significantly increase the number of online classes as well as other formal and informal internet based learning options, and Korea is now leading Asia in e-learning adoption.

A good example of a growing online learning project is the Korean Ministry of Education (KME) sponsored K-MOOC project. The K-MOOC project launched in October of 2015 with 27 open enrollment free online courses from ten Korean Universities. By 2016 the project had expanded to 140 course offerings and the program expects to offer 500 courses by 2018 (K-MOOC, 2017). One of the main motives for the project was to create alternative learning opportunities for Korean citizens that took advantage of the "advanced ICT infrastructure environment and e-learning technology" available in Korea. The project's webpage also proudly states that Korea has achieved the highest levels of technology development in the world and that this strong technology infrastructure is an important element of the project. Given these technology growth trends, it is logical to predict that Korea will consistently increase online class and program offerings over time across a range of credit and extended education settings.

ONLINE DISCUSSION AND ONLINE ANONYMITY IN KOREAN ESL COURSEWORK

Online discussions and online anonymity provide Korean university students with a different way to learn about English-speaking culture. It also adds some much-needed innovation and flexibility to the traditional teacher-focused ESL classrooms that are usually based around a textbook. The positive feedback that can be incorporated into online discussions when combined with the above average digital media usage of Koreans makes them good candidates for collaborative online learning. 81.6% of South Koreans access the internet on their mobile phone. "About 31 percent of South Korean smartphone users watch mobile video on a daily basis, and another 25 percent of them watch mobile videos weekly." (Internet usage, n. d.). Koreans also consume a large amount of their digital media through their social networking sites. This is reflected in 82% of South Koreans being active on social media in 2016, the highest in Asia (Internet usage, n.d.). Online discussions can be used as a tool to help facilitate student engagement, discourse, and discussion and to remove unauthentic feedback and the fear of embarrassment (Gosling, Augustine, Vazire, Holtzman, & Gaddis, 2011).

THE BENEFITS OF ONLINE DISCUSSION

Instructors use discussion boards as the primary source for student interaction. The majority of online discussion boards primarily involve reading and writing exercises. This appeals to Korean students because there is no speaking component and it helps to facilitate improvement in their reading and writing skills. Improved reading and writing skills boost their scores on the TOEIC, Test of English for International Communication. Doing well on standardized tests such as the TOEIC is expected of Korean students (Choi, 2008). Korean parents, educators, and peers place extreme emphasis on test scores and the Korean education system is test driven (Choi, 2008). Because the TOEIC exam places heavy emphasis on reading comprehension, vocabulary, and writing Korean students are motivated and eager to find ways to improve in these core testing areas.

The pressures that Korean students are under is immense (Choi, 2008; Wang, 2013). The strong Confucian influences of Korean society place an enormous amount of emphasis on education. 69 percent of South Korean high school students attend four-year university and this is the highest college going rate in the world (South Korean Education Reforms, 2016; Population with tertiary education, 2016). Another indicator of the Korean emphasis on education is that the Korean Ministry of Education's budget has increased six-fold since 1990.

Students are ranked from an early age and expected to consistently perform at the top of their class. Pressure comes from all directions with family, friends, and educational institutions all focused on test scores and competition for limited spaces at desirable universities (Choi, 2008). These pressures come to a boiling point when students take the Suneung, the Korean equivalent of the Scholastic Aptitude Test. This high stakes test is responsible for high school students studying 14 hours a day (Hu, 2015). This gateway test provides all Korean high school students with immense stress in the lead up to the test. The test is also linked to psychological depression and high suicide rates among Korean teenagers. Liang Wang (2013) researched the effects of high stakes testing and Korea's leading suicide rate within the OECD of 28.9 per 100,000 people (OCED, 2013). Wang's (2013) research on Korea finds that "suicide was the leading cause of death of Korean individual's ages 15-24 years in 2008" (p. 2).

Given these pressures and expectations, South Korean students are conditioned by society to conform. Pressures to excel in school and to conform to family expectations and cultural norms can start as early as kindergarten (Lee & Yi, 2002; Shin & Koh, 2005). This stunts in-class discussion as Korean students do not want to be seen as 'individuals.' The cultural norm and expectation is that Korean students are part of a collective social structure and that they defer to the elders who lead the group (Kashima et al., 1995). It is a deeply held Confucian belief within Korean society that the group should come first and that any form of individualism or standing apart from the group is not only contrary to the group concept but it is explicitly and strongly discouraged (Daehwan, 2003, Koh, 2003). This directly impacts students and their learning styles. Students become unwilling to participate in class discussions for fear of being seen as individualistic or saying something out of the norm (Kashima et al., 1995). They also worry constantly about how they are being perceived by instructors and peers and this anxiety negatively impacts their ability to participate in a wide range of learning activities (Arnau et al., 2007; Diener, Ng, Harter, & Arora, 2010). Given these constraints, teachers can use online discussions to promote student engagement and to create individual learning spaces where authentic teacher student interactions can take place without social conventions and cultural norms dictating behavior. The Gosling (2011) study supports this conclusion. The student's online self is more expressive than what you might find when speaking to the same person face to face (Gosling et al., 2011).

Korean students are already familiar with and comfortable with online interactions. This is evident by their vast amounts of data consumption, high internet speeds, and the extreme technology based connectivity within the country. Hoe, Oh, Subramanian, Kim, and Kawachi (2014) surveyed over 50,000 middle and high school students about their internet usage concluding that males primarily use the internet for online gaming and searching information while women primarily use the internet for searching information and blogging or updating their personal homepages. Over 75% the population in South Korea has a smart phone (The Global Media, 2012), and 73% of the population that use the internet in South Korea utilize KakaoTalk, a popular free mobile instant messaging application, on a daily basis (Internet usage, n. d.). These factors indicated widespread familiarity with technology and communicating online, indicating that most college age Korean students are already comfortable and competent in an online teaching and learning environment. Another benefit of online discussion is that students in the online chat or forum deal with only text and do not have to address various accents. Tauroza and Luk's (1997) study found that one of the most difficult aspects of English second language learning (ESL) is parsing different accents and diverse dialectic emphasis and pronunciations. In an online discussion, students also have more time to respond, and they can use the internet or the course text to aid their writing while they are working in the chat or discussion board (Campbell, 2007; Guardado & Shi, 2007; Hoon & Emerson, 2002).

Making the shift from in-class discussions to anonymous online discussions can dramatically improve ESL learning among the Korean college age population. The instructor can deliver positive feedback in the online classes or the online discussion without singling out the student or calling unwanted attention to them amongst their peers (Campbell, 2007; Chester & Gwynne, 2006). Given Korea's Confucian roots, online discussions are a good way to promote student participation and they are a good way to give focused positive feedback. They are also a good place for instructors to address individual student questions or issues. Further, students in an anonymous online discussion can ask questions or ask for clarifications or further help without worrying about appearing needy or slow. Given these important benefits, online discussions and chats can be more beneficial than in-class discussions especially in ESL classes (Chapelle, 2001; Warschauer, 1995, Zheng & Warschauer, 2015).

Anonymous online discussion can be used to help promote discussion and lessen the fear of public embarrassment and intimidation (Roed, 2003). The rules about online behavior and requirements regarding online conduct, post lengths, content, and deadlines should be clearly discussed to ensure student security and privacy (Bikowski & Kessler, 2002; Garrison & Cleveland-Innes, 2005). Students should be respectful of others and refrain from using inappropriate online etiquette, such as typing in all caps, disrespecting another student, stealing another student's alias, or impersonating another student (Chester & Gwynne, 2006). Internet etiquette and the rules for the discussion boards should stem from the code of conduct at the university. If done properly, the anonymous discussion boards can be a powerful tool that the teacher can use to promote authentic discourse and discussion.

CREATING ANONYMOUS DISCUSSIONS

Blackboard, Moodle, and Canvas are three popular learning management systems (LMS) for online course and program delivery. All three systems require students to input their real names and information so that teachers and other students can recognize each participant within the class. Within each learning management system, there are forums that the instructors can use to set up online discussions, and students can upload documents to the teacher or to the class. Blackboard, Canvas, and Moodle all currently allow or have allowed the creation of some level of anonymous discussion either within a course shell or as a separate discussion thread.

In Blackboard instructors can go into the Forum Settings and click a box to allow anonymous posts. Directions for anonymous discussion creation in Canvas can be found at https://community.canvaslms.com/thread/2591. And, Moodle used to have a 'plugin' that allowed for the setup of anonymous discussions (https://moodle.org/plugins/view/local_anonymousposting) but unfortunately, this plugin has not been updated and it should not be used in its outdated format. So, while Blackboard, Canvas, and Moodle all do or did allow an instructor or course designer to set up an anonymous discussion, doing so in a way that specifically meets teaching goals or that provides maximum student privacy might require the expertise of a seasoned instructional designer familiar with the platform. However, it is more important that current and future LMS's have functions to allow anonymity for students, either within an online course or in a separate anonymous online forum space, to help offset the Confucian ideas and ideals that have been implemented within the Korean education system.

AUTHENITIC FEEDBACK FLOURISHES IN ANONYMOUS ONLINE ENVIRONMENTS

As discussed above, the Confucian ideas and ideals of Korean culture result in a lack of authentic in class interactions and they significantly hinder the learner and the instructor. One example of this hindrance is that instructor feedback to students often has to be 'sugar coated' so as not to embarrass the student in front of classmates. Guardado and Shi (2007) discuss the issue of 'sugarcoated feedback' when students' were not given anonymous feedback when they said, "many participants expressed little confidence in peer commenting. Some chose to provide positive feedback or sugarcoat their negative feedback as reviewers or found peer feedback useless, and therefore, did not follow it in their own revisions as authors" (p.458). In contrast, anonymous discussion and postings allow and promote honest and authentic feedback in an environment where public and peer perceptions are not a concern. Removal of this cultural barrier to provide an authentic instructor and student feedback and dialog facilitates improved learning outcomes and improved test scores.

Without anonymity, students are likely to give only positive feedback to each other, out of fear of upsetting other students and in order to maintain their friendships and relationships with other students. Being anonymous helps students feel more at ease, and it also allows them to give honest peer to peer feedback that might not otherwise been given. While my Korean students clearly understand that giving inauthentic or overly positive feedback does not help them or their peers they almost always feel compelled to give only positive feedback in any public setting. Anonymity can provide Korean university students more comfort when they are in-front of others, allow them to worry less about feeling humiliated for their own perceived low proficiency levels and to break from conformity to cultural norms that might inhibit their actions (Campbell, 2007).

Online anonymity helps Korean students avoid societal rules that play major roles in everyday life in Korea, and it helps promote student-to-student discussions in addition to instructor-student discussions. One of the biggest issues in the Korean classroom is the above average power-distance that permeates Korean culture. Power distance is best defined by Geert Hofstede (2017) as, "the extent to which the less powerful members of institutions and organizations within a country expect and accept that power is distributed unequally." Korea's Confucian instilled society has a hierarchical structure that can be found across numerous aspects of Korea life and culture. Those on the lower end of the hierarchical structure in Korea are expected to be consistently respectful and agreeable and they are expected to accept their roles without question. This results in a person without power, in this case the student, unquestionably accepting the ideas from the person with the power, the teacher. Students are expected to accept the information and to not question the instructor and they are conditioned to be respectful at all times. Because of the large power-distance in Korean culture, classes within Korean universities are teacher-centered and this makes in class discussions or interactions with the instructor difficult, stilted and overly formal.

These cultural norms and classroom dynamics are evident in Warschauer's (1995) study of a class of ESL students in Hawaii composed of Asian students. He noted that the Japanese students were the most reserved, and this was probably because of their cultural upbringing and the education methodology within the Japanese school system. In Warschauer's (1995) study there were five Japanese students; four of the Japanese students contributed 1.8% of the comments in face-to-face discussions meaning that they were almost silent. The fifth student made 48% of the comments in face-to-face discussions. "This single student made five times as many face-to-face comments as all four other Japanese students combined" (Warschauer, 1995, p. 21). It is extremely interesting to note that the fifth Japanese student was married to an American. As a result, he had the opportunity to practice English and to learn about American culture, values, and beliefs. From his prolonged exposure to and extensive practice in English language and American culture, this Japanese student had learned to 'culture switch' and his behavior was consistent with the other American students in the class.

According to Cambell (2007), ESL students from Asian cultures tend to be reluctant to express views that disagree with the views or teachings of the instructor. "They feel more comfortable in a student-led, student-centered discussion" (p.41). Online anonymity can help Korean university students feel more comfortable during online discussions. Online discussions can increase participation dramatically. Several studies have reported increases of as much as 10 times more than in-class discussion participation levels, as well as increases in the student's comfort levels (Chester & Gwynne, 2006; Warschauer, 1995). Use of anonymous discussions and other online exercises can also help to change the peer-to-peer dynamics and shift the power balance in the classroom away from the instructor.

PARTICIPATION IN ONLINE DISCUSSIONS

Korean ESL students can be difficult to manage and to engage in the curriculum because of the Confucian constructs that lead them to exhibit low participation rates. The traditional style of teacher-centered classes is prominent throughout Korean universities. The Confucian ideologies present in Korea have made the role of the teacher, instructor and professor highly desirable and highly respected. Confucian ideas require that individuals seek harmonious relationships through respecting and following tradition and social hierarchy. These relationships focus around the elder having the power and status to which other younger persons must submit. The Confucian relationships, father and son, ruler and ruled, husband and wife, older brother and younger brother, and friend to friend, focus on each individual knowing his or her place within the social structure and hierarchy. Social structures and hierarchies are based on relationships of age and status and through individuals showing respect to those of higher status and older age. Koreans, following the teachings of Confucius, seek harmony in society through sincerity, tolerance, and forgiveness of others. Relational harmony, social hierarchical relationships, and traditional conservatism are core elements of Confucius' teaching as are having few items, being pure, non-competitive, showing moderation, and having few desires or individual wants and needs (Zhang, Lin Nonaka, & Beom, 2005).

Based on Confucian teachings the instructor-student dynamic results in the instructors being seen as a ruler. The student must defer to and show complete respect to the instructor and by extension all teachings of the instructor. Add to this established relationship the large age difference between the instructor and the students and the Confucian focus on education, and the result is that the instructor holds immense power and commands complete respect in the classroom. These ideas have contributed to a power-distance dynamic in the Korean classroom. This dynamic ensures that students are obedient and quiet and that they never challenge the instructor. Lee, Fraser, and Fisher's 2003 study of teacher-student interactions in a high school support this when they said, the "present study revealed that teacher–student interactions in Korean senior high school science classrooms reflect the general image of the youth–elder relationship in society of 'directing teachers and obeying students" (p. 83). The authors (2003) continue to describe Korean high schools as "directive, controlling and not supportive of students' self-activities" (p. 82).

Korea has deep-rooted and unaltered ideas that ensure Korea's traditional views and orientations endure. Geert Hofstede (2017) defines long term orientation as, "how every society has to maintain some links with its own past while dealing with the challenges of the present and future." Geert Hofstede (2017)'s research on culture describes Korea as, "one of the most pragmatic, long-term oriented societies". Korea's desire to maintain links with the past when dealing with current and future challenges gives Korea its cultural identity. This desire to hold onto established ideas and tradition makes it difficult to implement new concepts and ideas if these newer ideas conflict with long term, previously held beliefs.

With Korea's rapid rate of technical and scientific advances, Korea will be looking to integrate technology advances into the classroom. But given the power of traditionalism, this integration will present unique challenges throughout the Korean education system. Korea's reliance on older ideas when using present concepts presents unique issues that seriously and consistently limit learning and student success (Geert Hofstede, 2017). Anonymous online learning can provide students with the ability to break out of the confines of teacher-centered classroom through the use of online discussions and other anonymous online activities (Warschauer, 1995, Zheng & Warschauer, 2015). However, anonymous online learning as well as the role of the teacher as a facilitator in online course work will go against deeply ingrained and long-standing ideas and beliefs around how education should occur. It will be fascinating to see how these strong and opposing forces interact and enable or constrict change and innovation within the Korean education system.

Anonymous online discussions can provide students increased equality and can lead to higher participations rates (Collins & Berge, 1995; Hartman, Neuwirth, Kiesler, Cochran, Palmquist, & Zubrow, 1995; McComb, 1994; Ruberg & Taylor, 1995). Anonymous online forums give students the opportunity to participate synchronously or asynchronously. Students who are typically quiet or voiceless in class have the opportunity to engage in discussion with their classmates. In contrast to a live conversation that requires the student to process the words and information at faster processing speeds, online forums and discussions provide students the time to fully process the conversation and information.

A study conducted by Polat, Mancilla, and Mahalingappa (2013) discusses participation in L2 (Second-Language Learning) learning. But it is not clear if these findings help us to better understand Korean students. The Polat, Mancilla, and Mahalingappa (2013) study found that there was a significantly higher number of posts and a higher word count in the nonanonymous forums than the anonymous ones (p. 67). The Polat, Mancilla, and Mahalingappa (2013) study focused on American students studying

Spanish. As discussed above, Confucian based Korean culture is vastly different, and these differences are significant as they relate to online learning versus instructor lead classroom based learning.

Koreans enjoy using Korean made products and they are proud of Korean companies and how they rank against foreign competitors. KakaoTalk is a good example of the loyalty displayed by South Koreans. As of 2015 KakaoTalk boasts almost 29 million unique users and reaches nearly 95% of the Korean population. Naver, Korea's main website and search engine, since its inception in 2015 has 15 million unique daily visitors and reaches just over half the population while Google can claim only 11 million unique daily visitors and a 37% impact rate (Nielsen Korea, n. d.). If Polat, Mancilla, and Mahalingappa's research was conducted in Korea, given the tight regulations requiring South Koreans to be identifiable online, we can suggest that Koreans would participate more in anonymous online discussions since this would be their chance to voice their opinion without direct repercussions and concerns about not following long held beliefs and societal rules (Guardado & Shi, 2007; Lee & Yi, 2002; Shin & Koh, 2005; Warschauer, 1995). Their research also suggests that when considering online forums both anonymous and identifiable options should be considered when looking for participation in discussions in the online forums.

Koreans traditional educational style as influenced by its strong Confucian roots has made participation difficult for students in the classroom. Online learning gives students a platform to participate in robust and engaged discussions and learning activities compared to the traditional Korean classroom where students are expected to accept the one-way communication of the traditional lecture style class in which they must adhere to a large set of societal norms.

MOTIVATION FACTORS IN ANONYMOUS ONLINE LEARNING

Motivation is one of the biggest factors in determining whether or not a student will succeed at learning online (Dembo & Seli, 2012; Rueda, 2011). A student's personality and the details of the course drive the motivation for students to participate in the class and to become engaged with the course material. For online classes, students need to develop a different skill set and approach as opposed to learning in the traditional classroom with a teacher leading the class, usually through a lecture. Whether the class is synchronous or asynchronous, having to use technology to learn is different than physically going to a location to learn with other students and the teacher (Clayton, Blumberg, & Auld, 2010; Wighting, Liu, & Rovai, 2008).

Each student's unique personality determines their own motivational factors when considering online learning (Rueda, 2011). In Korea, motivation can be an issue since online learning is not common and students may not have experience with online learning and they may be unsure of how to approach online coursework. When comparing South Korean students to American students, Korean students show significantly less motivation to learn online (Lim, 2004). Lim's (2002) study found that Korean students listed course relevancy, learning control, and course interest as the most important motivational factors. Self-efficacy is listed as the least important motivational factor signaling that students see courses as a compulsory requirement.

Comparatively, the American students listed course relevancy, self-efficacy, and reinforcement as the top motivational factors and affect and emotion as the least motivation factor. The only factor that Korean students scored significantly higher on is for learner control when compared to Americans (Lim, 2004). This suggests that American students strive for mastery of the course content to increase their self-efficacy. While Korean students seek to attain high performance levels, high test scores and high grades rather of mastery of the subject matter. Korean students displayed low self-efficacy and low levels of learner control and this affirms their cultural preference for performance over mastery. This distinction is likely due to the testing emphasis within the Korean education system.

Once a student enrolls in an online course to improve their second language acquisition, the student can reap quick and significant benefits. The typical asynchronous online forums give students the time to better understand the course material at a pace that suits them. Replaying or revisiting content is easy and the student is in control of the pace of learning. This lessens the fear of difficult social interactions and embarrassment. It also fosters a higher level of comfort through the more manageable and slower paced online discussions as compared to a more dynamic real-time classroom interaction (Bradley & Lomicka, 2000; Campbell, 2007, Roed, 2003). This alleviation of stress gives Korean students the opportunity to be more engaged and to gain control of their learning, something that Koreans rank as a highly desirable element when taking online classes.

ENGLISH LANGUAGE LEARNING SURVEY AT NAMSEOUL UNIVERSITY

In 2014, the General Education Department of Namseoul University conducted a survey of 141 Korean university students to understand student's preferences for English language learning. The survey also captured data about course delivery formats and student preferences for online or face to face learning. The survey was written in English, and a Korean staff member gave verbal instructions in Korean. Students responded to all the answers that applied to them for each question, and the students were able to comment on all parts of the survey in order to help the researchers gain better insight into and more information about their answers.

Survey Results and Findings

Only 18 of the 141 participants said that they had any experiences learning English online. This is not a surprise as South Korean universities rarely offer online classes. At Namseoul University there were only 11 online classes offered in the spring 2014 semester, and all of them were taught by Korean professors in Korean. Within the survey, students could also comment on their answer choices and a recurring theme amongst the 141 survey students was they wanted to study with and communicate with friends in English with 86 students saying that studying with friends was important. Even though many students mentioned that they preferred learning in class, the comments about learning with friends were the same as students who wanted to learn in a group environment. These data points strongly suggest that students want to learn with their friends regardless of the class delivery format. Online discussions and online classes can help students stay connected and build relationships with their peers while learning as a group. Online courses can also facilitate individual learning, depending on the assignments given.

However, the most important takeaway from the survey was that online courses give students a voice and a place to learn and to interact with friends. 72% of students who took online classes to learn English said that they enjoyed it. While the total number of students who had taken online class was small, the interest in online classes was measurable and consistent with the small number of students who were exposed to online learning.

Survey Results

The lack of interest in online learning in Korean stems from the cultural idea that learning should be done in a classroom with a teacher. With all of the online classes at Namseoul University being taught by Korean professors in Korean, students have very few opportunities to learn from English language teachers through online course mediums. Yet, there are signs that online learning, especially in the L2 area, could be beneficial and of interest to Korean university students. Over 70% of the students who took online classes said that they enjoyed studying online and 43 of 141 students said that they would try online or hybrid classes. Combining this interest level with a review of the literature strongly suggests that Korean students would enjoy online learning that is student-centered.

The literature shows that ESL and English as a Foreign Language students who study online can improve their English proficiency and their relationships with other students in the class (Campbell, 2007; Chester & Gwynne, 2006; Duff, 2002). The large amount of positive feedback regarding online learning and the strong personal connections that are built through online interactions make online learning a natural choice for Korean students wanting to learn with their friends in English. This relationship with friends and group learning was extremely important to the students who were surveyed. If students are able to take classes with their peers this can increase the feelings of and perspective of control that a student has when learning online. This sense of increased control and communication leads to increased motivation to learn and increased levels of student engagement (Rueda, 2011).

The survey shows that not many Korean students have learned online, and this appears to stem from the long-term orientation culture that Korea has developed. While the amount of online degree programs and online classes that are currently offered are limited, Lee, Yoon, and Lee's 2009 study found, "Korea is one of the fastest growing countries in e-learning" (p. 1327). This is because of Korea's rapid growth of its information and communications technology industry. This is increase in e-learning is confirmed by K-MOOC's creation in 2015 and its rapid and ambitious expansion.

ANONYMOUS ONLINE LEARNING OUTSIDE OF ESL

Anonymity is a double-edged sword that can liberate students in both positive and negative ways. Anonymity can provide the freedom necessary to participate freely and without worry of receiving direct negative feedback in public. But it can also provide students with a veil that allows them to be more aggressive and less empathetic of others (Christopherson, 2007; Zimbardo, 1969). The code of conduct that is enforced at most universities helps ensure that students who choose to violate these interpersonal and netiquette rules are punished appropriately. Grades and class participation requirements in group discussions, discussion threads and other group work should also be required and each of these online or face to face course sections should clearly outline class conduct and netiquette expectations. These point allocations tied to proper and appropriate class interactions combined with the code of conduct should be sufficient to positively shape student's online behaviors while providing students with a safe place to voice their opinion in a digital classroom setting regardless of the subject matter.

FUTURE RESEARCH DIRECTIONS

The survey of Namseoul University students showed that most students were unaware of online learning options but that studying with friends was an important factor in course and program selection. These results show that students want to be in more collaborative and less rule binding learning environments. This small single institution survey provided only a cursory understanding of student perceptions of online learning and the factors that contribute to motivation to learn and to participate in the learning process. Future research in this area should include a comprehensive survey examining student's motivational factors and the factors that hinder and promote participation in the online and face to face classroom. The survey should be conducted across a range of South Korean post-secondary institutions. Looking at private and public universities ranging from large elite universities to small regionally accredited colleges will provide a more complete understanding of the factors that affect motivation and participation in South Korean post-secondary institutions and it will further inform the utilization of anonymous online learning as an element of both online and face to face courses.

CONCLUSION

Given Korea's deep Confucian roots, online discussions offer a good way to promote student participation and they are a good way to give focused positive feedback. They are also a good place for instructors to address individual student questions or issues. Further, students in an anonymous online discussion can ask questions or ask for clarifications or further help without worrying about appearing needy or slow. Given these important benefits, online discussions and chats can be more beneficial than in-class discussions especially in ESL classes.

Without the opportunity for Korean students to engage in online discussions, students will continue to be held to the Confucian rules that govern and shape Korean society. While the Namseoul University ESL survey did not show much participation in online learning, those who did participate enjoyed online classes more than traditional classes. This suggests that there needs to be more of an opportunity for students to learn online and for students to use some elements of anonymous learning to express themselves without the fear of embarrassment or fear of violating long held cultural norms and ideals.

The K-MOOC project is changing education in Korea by opening up online classes and online learning to the general population. With Korea's ICT industry booming and the country boasting the world's top internet speed, online learning will surely continue to increase in Korea. This increase in online classes gives students an opportunity for alternative methods of education. Online course and learning options will give instructors the opportunity to increase participation and motivation to learn while decreasing cultural barriers to learning in the classroom. It is imperative that teachers and students not only adjust to online learning but that they maximize its benefits when living within a Confucian society.

Understanding the societal rules that Korean students must adhere is paramount to delivering impactful online classes and to providing improved education opportunities for ESL in South Korean post-secondary institutions. By implementing anonymous online learning in South Korea, students are freed from the binding rules that govern their language and society. With properly regulated anonymous online discussion boards, instructors and students can create a harmonious learning environment that will help them succeed on their educational journey while utilizing the vast technological advantages that are available within South Korea.

REFERENCES

Arnau, R. C., Rosen, D. H., Finch, J. F., Rhudy, J. L., & Fortunato, V. J. (2007). Longitudinal Effects of Hope on Depression and Anxiety: A Latent Variable Analysis. *Journal of Personality*, *75*(1), 43–64. doi:10.1111/j.1467-6494.2006.00432.x PMID:17214591

Asian Society. (2016). South Korean Education Reforms. Retrieved July 7, 2016 from http://asiasociety.org/global-cities-education-network/south-korean-education-reforms

Bang, H., & Kim, J. (2016). Korean and American teachers' praising styles and teaching practices. *Mid-Western Educational Researcher*, *28*(1), 28–54.

Bikowski, D., & Kessler, G. (2002). Making the most of the discussion boards in the ESL classroom. *TESOL Journal*, *11*, 27–30.

Bradley, T., & Lomicka, L. (2000). A case study of learner interaction in technology-enhanced language learning environments. *Journal of Educational Computing Research*, *22*(3), 347–368. doi:10.2190/TCUA-3YUV-B1P5-26P3

Campbell, N. (2007). Bringing ESL students out of their shells: Enhancing participation through online discussion. *Business Communication Quarterly*, *70*(1), 37–43. doi:10.1177/108056990707000105

Chapelle, C. (2001). *Computer applications in second language acquisition*. Cambridge University Press. doi:10.1017/CBO9781139524681

Chester, A., & Gwynne, G. (2006). Online teaching: Encouraging collaboration through anonymity. *Journal of Computer-Mediated Communication*, 4.

Choi, I. C. (2008). The impact of EFL testing on EFL education in Korea. *Language Testing*, *25*(1), 39–62. doi:10.1177/0265532207083744

Christopherson, K. M. (2007). The positive and negative implications of anonymity in Internet social interactions: On the Internet, nobody knows you're a dog. *Computers in Human Behavior*, *23*(6), 3038–3056. doi:10.1016/j.chb.2006.09.001

Clayton, K., Blumberg, F., & Auld, D. (2010). The relationship between motivation, learning strategies and choice of environment whether traditional or including an online component. *British Journal of Educational Technology*, *41*(3), 349–364. doi:10.1111/j.1467-8535.2009.00993.x

Collins, M., & Berge, Z. (1995). Introduction: Computer-mediated communications and the online classroom in higher education. In Z.L. Berge & M.P. Collins (Eds.), Computer mediated communication and the online classroom (Vol. 2, pp. 1–10). New Jersey: Hampton Press.

Worldatlas.com. (2016, November 24). Countries Where People Spend the Most Time Online. Retrieved March 20, 2017 from http://www.worldatlas.com/articles/top-countries-which-spend-the-greatest-amount-of-time-online.htm

Daehwan, N. (2003). The eclectic development of the Neo-Confucianism and statecraft from the 18th to the 19th century. *Korea Journal*, *43*(4), 87–112.

DeCapua, A., & Wintergerst, A. (2004). *Crossing cultures in the language classroom*. Ann Habor, MI: The University of Michigan Press.

Dewar, R. (2002). Case study: Space station robot embeds Ada. *COTS Journal, 3*. Retrieved from http://www.adacore.com/uploads/customers/CaseStudy_SpaceArm.pdf

Diener, E., Ng, W., Harter, J., & Arora, R. (2010). Wealth and happiness across the world: material prosperity predicts life evaluation, whereas psychosocial prosperity predicts positive feeling. *Journal of Personality and Social Psychology, 99*(1), 52–61. doi:10.1037/a0018066 PMID:20565185

Duff, P. (2002). Pop culture and ESL students: Intertextuality, identity, and participation in classroom discussions. *Journal of Adolescent & Adult Literacy, 45*(6), 482–487.

Education First. (2013). South Korea. Retrieved from http://www.ef.edu/epi/spotlight/south-korea/

Garrison, D. R., & Cleveland-Innes, M. (2005). Facilitating cognitive presence in online learning: Interaction is not enough. *American Journal of Distance Education, 19*(3), 133–148. doi:10.1207/s15389286ajde1903_2

Geert Hofstede. (2017). Retrieved March 20, 2017, from https://geert-hofstede.com/south-korea.html

Gosling, S. D., Augustine, A. A., Vazire, S., Holtzman, N., & Gaddis, S. (2011). Manifestations of personality in online social networks: Self-reported Facebook-related behaviors and observable profile information. *Cyberpsychology (Brno), 14*(9), 483–488. PMID:21254929

Guardado, M., & Shi, L. (2007). ESL students experience of online peer feedback. *Computers and Composition, 24*(4), 443–461. doi:10.1016/j.compcom.2007.03.002

Hartman, K., Neuwirth, C. M., Kiesler, S., Cochran, C., Palmquist, M., & Zubrow, D. (1995). Patterns of social interaction and learning to write: Some effects of network technologies. In Z.L. Berge & M.P. Collins, (Eds.), Computer mediated communication and the online classroom (pp. 47–78). New Jersey: Hampton Press.

Heo, J., Oh, J., Subramanian, S. V., Kim, Y., & Kawachi, I. (2014). Addictive Internet Use among Korean Adolescents: A National Survey. *PLoS ONE, 9*(2), e87819. doi:10.1371/journal.pone.0087819 PMID:24505318

Hinkel, E. (1990). *Culture is second language teaching and learning*. New York, NY: Cambridge University Press.

Horvath, Z. (2014). History of the South Korean flag. Retrieved April 10, 2014 from http://www.crw-flags.com/fotw/flags/kr_hist.html#1856

Hu, E. (2015, April 15). The All-Work, No-Play Culture Of South Korean Education. Retrieved March 27, 2017 from http://www.npr.org/sections/parallels/2015/04/15/393939759/the-all-work-no-play-culture-of-south-korean-education

ICEF. (2012, June 288 countries leading the way in online education. Retrieved from http://monitor.icef.com/2012/06/8-countries-leading-the-way-in-online-education/

James, M. (2000). Culture in ESL instruction: An analytic framework. *TESOL Canada Journal, 17*(2), 36–49.

Jeon, B., & Kim, J. (2012). Development and validation of a teachers praise disposition scale. *Journal of Fisheries and Marine Sciences Education, 24*(5), 627–642. doi:10.13000/JFMSE.2012.24.5.627

K-MOOC. (2017). Retrieved March 27, 2017, from http://www.kmooc.kr/

Kashima, Y., Yamaguchi, S., Kim, U., Choi, S., Gelfand, M. J., & Yuki, M. (1995). Culture, gender, and self: A perspective from individualism-collectivism research. *Journal of Personality and Social Psychology, 69*(5), 925–937. doi:10.1037/0022-3514.69.5.925 PMID:7473038

Kim, J. (2013, November 12). How to cure South Korea's English fever? *The Wall Street Journal.* Retrieved from http://blogs.wsj.com/korearealtime/2013/11/12/how-to-cure-south-koreas-english-fever/

Koh, Y. J. (2003). Neo-Confucianism as the dominant ideology in Joseon. *Korea Journal, 43*(4), 59–86.

Lee, B., Yoon, J., & Lee, I. (2009). Learners acceptance of e-learning in South Korea: Theories and results. *Computers & Education, 53*(4), 1320–1329. doi:10.1016/j.compedu.2009.06.014

Lee, J. K., & Yi, C. G. (2002). *Korean higher education: A Confucian perspective. 22.* Jimoondang Publishing Co.

Lee, S. U., Fraser, B. J., & Fisher, D. L. (2003). Teacher-student interactions in Korean high school science classrooms. *International Journal of Science and Mathematics Education, 1*(1), 67–85. doi:10.1023/A:1026191226676

Lee, Y. H. (2016). ICT as a key engine for development: good practices and lessons learned from Korea. Retrieved April 13, 2017 from siteresources.worldbank.org/INTEGOVERNMENT/Resources/NoteKoreaICT.doc

Lim, D. H. (2004). Cross cultural difference in online learning motivation. *Educational Media International, 41*(2), 163–175. doi:10.1080/09523980410001685784

McComb, M. (1994). Benefits of computer-mediated communication in college courses. *Communication Education, 43*(2), 159–170. doi:10.1080/03634529409378973

Nguyen, L. V. (2008). Technology-enhanced EFL: Syllabus design and materials development. *English Language Teaching, 1*(2), 135–142. doi:10.5539/elt.v1n2p135

Nielsen Korea. (n. d.). Retrieved March 20, 2017 from http://www.koreanclick.com/english/top10/mobile_web_ranking.html

OECD. (2013). *Suicide rates* [Data file]. Retrieved from https://data.oecd.org/healthstat/suicide-rates.htm

OECD. (2016). *Population with tertiary education.* Retrieved July 6, 2016, from https://data.oecd.org/eduatt/population-with-tertiary-education.htm

Polat, N., Mancilla, R., & Mahalingappa, L. (2013). Anonymity and motivation in asynchronous discussions and L2 vocabulary learning. *Language Learning & Technology, 17*(2), 57–74.

Roed, J. (2003). Language learner behaviour in a virtual environment. *Computer Assisted Language Learning*, *16*(2-3), 155–172. doi:10.1076/call.16.2.155.15880

Rowsell, J., Sztainbok, V., & Blaney, J. (2007). Losing strangeness: Using culture to mediate ESL teaching. *Language, Culture and Curriculum*, *20*(2), 140–154. doi:10.2167/lcc331.0

Ruberg, L. F., & Taylor, C. D. (1995). *Student responses to network resources: Formative evaluation of two classes.* Paper presented at the Annual Meeting of the American Educational Research Association, San Francisco, CA.

Rueda, R. (2011). *The three dimensions of improving student performance: finding the right solutions to the right problems*. New York: Teachers College.

Shin, S., & Koh, M. (2006). Korean education in cultural context. *Essays in Education*, *14*, 1–10.

Skinner, B., & Austin, R. (1999). Computer conferencing: Does it motivate EFL students? *ELT Journal*, *53*(4), 270–279. doi:10.1093/eltj/53.4.270

Statista. (n. d.). Internet usage in South Korea. Retrieved April 10, 2017, from https://www.statista.com/topics/2230/internet-usage-in-south-korea/

Worldatlas.com. (2017, February 21). Suicide rates by country. Retrieved March 27, 2017 from http://www.worldatlas.com/articles/countries-with-the-most-suicides-in-the-world.html

Tauroza, S., & Luk, J. (1997). Accent and Second Language Listening Comprehension. *RELC Journal*, *28*(1), 54–71. doi:10.1177/003368829702800104

Campaign Brief Asia. (2012). The Global Media Intelligence Report. Retrieved from http://www.campaignbriefasia.com/2012/10/10/eMarketer%20Global%20Media%20Intelligence%20Report%20Exec%20Summary.pdf

The World Factbook. (2014). South Korea. Retrieved from https://www.cia.gov/library/publications/the-world-factbook/geos/ks.html

Warschauer, M. (1995). Comparing face-to-face and electronic discussion in the second language classroom. *CALIO Journal*, *13*, 7–26.

Weidman, J. C., & Park, N. (2000). *Higher education in Korea: tradition and adaptation*. New York: Falmer Press.

Wighting, M. J., Liu, J., & Rovai, A. P. (2008). Distinguishing sense of community and motivation characteristics between online and traditional college students. *Quarterly Review of Distance Education*, *9*(3), 285–295.

Williamson, L. (2014, February 4). Kimchi: South Korea's efforts to boost its national dish. *BBC*. Retrieved from http://www.bbc.com/news/world-asia-25840493

Yun, C., Kim, K., & Kim, J. (2009). Praise types perceived by pre-service secondary teachers. *The Journal of Korean Teacher Education*, *26*(4), 55–78. doi:10.24211/tjkte.2009.26.4.55

Zhang, Y. B., Lin, M., Nonaka, A., & Beom, K. (2005). Harmony, Hierarchy and Conservatism: A Cross-Cultural Comparison of Confucian Values in China, Korea, Japan, and Taiwan. *Communication Research Reports*, *22*(2), 107–115. doi:10.1080/00036810500130539

Zheng, B., & Warschauer, M. (2015). Participation, interaction, and academic achievement in an online discussion environment. *Computers & Education*, *84*, 78–89. doi:10.1016/j.compedu.2015.01.008

Zimbardo, P. G. (1969). The human choice: Individuation, reason, and order vs. deindividuation, impulse, and chaos. In W. J. Arnold & D. Levine (Eds.), *Nebraska symposium on motivation* (Vol. 17, pp. 237–307). Lincoln: University of Nebraska Press.

KEY TERMS AND DEFINITIONS

Confucianism: Philosophy and teachings attributed to Confucius relating to norms, customs, ethics, education, relationships, and harmony within the society.

K-MOOC: Collaborative Korean Massive Online Open Course project with about thirty Korean universities contributing at least online course offering. As of 2017, the project offered just over 200 open enrollment online courses, mostly in Korean (http://www.kmooc.kr).

Power-Distance: The chapter utilizes the definition and description of power distance from Geert Hofstede (2017). Hofstede explains power-distance as "the extent to which the less powerful members of institutions and organizations within a country expect and accept that power is distributed unequally."

Self-Efficacy: The chapter utilizes Albert Bandura's definition of self-efficacy from his seminal 1977 work, Self-Efficacy: Toward a Unifying Theory of Behavioral Change. Bandura's definition can be summarized as an individual's confidence in their ability to achieve desired goals and intended outcomes.

The Suneung: The Suneung is the name of the Korean national college entrance exam that is accepted by South Korean colleges and universities and that determines college placement. The test is offered once a year in November. The Suneung is the Korean equivalent of the ACT or SAT exams.

Chapter 8
Supporting At-Risk, Underrepresented, and Female Undergraduate Students to Advance Directly to an MBA

Helen Eckmann
Brandman University, USA

ABSTRACT

This Chapter explores the reasons for and provides a range of ways for faculty members, staff and administrators to build a bridge that will encourage and help a higher number of at-risk, underrepresented and female students matriculate directly into the MBA program upon completion of an undergraduate degree. First, the chapter will provide data driven reasons why it is in the best interest of at-risk, underrepresented minority and female students to pursue a graduate degree. Second, it will provide concrete and proven examples of how to help these student's move directly from undergraduate degrees into the MBA program. The examples of how to encourage this direct pathway fall into three categories: faculty focus, curriculum design and university-wide systems.

INTRODUCTION

In "The Atlantic", Anthony Carnevale, who directs Georgetown University's Center on Education and the Workforce stated that graduates enrolled in college have limited information as postsecondary education in the U.S. is "like a big computer with no operating system. Americans have a chaotic understanding of the role of college in the economy. Students are lost; they don't know how to make these connections between the value of a college degree and their position within the economy" (Wong, 2016). This is a description of undergraduate college selection but it could also apply to graduate education, especially for underrepresented minority students and other at-risk students. Underrepresented minorities (URM) began to be mentioned in literature in relation to STEM disciplines but the term can also be used to describe graduate degree achievement (Baron, 2010). Although there is a trend to develop and offer op-

DOI: 10.4018/978-1-5225-2998-9.ch008

portunities for URM students in undergraduate degrees, the opportunity for graduate education remains confusing and out of reach for some females, especially those from low socio-economic backgrounds, and significantly for URM students.

Specifically, this Chapter will address and outline the ways that members of the university administration and faculty members can work together to develop a bridge to assist students through the chaotic world of postsecondary education providing a "bridge over troubled waters" in the form of a direct and highly supported pathway to successful MBA matriculation.

REVIEW OF THE LITERATURE

According to a recent study by Georgetown University, by 2020, 65% of all U.S. jobs will require post-secondary education, (Georgetown, 2015). Since 1940 the gap between the genders in undergraduate degree attainment has narrowed. 2015 was the first year in the history of the US that more women received a college degree than men (Stastica, 2016). In contrast, in 1940 almost twice as many men received college degrees than women. The trend moved significantly during the 1960s; but, the most notable changes came during and after 1963 when the Equal Pay Act was passed (EPA, 1963).

The 2016 Lumina Report examines the trends for degree-attainment rates for 15-64 year olds in the U.S. The lowest attainment rates are for Hispanic, Native American, and African American students. All of these students are part of the Lumina Report's underrepresented minorities (URM) statistic (Aranda, 2015). Women and students in at-risk categories, such as non-first time (NFT) enrollment and online degree completion program students often present the same completion rates as URM populations. Notably, most of these students are not seeking to achieve graduate degrees, in part because no one is showing them a meaningful and viable pathway to make and support the journey to and through graduate education.

The 2016 Lumina Foundation reports that only 11.47 percent of the general population has earned a graduate degree (Lumina, 2016). Given that a graduate degree is often the key for attaining a professional standard in the workplace along with a high income this statistic seems low. The median weekly earnings for the education attainment categories are: less than a high school diploma, $471; high school diploma, $652; some college but no degree, $727; associate degree, $785; bachelor's degree, $1,066; master's degree, $1,300; doctoral degree, $1,624; and professional degree, $1,735 (Review Journal, 2015). This data combined with the fact that URM females have few role models of successful graduate degree holders in their workplaces and communities illustrates that graduate school may not be even a remote idea or concept for this student population. This Chapter will offer a strategy to increase the number of students who proceed directly from undergraduate degree to graduate work, with an emphasis on and examples of how to help female, and at-risk students enter an MBA.

AN OVERVIEW OF WOMEN IN THE WORKFORCE

Women's share of the total U.S. labor force is 46.8%. According to the 2016 Catalyst Report, 56.7% of all women 16 years and over are in the labor force, compared to 69.1% of all men. Women currently hold 20 or only 4% of CEO positions at S&P 500 companies. Even, 61% of all mothers with children under the age of three are in the labor force (Catalyst, 2017). While women are participating in the work force, they predominately work in low wage and non-professional job classifications; and, their earning power is significantly less than men of similar background and education (Catalyst, 2017).

Women of color have significantly fewer professional opportunities (Catalyst, 2017). One way to view workplace parity is to consider the number of women and minorities who hold board positions. The graph below depicts the percentages of those in board positions held by minority vs majority females as of 2014:

As the above Figure indicates the numbers of women who are sitting on public company boards is very small, and the number of board positions that are held by minority females is even smaller. Participation on public and large company governing boards is important because the board room is the place were policies are changed and constructive road maps can be made to create more equality and parity in the workplace (Catalyst, 2017). Women of color occupy only 11.9 percent of managerial and professional positions. And of those women, 5.3 percent are African American, 2.7 percent are Asian American, and 3.9 percent are Latina. More than two-thirds of Fortune 500 companies have no women of color as board directors, according to statistics from the Center for American Progress (Warner, 2014).

Figure 1. Women holding board positions by race and ethnicity

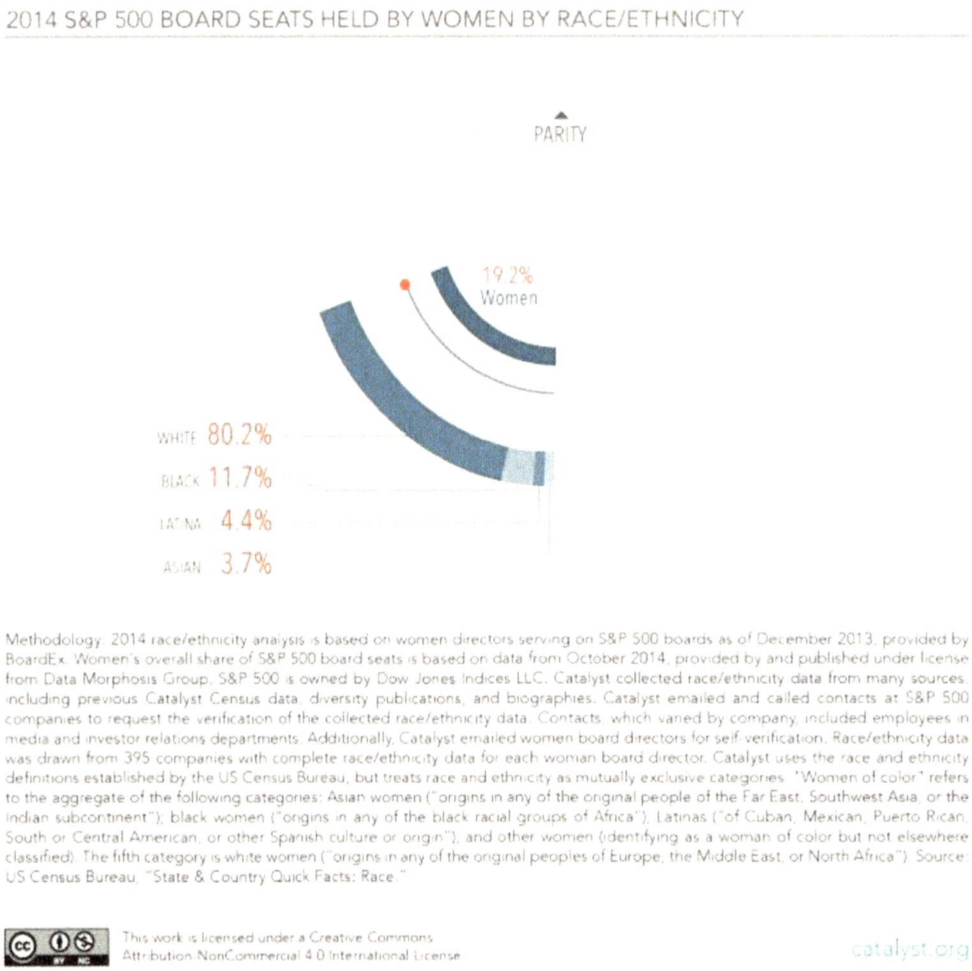

GRADUATE DEGREES PROVIDE A CONCRETE WAY TO CLOSE THE ACHEIVEMENT GAP AND THEY OPEN UP PATHWAYS TO EQULITY AND PARITY FOR WOMEN IN THE WORKFORCE

One way to close the achievement gap, is to develop and support pathways for females and at-risk student communities to quickly and successfully move into graduate work at much higher rates. The majority of non-first-time women students are dealing with dependent children, aging parents and the rigors of full time employment while trying to complete their undergraduate education programs (Black, 2016). These women are often in lower level administrative positions and they receive little or no support from their families and employers for the educational efforts. These women and at-risk students report being exhausted after completing the undergraduate degree and are often unaware of, the following:

- Uninformed that a graduate degree is approximately one third as long as the undergraduate degree (BBA: 120 – 130 units versus MBA 36- 48 units).
- Unacquainted with the requirements or preferences of employers to interview individuals with graduate education, and that employers are more likely to hire and promote those with graduate degrees.
- Lack the vision to see themselves as successful contributors at the higher ranks of business. They have fewer role models in the workplaces and the community of people like them (single female with children and/or non-white) who have successfully moved from situations similar to theirs to leadership roles in business or the community.

To address these issues universities and faculty members who teach these types of students need to build and then systematically support bridges and pathways for success for these at-risk and female students to become successful graduate students where they will have increased opportunities for success in the marketplace and within their communities.

BUILDING AND SUPPORTING BRIDGES AND PATHWAYS TO SUCCESS FOR URM, AT-RISK, AND FEMALE STUDENTS

This section of the paper offers twelve strategies to increase the number of students who proceed directly from an undergraduate degree to graduate work, with an emphasis on and examples of how to help female, and at-risk students enter an MBA. With comprehensive university-wide support from curriculum design, systematic instructor engagement and university administration support throughout the student life cycle (from admissions throughout the coursework and to graduation and beyond), these URM and female students can be thoughtfully and comprehensively supported to transition quickly and seamlessly from undergraduate to graduate completion. These subordinated groups need to be encouraged and supported to compete to earn living wages so that they can care for themselves and others, and so that they can be recognized as individuals with power and authority in business and in their communities.

Most undergraduate students at Brandman University will be the first in their family to graduate from college. Additionally, many Brandman undergraduate students are returning to college after one or more major interruptions in their course of study in an effort to finally complete their bachelors' degree. Almost all of these students will fail to complete an undergraduate degree within the six-year reporting cycle of the regional accrediting agencies and the IPEDS data. These non-traditional students often do not have the vision or mindset to see themselves with a graduate degree. They often lack role models and a mind set for continuing school. Most of these students had set their sights set from the beginning on completing the bachelor's degree. For the vast majority of these students, bachelor degree completion was such an enormous goal that thinking beyond undergraduate work is simply a bridge too far for many of them. It is critical that these students have a solid and consistent support system in order to be successful in undergraduate completion. However, a solid and systematic set of student supports is more important for graduate degree completion and a program to develop student thinking about the importance of continuing their education needs to begin early in their undergraduate program.

The following examples will provide proven strategies to show increased matriculation from a BBA to MBA programs. These examples will highlight pathways and processes can be applied across other schools and academic disciplines. Developing a multi-stage program such as this should be a key element of the mission and vision of universities serving URM and female students. Universities focused on serving this student community should provide a meaningful student pledge that URM female students can find bridges and pathways to success in graduate level coursework and that they can become important leaders and role models in business and within their communities.

The section of the paper will highlight and give examples of specific ways to be able to encourage students to continue their education after undergraduate work and stay in school to complete a graduate degree. These steps need to span the student life-cycle to ensure that they have an expanded vision and the necessary confidence to seek and complete a graduate degree immediately upon completing their undergraduate work. The suggested steps and strategies are offered in three categories as each category will offer concrete and proven examples of ways to assist students in matriculation from undergraduate to graduate programs:

Category One: Direct and Consistent Faculty Engagement
Category Two: Using Curriculum Design to Support Onward Matriculation
Category Three: Consistent and Supportive Communications from the Institution

CATEGORY ONE: DIRECT AND SYSTEMATIC FACULTY SUPPORT FOCUSED ON MBA MATRICULATION FROM THE START OF THE STUDENT'S PROGRAM

Many of us remember a teacher or an administrator that encouraged us at a time when we needed it the most. In my own High School experience at Homestead High School in Cupertino, California the only encouragement I remember came from Mrs. Liska, who ran the school office.

My own father had not graduated from High School and he had to leave school to work in order to support his family. My parents did not think that the grades girls made in high school were important and they did not even encourage me to attend the local community college. However, when I was working in the school office during my senior year, Mrs. Liska, always encouraged me to continue my studies. She repeatedly told me that "I was very strong at performing office skills." I remember her consistent encouragement and even though I was getting C's and D's in High School she helped me to develop a vision and a belief that I could find a way to be successful in life. My only A in High School came from my work experience course with Mrs. Liska. Mrs. Liska provided a single bright light about my future that was appearing dim and devoid of achievement and opportunities to excel. From personal experience, I know that sometimes all it takes to set a student on a pathway to success is one authentic and personalized intervention.

The examples below are specific ways to encourage URM female students to be retained in school and to proceed directly on to a graduate degree. These steps should be developed into a strategic plan for the program or institution so that the efforts can become part of the culture of support and encouragement within the university.

Clear and Consistent Faculty Feedback About Success in Graduate Level Coursework

Faculty should provide clear feedback to students in their classes when their students are successful in upper division course work at the bachelor level. These faculty comments will remind students that the upper division work being performed closely mirrors the style and rigor of the work they will be doing at the graduate level. Students can and should be reminded early and often as they excel. An example of a constructive comment that a faculty could make to a successful student in an undergraduate class might be:

If this paper was submitted in my graduate level class it would an A so you are already doing graduate level work and you are doing it successfully. Please remember you currently have all of the learning skills you need to move up to and be successful in graduate level work and perform at the graduate level in the workplace.

Faculty members should develop a PowerPoint or other type of student communication that is embedded into their upper-division courses in the undergraduate program. This presentation can be embedded into as many as 5 or 7 upper division courses to clearly and consistently remind students how close they are to doing graduate level work. The below slide from a PowerPoint provides an example of the type of communications that can be effective.

It is important that administrators ensure that both full time and adjunct faculty clearly understand that it is an important part of their role and responsibility to encourage students to move from the undergraduate program directly to the graduate program. As part of faculty training it is helpful to create templates for grading, course discussions and other course elements that focus on noticing the at-risk and female students, and then, concentrating support and resources on them and their needs, especially as it relates to their prompt progression to graduate education.

Faculty who consistently and positively provide encouragement should be recognized by the program chair and the dean at faculty in-services and other faculty retreats and conferences. Administrators should also review and reward faculty efforts and comments designed to encourage URM female students. Institutions should also work on attracting, developing and retaining faculty and staff that mirror the demographics of their student body. If the student population is 10% African American, then the institution should set a goal for 10% of its faculty be African American. Ohio State University has shown marked success in their recruitment efforts for minority faculty as part of their diversity and inclusion efforts (OSU Diversity Action Plan, 2007-12).

The program and the institution should create an environment that delivers the undergraduate curriculum using "Mindset" research and the best practices that Dweck and Yeager (2012) discuss in their seminal work, "Mindsets That Promote Resilience: When Students Believe that Personal Characteristics

Figure 2. Sample PowerPoint slide encouraging MBA consideration and matriculation

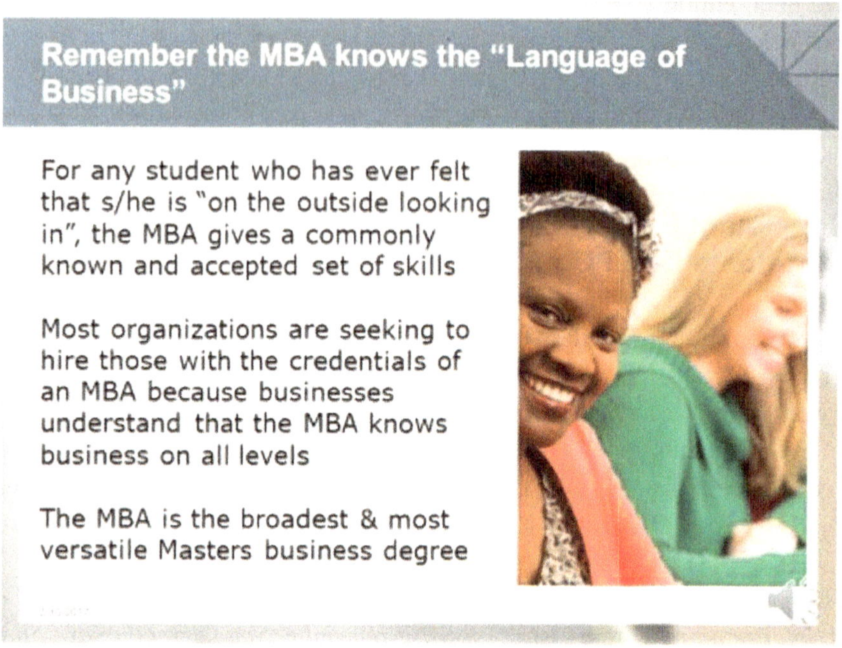

Can Be Developed." Leveraging the work of Dweck and Yeager (2012) in the "Mindset" area would include discussion questions and assignments that are directed toward encouraging students toward applying what they are learning to their current or previous personal or professional experiences. A growth mindset is one where students are encouraged to question, to try new things and to even make mistakes so that they can understand and work through failure. This could even include some assignments that are graded "A, B and anything below a B would be not yet." The growth mindset thrives in an environment where every misstep is considered "data" rather than blame.

CATEGORY TWO: CURRICULUM DESIGN

Faculty should design curriculum that establishes and supports a strong connection between what students are doing in class and what they will soon be doing in the marketplace. Developing directed assignments that have the students read, work on assignments, and complete projects that demonstrate the value of continuing their education is an integral part of the process of streamlining a transition directly to graduate enrollment. An example of this type of assignment might be:

An assignment was created for students to research and post their findings about how completing their undergraduate degree in business would increase their pay. Specifically, students were asked to review five possible jobs that they would be qualified for using the titles of the courses in their business program. Students then put their findings into the discussion thread and commented not only their own findings, but the findings of other students who lived in a variety of cities.

These types of targeted assignments and exercises when embedded in courses and across the curriculum ensures that students are regularly encouraged to think about continuing with their education and that they understand that continuing through to a graduate program has a positive and a significant pay-off. Students need to make strong and regular connections between what they are doing in class today and the positive impact for their immediate and long-term futures if they attain a graduate degree.

Ask local employers to create PowerPoints encouraging students to complete a graduate degree and to offer examples of jobs for individuals with recent graduate degrees. These could include vignettes extolling the benefits of a graduate degree within their organization that can be recorded and placed in selected courses for students to view and comment on. One suggestion is that the employers have the PowerPoints narrated by managers and directors who are female and from minority groups.

Another important element of creating a bridge to the graduate program is to identify the points within the curriculum where URM female students are most likely to "stop out." Faculty could redesign these courses putting more emphasis on assignments that encourage student success, focusing on inspiration for continued enrollment and graduation. The faculty who teach these courses could also be trained to provide the messaging of "going the distance" and "completing a master's degree." Faculty should work to create curriculum and course design that mitigates drop out factors (Street, 2010). The curriculum could include assignments that are a combination of student research and their personal stories, so that students understand that their personal stories are welcomed and encouraged. Another course design approach would allow the student to understand within each course what they are learning and how to apply this information to their lives leading to a possible graduate program.

CATEGORY THREE: PROGRAMS AND COMMUNICATION FROM THE INSTITUTION AND SENIOR ADMINISTRATORS TO THE STUDENT

In the development of a bridge program or direct on-ramp program for an MBA, it is important for the Academic Advisors and student-facing staff members to acquire the knowledge about degree requirements and to communicate with their students the difference in the number of courses and time necessary to complete a graduate degree compared with the time it takes to complete the undergraduate degree program. Academic Advisors and other student facing staff need to regularly remind students that completing a graduate degree takes approximately one-third of the time and financial commitment as an undergraduate degree.

Another important element of the bridge or direct on-ramp program would be for the Dean of the School of Business to work with faculty and the university marketing department to create a letter, email, and/or phone call that regularly goes out to selected successful URM and female undergraduate students. The communication should go out to URM and female students who have earned a B or better in one or more of the core upper division undergraduate classes. Each student could be invited to meet with the Dean as a group to discuss the graduate program and the pathway and process to continue directly to the MBA. The meeting should be marketed as a celebratory event and perhaps include a luncheon or coffee hour with the Dean. The communication to the selected students can also include information on higher salary ranges available to students who attain a graduate degree, and perhaps, include information on recent URM students who have moved into new positions or attained promotions after graduating with their MBA.

Programs should also explore dual credit opportunities between the BBA and MBA. For example, one option that Brandman University is currently exploring is to offer an accelerated program that includes four MBA classes within the undergraduate program curriculum, which would reduce the MBA core requirements making the degree path from BBA to MBA up to twelve credits shorter. If students are successful in the four designated BBA classes, credit for each of the four classes will count toward both the BBA and the MBA degree. Because of the direct transition to the graduate program, obtaining the MBA would become less expensive and time consuming for the student.

The bridge program should also create in-service programs with local and regional employers. The purpose of the program would be to encourage the employees of these local and regional businesses to progress from the BBA to the MBA. The university could then work in tandem with these employers to offer financial, promotional and/or time incentives for obtaining a degree. One particularly appealing element is to try to get key local and regional employers to designate up to four to five hours of work time to studying. Brandman University is currently working with a group of employers called "Premiere Partners." Students employed at Premiere Partners receive a reduced tuition rate and also enjoy some tuition matching educational benefits offered by the employer. This significantly reduces the cost for employees to obtain their degree and provides an access point for the university to work with the employer to encourage a direct transition to the MBA. This type of educational benefit offering can also be a part of the company wide effort to increase diversity and inclusion.

The following "Ten Cool Things" table is an example of a document used at Brandman University to encourage direct enrollment in the MBA upon graduation from the undergraduate program.

FURTHER RESEARCH DIRECTIONS

While there is clear support in the literature about the benefits of direct progression to the MBA, there is little literature available on what specifically motivates students to make these types of enrollment decisions and what assessment processes the students use to facilitate their decision making. Faculty and administrators working in this area would benefit from studies that explore and help to define what types of messaging are most encouraging and what message mediums are most effective with various types of students or student groups. Similarly, what number of messages throughout the term are optimal to encourage and not detract from student decision making in this area would be useful data in designing an optimized approach.

Additionally, while this study puts forward a range of communication methods and approaches to encourage direct matriculation into an MBA, studies ranking the efficacy of the various approaches would be beneficial and would help administrators and faculty members assign financial and human resources to the most effective and efficient approaches. Likewise, if there are approaches within the range of suggestions that are neutral or even detracting from the goal of direct matriculation for all or certain student groups this would be useful information as well.

Table 1. Ten cool things about the MBA and why you should consider enrolling in your MBA right away

Cool Thing	Ten Cool Things About the MBA	What this Means for You (as a Student)
1	Brandman has one of the highest student retention rates in the country in its graduate business school programs	You will be in class with a group of students who are working hard and succeeding. Therefore, you will have a stronger opportunity for success when you go to class with others who are going to be successful
2	The classes are designed to support success in the "soft" as well as technical skills. You will know how to BOTH lead people and to make solid economic decisions.	Four of the 12 Core Classes in the MBA focus on leadership, creativity and Human Resource skills
3	Help is "one click away" if you struggle with math	We have a free online math help center that will get you up to speed in no time. It does not matter if math is your favorite skill with our help you will be able to be taken seriously in terms of making good economic decisions (at work and at home). When someone hands you a financial statement you will sound so smart when with one glance you will be able to comment, "looks like we are up 5% from last year."
4	Getting an MBA means that no matter what people have to take your ideas seriously	You will understand and appreciate all of the key cornerstones of business (from marketing to operations) and you will be able to create and help everyone around you with strong options and opportunities
5	The skills learned for the MBA apply to all businesses whether they are for profit or not-for-profit	Students who want to create or work in businesses that are not-for-profit will be able to perform their coursework around their not-for-profit interest areas.
6	With an MBA people will see you as an Entrepreneur	Two of the 12 core classes focus on creativity and innovation. Students learn how to become creative problem solvers.
7	With an MBA you become a more critical or complex thinker	Students who graduate with an MBA become better at making decisions for both their professional and personal lives after practice in the skills of decision making in turbulent times
8	Fantastic options for Emphasis areas	Students can choose from a wide variety of emphasis areas; including Marketing, Health, Human Resources, Leadership and more......
9	With an MBA there is more money to be made	"*Post-MBA Starting Pay* was up 50% over *Pre-MBA Pay* for full-time students" "By the end of five years post-MBA, *Fifth Year Pay* was up 80% over *Post-MBA Starting Pay* for full-time students" (Source: Yeaple, 2014. MBA pay. Retrieved from http://www.forbes.com/sites/ronaldyeaple/2014/03/17/mba-alumni-profile-2/#7f92ac6e503e)
10	MBA's are taken seriously world wide	Once completing the MBA every business in the world knows that you know the entire scope of business and that you are a hard worker. This combination leads to continued success.

CONCLUSION

Universities who want to go the distance to improve at-risk and female student success should consider the offered multiple steps and approaches to supporting direct enrollment into the MBA program upon graduation with a bachelor's degree. It is in the best interest of both the students and the university to increase the number of women and at-risk students who successfully complete a graduate degree. There are three areas that universities can focus on to build a bridge for at-risk and female students so that they can become informed, acquainted with and understand the "chaos" of university education. These steps focus on faculty engagement, curriculum design and university-wide programs. Through these programs URM female students can develop a clear understanding of the value and importance of a graduate degree. These program elements will allow URM female students to understand graduate education, even if it is unusual in their community or if they do not know anyone who has a graduate degree. Attainment of a graduate degree can catapult them into a better paying job with more rewarding work, setting them on a path to distinction and leadership opportunities in the workplace and in their communities. Female URM students will then have an opportunity to become role models for others who are just beginning their educational journeys.

REFERENCES

Aranda, P. (2015, September). In most MBA programs, the diversity trend is down. *The Consortium*. http://cgsm.org/in-most-mba-programs-the-diversity-trend-is-down/

Baron, N. (2010). The forgotten underrepresented minority. *Huffington Post*. http://www.huffingtonpost.com/noah-baron/the-forgotten-underrepres_b_620995.html

Black, R. (2016). Coaching to increase retention in online degree completion programs. *USC*. Retrieved from, http://digitallibrary.usc.edu/cdm/compoundobject/collection/p15799coll40/id/298507/rec/2

Borkowski, J. G., Carr, M., Relliger, E., & Pressley, M. (1990). Self-regulated cognition: Interdependence of metacognition, attributions, and self-esteem. In B. Jones & L. Idol (Eds.), *Dimensions of Thinking and Cognitive Instruction* (Vol. 1). Hillsdale, NJ: Erlbaum.

Brahm, G. (2006). Students with staying power. *Nacubo*. Retrieved from http://www.nacubo.org/Business_Officer_Magazine/Magazine_Archives/June_2006/Students_With_Staying_Power.html

Carnevale, A. P. (2012, July 2). *The great sorting*. Retrieved from http://chronicle.com/article/The-Great-Sorting/132635/

Carnevale, A. P., Jayasundera, T., & Cheah, B. (2012). *The college advantage: Weathering the economic storm*. Retrieved from https://cew.georgetown.edu/wp-content/uploads/2014/11/CollegeAdvantage.ExecutiveSummary.081412.pdf

Carnevale, A. P., & Rose, S. J. (2015). *The economy goes to college: The hidden promise of higher education in the post-industrial service economy*. Retrieved from https://cew.georgetown.edu/wp-content/uploads/EconomyGoesToCollege.pdf

Carnevale, A. P., Smith, N., & Strohl, J. (2013). *Recovery: Job growth and education requirements through 2020*. Retrieved from https://cew.georgetown.edu/wp-content/uploads/2014/11/Recovery2020.FR_.Web_.pdf

Carnevale, A.P., Smith, N., & Strohl, J. (2014). Recovery: Job growth and education requirements through 2020 (executive summary). Georgetown University. Retrieved from https://cew.georgetown.edu/wp-content/uploads/2014/11/Recovery 2020.ES_.Web_.pdf

Covington, M. V. (2000). Goal theory, motivation, and school achievement: An integrative review. *Annual Review of Psychology*, *51*(1), 171–200. doi:10.1146/annurev.psych.51.1.171 PMID:10751969

Dembo, M. H., & Seli, H. (2008). *Motivation and learning strategies for college success: A focus on self-regulated learning*. New York, NY: Routledge Taylor.

Dweck, C. (2006). *Mindset: The new psychology of success*. New York, NY: Ballentine Books.

Eccles, J. S., & Wigfield, A. (2002). Motivational beliefs, values, and goals. *Annual Review of Psychology, 53*(1), 109–132. doi:10.1146/annurev.psych.53.100901.135153 PMID:11752481

Elliot, E. S., & Dweck, C. S. (1988). Goals: An approach to motivation and achievement. *Journal of Personality and Social Psychology, 54*(1), 5–12. doi:10.1037/0022-3514.54.1.5 PMID:3346808

Equal Employment Opportunity Commission. (1963). The Equal Pay Act of 1963. Retrieved from https://www.eeoc.gov/laws/statutes/epa.cfm

Farkas, K. (2014, October 8). Adults who re-enroll in college often don't graduate. *Northeast Ohio Media Group*. Retrieved from http://www.cleveland.com/metro/index.ssf/2014/10/ adults_who_re-enroll_in_colleg.html

Hidi, S., & Harackiewicz, J. M. (2001). Motivating the academically unmotivated: A critical issue for the 21st century. *Review of Educational Research, 70*(2), 151–180. doi:10.3102/00346543070002151

Karim, M. N., & Behrend, T. S. (2013). Controlling engagement: The effects of learner control on engagement and satisfaction. In C. Wankel & P. Blessinger (Eds.), *Increasing Student Engagement and Retention in e-learning Environments: Web 2.0 and Blended Learning Technologies (175 – 197)*. Bingley, UK: Emerald Group Publishing Limited. doi:10.1108/S2044-9968(2013)000006G005

Kelly, A. P., & Schneider, M. (2012). *Getting to graduation: The completion agenda in higher education*. Baltimore, MD: John Hopkins University Press.

Las Vegas Education Review Journal. (n. d.). 60 percent of job openings will require a college education. Retrieved from http://www.reviewjournal.com/news/education/2018-60-percent-job-openings-will-require-college-education

Lotkowski, V. A., Robbins, S. B., & Noeth, R. J. (2004). The role of academic and non-academic factors in improving college retention. *ACT.org*. Retrieved from https://www.act.org/research/policymakers/pdf/college_retention.pdf

Lumina Foundation. (2016). Report. Retrieved from http://strongernation.luminafoundation.org/report/2016/

Müller, T. F. (2007). *Factors affecting persistence of women in online degree-completion programs* [Doctoral dissertation].

NASPA Student Affairs Administrators in Higher Education. (2014). *National study of non-first-time students shows disturbing completion rates*. [News Release]. Retrieved from NASPALowestGradRates-NonFirstTimeOct2013.pdf

National Center for Education Statistics. (2014, May). *The condition of education. Institutional retention and graduation rates for undergraduate students*. [Data file]. Retrieved from http://nces.ed.gov/programs/coe/indicator_cva.asp

National Center for Higher Education Management Systems. (2016). *Persistence and completion: Graduation rates.* [Data file]. Retrieved from http://www.higheredinfo.org/dbrowser/index.php?submeasure=27&year=2009&level=nation&mode=data&state=0#/-1/

O' Shea, S., & Stone, C. (2014). The hero's journey: Stories of women returning to education. *The International Journal of the First Year in Higher Education, 5*(1), 79. Retrieved from http://dx.doi.org.libproxy1.usc.edu/10.5204/intjfyhe.v5i1.186

Ohio State University Diversity Action Plan 2007-2012. (2012). Retrieved from https://www.osu.edu/diversityplan/index.php

Paradise, A. (2008). *Informal learning: Overlooked or overhyped?* Alexandria: American Society for Training & Development, Inc.

Pintrich, P. R., & De Groot, E. V. (1990). Motivational and self-regulated learning components of classroom academic performance. *Journal of Educational Psychology, 82*(1), 33–40. doi:10.1037/0022-0663.82.1.33

Rheinberg, F., Vollmeyer, R., & Rollett, W. (2000). Motivation and action in self-regulated learning. In M. Boekaerts, P. R. Pintrich, & M. Zeidner, (Eds.), Handbook of Self-Regulation (pp. 503-529). San Diego, CA: Academic Press. doi:10.1016/B978-012109890-2/50044-5

Richardson, M., Abraham, C., & Bond, R. (2012). *Psychological correlates of university students' academic performance: A systematic review and meta-analysis.* Retrieved from http://www.ncbi.nlm.nih.gov/pubmed/22352812

Rueda, R. (2011). *The three dimensions of improving student performance: Finding the right solutions to the right problems.* New York, NY: Teachers College Press.

Sansone, C., & Harackiewicz, J. M. (2000). *Intrinsic and extrinsic motivation: The search for optimal motivation and performance.* San Diego, CA: Academic Press.

Schunk, D. H., & Ertmer, P. A. (2000). Self-regulatory and academic learning self-efficacy enhancing interventions. In M. Boekaerts, P.R. Pintrich, & M. Zeidner, (Eds.), Handbook of Self-Regulation (pp. 631-649). San Diego, CA: Academic Press.

Scrivner, S., & Weis, M. J. (2009). More Guidance, Better Results? Three-Year Effects of an Enhanced Student Services Program at Two Community Colleges. Retrieved from http://www.mdrc.org/publication/more-guidance-better-results

Stastista. (2017). Percentage of the U.S. population who have completed four years of college or more from 1940 to 2015, by gender. Retrieved from https://www.statista.com/statistics/184272/educational-attainment-of-college-diploma-or-higher-by-gender/

Street, H. (2010, Winter). Factors influencing a learner's decision to drop-out or persist in higher education research. *Online Journal of Distance Learning Administration, 13*(4).

United States Department of Education. (2012, February). *Pathways to success, integrating learning with life and work to increase national college completion.* Retrieved from http://www2.ed.gov/about/bdscomm/list/acsfa/ptsreport2.pdf

Wankel, L. A., & Blessinger, P. (Eds.). (2013). *Increasing student engagement and retention using multimedia technologies: Video annotation, multimedia applications, videoconferencing and transmedia storytelling*. Bingley, UK: Emerald Group Publishing Limited.

Warner, J. (2014). *Fact sheet: The women's leadership gap*. Center for American Progress. Retrieved from: https://www.americanprogress.org/issues/women/reports/2014/03/07/85457/fact-sheet-the-womens-leadership-gap/

Witherspoon, R., Goldsmith, M., Lyons, L., & Freas, A. (2000). *Coaching for leadership: How the world's greatest coaches help leaders learn*. San Francisco, CA: Jossey-Bass/Pfeiffer.

Wlodkowski, R. J. (2008). *Enhancing adult motivation to learn: A comprehensive guide for teaching all adults*. San Francisco, CA: Jossey-Bass.

Wong, A. (2016). The Atlantic. Where are all the high school grads going? Retrieved from https://www.theatlantic.com/education/archive/2016/01/where-are-all-the-high-school-grads-going/423285/

Yeager, D. C., & Dweck, C. S. (2012). Mindsets That Promote Resilience: When Students Believe That Personal Characteristics Can Be Developed. *Educational Psychologist*, *47*(4), 302–314. doi:10.1080/00461520.2012.722805

Yeaple, R. (2014). MBA Pay. *Forbes*. Retrieved from http://www.forbes.com/sites/ronaldyeaple/2014/03/17/mba-alumni-profile-2/#7f92ac6e503e

KEY TERMS AND DEFINITIONS

EPA: The Equal Pay Act of 1963.

Growth Mindset: A growth mindset is one where students are encouraged to question, to try new things and to even make mistakes so that they can understand and work through failure in a positive and confidence building way. The chapter leans on the work of Carol Dweck and the definitions of growth and fixed mindset that are explored in her work.

Mentor: Within the context of this chapter mentor means being supported and encouraged by someone who looks like you and who comes from similar circumstances as you and who has already achieved a high level of success in one's desired professional field.

NFT: Non-first-time students and students returning after one or more interruptions in their studies.

Role Model: Within the context of this chapter role model means seeing someone who looks like you and who comes from similar circumstances as you achieve what it is that you are seeking to achieve.

STEM: The STEM disciplines are science, technology, engineering and mathematics.

URM: Underrepresented minority students.

Chapter 9
Coaching Based on the Hope Model to Address Student Success in South Korean Educational Institutions

Alan Cromlish
Namseoul University, South Korea

Ruth Claire Black
Namseoul University, South Korea

ABSTRACT

South Korea is one of the highest achieving education countries in the world. However, there are several factors that inhibit South Koreans from realizing their full potential. South Korean students exhibit low scores on a number of learning and success measures, including low satisfaction with education and career prospects, low numbers of positive feelings and high numbers of negative feelings on a range of measures such as, poor family connections and poor social connections. These factors contribute to high depression rates and high rates of suicide among young adults. The depressive states that are common among young South Koreans can be addressed using a Hope Theory based professional coaching intervention to generate and then support more positive feelings about education and personal and professional goals. The combined approach can significantly and positively impact depressive issues and address student confidence, motivation, and learning ability. Thereby, leading to improved metrics on a wide range of life and educational satisfaction indexes.

INTRODUCTION

South Korea is one of the highest achieving education countries in the world yet there are some key factors that inhibit South Koreans from realizing their full potential as students and young adults. South Korean students exhibit low scores in several learning and success measures, including low satisfaction with education and career prospects, poor family connections, and poor social connections. In addition,

DOI: 10.4018/978-1-5225-2998-9.ch009

on a wide range of educational, personal, and professional measures, the results show low positive feelings about their future, and correspondingly, high negative feelings. These factors contribute to a high overall depression rate among young South Koreans, with a high rate of suicide specifically among high school and college age South Koreans. The depressive states that are common among young South Koreans could be addressed using a "hope theory-based professional coaching intervention" to generate and then support more positive feelings about education, and as a result, higher levels of self-reported student success and better attainment of personal and professional goals.

A comprehensive study of coaching as a promising practice for retention and student success took place in 2015 and 2016 as part of a U.S. based dissertation study. To determine how the results of the study might be instructive in developing a specialized coaching-based intervention to support South Korean post-secondary students, this paper re-examines the coaching findings from the 2015-16 study. The goal of this paper is to help identify best practices, innovative approaches, and viable alternatives to existing approaches for these significant student success and life issues and to improve the South Korean post-secondary student experience on student satisfaction and student success indexes. By significantly and positively impacting these depressive and performance issues, along with promoting student confidence, motivation and learning ability, it is possible to improve life optimism and eventual success in South Korea.

SOUTH KOREAN COMMITMENT TO EDUCATION

South Korea is consistently ranked as one of the leading educational countries in the world. The Korean devotion to post-secondary education is unparalleled. The country devotes immense financial and human capital resources to higher education. Since 1990 the national South Korean educational budget has risen from $1.4 billion to $29 billion, a six-fold increase in less than thirty years (High Performance, 2015). Teaching in Korea is a highly competitive and respected career choice. Only 20% of Korean students that enter post-secondary teacher preparation programs pass the highly selective teacher examination and go on to become secondary school teachers (High Performance, 2015).

In addition to South Korea's strong commitment to school funding, students are highly incentivized to obtain a baccalaureate degree because the earning power of a baccalaureate degree holder is up to two and half times that of high school graduates (South Korean Education Reforms, 2016). South Korea's rapid economic and technological development has students rushing to keep up with the technological and degree-driven demands in the labor market (South Korean Education Reforms, 2016). A 2016 OCED report on education attainment found that South Korea has the largest percentage of college graduates among 25 to 34 year old's in the world (Population with Tertiary Education, 2016).

PRESSURE TO EXCEL

While Korea is leading the world in many education metrics and measures while simultaneously demonstrating high education attainment rates, high school and post-secondary students are exhibiting some alarming emotional issues. These issues are a symptom of significant and deep-rooted social challenges. Korean students are under immense pressure to do well on the national university admission test, the "Suneung." The Suneung test is the sole measure of a student's ability to enter a Korean university, which makes it a major contributor to stress, psychological depression, and suicide among teenagers and young adults. "Suicide was the leading cause of death for Koreans, ages 15 - 24 years, in 2008…" (Wang, 2013, p. 2). In addition to the alarmingly high rates of stress and suicide among high school and young adults, Koreans also present with high degrees of negative feelings and difficulties with social connections, and correspondingly, low mental and social health scores. Young Koreans and Korean college students regularly exhibit difficulties in the areas of self-anchoring and general life satisfaction (Diener, Ng, Harter & Arora, 2010).

A comprehensive study of coaching as a promising practice for retention and student success took place in 2015 and 2016 as part of a US based dissertation study. This paper re-examines the pertinent findings from the comprehensive study as they relate to a specialized coaching based intervention to try to offset these societal pressures, performance and student satisfaction issues among Korean college students. The recommendation of the paper is to develop a specialized coaching intervention based on "Hope Theory" to tackle these pervasive social pressure, success and satisfaction issues as part of the college experience. The combination of the Hope Theory approach embedded across a comprehensive coaching program is a targeted intervention specifically designed to support Korean post-secondary students as they grapple with social, peer and educational pressures and the depressive symptoms that result from these significant and ongoing pressures. Re-examining the coaching findings from the 2015-16 study, within this distinct but comparable educational context, can help to identify best practices, innovative approaches, and viable alternatives to existing approaches to these difficult student success issues and challenges in Korea. By significantly and positively impacting these issues, along with promoting student confidence, motivation and learning ability, it is possible to improve life optimism and self-perceptions of success for young South Koreans.

REVIEW OF THE LITERATURE

While South Korean ranks high in many happiness indexes, those indexes are based on items such as per capita GDP, physical health, life expectancy, positive feelings toward the government, education, safety and security, personal freedom, and social capital (Diener, Ng, Harter, & Arora, 2010). However, these often-cited Korean metrics and measures fail to address personal feelings, interpersonal relationships, and social support perceptions. A review of the literature regarding emotions and interpersonal relationships among South Koreans indicates that they score low on measures of personal and interpersonal concepts and self-perception.

Diener, Ng, Harter, and Arora (2010) compiled the results of selected nations and then ranked them by per capita GDP, superimposing each nation's social psychological prosperity, self-anchoring, positive feeling, and low negative feelings over the GDP data to provide a more comprehensive analysis of the selected nations. The authors' research analyzed the first representative sample of respondents' reasons for happiness across the globe using the "Gallup World Poll" (Diener, Ng, Harter, & Arora, 2010). Some research studies indicated that income is a strong predictor of happiness and well-being; therefore, the authors reviewed raw income and its comparison to happiness.

While Diener's et al. work concludes that income was a moderately strong predictor of one's own life evaluation, income was seen as a weak predictor of positive and negative feelings. Their study discovered that, while income is a reliable predictor of a nation's social-psychological prosperity and other attributes, such as self-anchoring and positive and negative feelings, there are other important factors that need to be considered. South Korea is an outlier and is an example of a country deserving further study and evaluation because it ranks 35th out of 89 countries for GPD per capita, yet it reports significantly low social-psychological prosperity (83rd out of 90), low self-anchoring (53rd out of 90), low positive feelings (58th out of 90), and high negative feelings (77th out of 90) scores. These combined, multi-level metrics indicate that they have significant issues with self-perception on key happiness and life indices.

South Koreans have a high suicide rate. As of 2012, South Korea ranked first in suicide among OECD countries with 29.1 suicides per 100,000 people (OECD, 2016). A 2016 report of the World Health Organization lists South Korea at 36.8 suicides per 100,000 people as of 2012, which is the highest suicide rate in the world and could indicate the high levels of underlying emotional distress that present as a range of depressive disorders. These emotional problems are being produced within South Korean culture generally and within the high school and post-secondary education system specifically.

A number of studies explore and address issues that affect suicide. Arnau, Rosen, Finch, Rhudy, and Fortunato (2007) studied depression in a seminal longitudinal study. This study identified a number of key factors that contribute to depression and elevated suicide rates. "The number of recent stressors, major life events, coping style, and social support could all have direct effects on depression" (Arnau, et al., 2007, p.59). These factors are consistent with the 2010 study by Diener et al. regarding Korean society. Diener found a clear link between the data and information that Koreans report in surveys and the significant levels of long term depression that are present within Korean society. Cho, Nam, and Suh (2005) reiterated this conclusion through a survey of 3,711 respondents in South Korea that found that 34.3% of men and 47.5% of females exhibited depressive symptoms.

When these long-term depression rates are compared to North America and Western Europe the comparison is alarming. In North America and Western Europe only 2% and 6% respectively of the adult general population had depression (Ohayon & Hung, 2006). A 2011 article in *The Economist* discussed the findings of a study done in South Korea that concluded that the obsession for education within South Korea was becoming an emotional stressor for children. Citing a study by Seoul's Yonsei University, the article also reported that Korean teenagers are by far the unhappiest in the OECD. Given the large percentage of the South Korean population showing depressive symptoms, the number of individuals that seek and get help is correspondingly low. Ohayon and Hung (2006) researched the prevalence of depression disorders in 3,719 non-institutionalized individuals aged 15 years or older in South Korea. Only 15.7% of the participants had consulted a physician in the last 12 months and only 7.4% of those showing depressive symptoms were taking antidepressant treatments or in active therapy.

Depression association with negative emotions can lead to low vigilance, restlessness, being easily distracted and having poor focus, lower processing speeds, memory encoding problems and slower cognition (Ai, Cascio, Santagelo, & Evans-Campbell, 2005). That Korea is one of the top educational countries in the world is an amazing feat for a country that appears to have wide spread, recurring emotional issues that directly affect the population's ability to learn and to experience personal concepts of success and satisfaction. This long-term, untreated depression could result in deficiencies in the classroom that affect learning and memory functions.

IMPACT OF DEPRESSION ON LEARNING AND MEMORY

Mental state is impacted by multiple stimuli and events. Long-term memory is crafted over time through holistic events that slowly shape thoughts and perceptions of future events (Compton & Hoffman, 2013).

Memory can also be impacted by current thoughts, perceptions and judgements that immediately impact conscious thoughts and feelings. Emotions and mood are the driving forces behind mental state. Mood is a general, wide-spread feeling while emotions are feelings that can rapidly appear or disappear due to recent events (Compton & Hoffman, 2013). Emotions can affect mood; constant negative emotions to stimuli and events will lead to a prolonged negative mood resulting in depression.

The American Psychological Association website (2016) defines depression as, "[p]eople with depression may experience a lack of interest and pleasure in daily activities, significant weight loss or gain, insomnia or excessive sleeping, lack of energy, inability to concentrate, feelings of worthlessness or excessive guilt and recurrent thoughts of death or suicide." The findings of Diener, et al. (2010) suggest that these common negative feelings, low levels of positive feelings, and lack of social support in Korea are a long-term issue facilitating and supporting negative emotions and thus leading to high rates of depression and finally high suicide rates. The cultural aspects explored by Diener et al. (2010) showed that the depression and hopelessness experienced by Korean students will have lasting effects on students and their ability to learn. However, the culture is difficult to change.

THE ROLE OF PROCESSING SPEED IN LEARNING AND STUDENT SUCCESS

An important element of learning is processing speed. The lower a student's processing speed, the more time the student needs to take tests, absorb information and perform tasks in the classroom. Research has been done linking the presence of depression and processing speed. Research conducted by Tsourtos, Thompson and Stough (2002) found that unmedicated depressed patients' processing speed was lower than both the medicated and the control group. "Our results indicated that speed of information processing, as measured by [inspection time], is impaired in young, unmedicated, unipolar depressed patients. This finding is consistent with the hypothesis that young depressed individuals do show cognitive slowing" (Tsourtos et al., 2002, p. 262). The research by Egeland et al. (2003) compared the cognitive and attentive profiles of schizophrenic and depressed patients to a control group. Their finding of a "speed reduction in both clinical groups is in accordance with previous research" (p. 281) and supports the results of the study by Tsourtos, et al. (2002) that depression inversely affects processing speed and directly affects learning and student success. A study conducted by Calhoun and Mayes (2005) looked at processing speeds in children with depression, "As expected, children with depression performed poorly on [Processing Speed Index] relative to other scores" (p. 339).

Student motivation becomes an issue when students are constantly trying to catch up with the material due to low processing speeds. This issue is compounded when students move into higher grade levels and study more complex ideas and topics. Students can lose vigilance and motivation when they require more effort and time to understand what their peers are able to quickly process. Tarkbuck and Paykel (1995) explain that "the changes associated with depression affect particularly 'complex' tasks that require a greater degree of internal processing and conscious effort" (p. 293). While students may be capable of completing a learning task while in a depressed state, the lack of vigilance brought on by depression inhibits a student's ability to learn. Egeland et al. (2003) discussed low vigilance and its correlation to depression, concluding that, "within the depressed group decreased vigilance correlated with increased level of depression" (p. 282). Their work built upon Tarbuck and Paykel's (1995) research that found that depression reduces a person's central drive and motivation. The apathetic state brought on by depression hinders students from learning and gaining the necessary educational skills to succeed in school and then later within the highly competitive South Korean job market. This apathetic state leads to a lack of hope for the student's future and lower positive feelings of student success.

DEPRESSION AND GOAL PURSUIT

The dispirited and lethargic condition that is brought on by depression not only affects memory, it also has a direct and important impact on an individual's ability to pursue goals and to develop the motivation and confidence necessary to achieve these goals. Depression hinders goal setting and goal pursuit, motivation, confidence and non-cognitive skill development. Students and young adults in South Korea experience educational and social pressures as a result of goals and expectations that are set by family members and by the government through the once a year high-stakes admission test that is required for college and university placement and for academic success.

The early work of Brandtstadter and Renner (1990) established that a low level depressive state leads the subject to a stronger belief in the importance of internal control. In subsequent work Brandstadter and Rothermund (2002) then established that "depressive symptoms point to difficulties in relinquishing blocked goals"; however, they also suggest that "feelings of depression may play a functional role in promoting accommodative revisions of the person's system of preferences through destroying the cognitive sets and biases that maintain assimilative efforts" (p. 140). Students need to learn how to properly assimilate and how to accommodate and overcome obstacles and challenges to achieve their goals. It is clear that the depressive state hinders this process and that students will need significant help and interventions in order to move past these major issues. A more positive mood and higher levels of hope leading to increased confidence and greater goal pursuit will allow students to endure and work through the difficulties they encounter within the Confucian foundations of Korean society generally and the highly competitive, intensive education system in particular (Johnson, 2010; Street, 2002). A blended intervention approach to the stated issues through the use of Hope Theory and focused professional coaching in the student's junior or senior year of undergraduate study is proposed as a solution to the issue of high rates of depression and low levels of self-reported student success among young Koreans.

HOPE THEORY AS AN APPROACH TO OVERCOMING DEPRESSION AND FEELINGS OF HOPELESSNESS

Snyder (1994) adopts the most prevalent definition of hope which he expresses as "a construct of pathways and agencies, with pathways being the ability to find ways to reach one's goals, and agencies being the motivation to reach one's goals" (p. 355). Hope Theory provides a means and methodology for improving goal identification and support, as well as, the development of pathways and incremental steps along a continuum to enable reaching these goals. Through professional coaching built around Hope Theory, students can construct healthy goals and begin to create meaningful pathways to success (Black, 2016). One of the keys to coaching as a promising practice is the "combination of a highly individualized and targeted support with the caring, empathetic and non-judgmental counseling of a "missional" professional who authentically desires to help the student. It is this intentional two-part approach that leads to meaningful improvement in retention and persistence as well as improvement in the individual student's feelings and perspective about their successful degree completion goals and future career success" (Black, 2016, p. 153).

Depression and hope are inversely correlated. Ai, Cascio, Santangelo, and Evans-Campbell (2005) evaluated the relationship between hope and spiritual meaning after 9/11 and established that higher levels of education and hope are directly correlated to lower levels of depression and anxiety. It was also asserted that "depression was positively associated with negative emotion" (Ai et al., 2003, p. 533). Arnau et al. (2007) confirmed an inverse relationship between hope and depression through their longitudinal study and concluded that "high levels of hope predicted decreased levels of depression" (p.57). Han, Park, Im, Ju, Bae, and Kang (2010) researched the relationship between hope, depression, and suicidal ideation in chronically mentally ill patients. They found significant correlations between these factors and concluded that patients should focus on increasing hope and decreasing depression. Focusing on improving hope will lead to better cognition through improvements of pathways and agencies (Ai et al., 2003; Arnau et al., 2007; Hedayati & Khazaei, 2014; Mascaro & Rosen, 2005, Synder, 1994).

Improving hope could help reduce depression and restore the ability to succeed within educational institutions for students experiencing low satisfaction, low positive feelings, high negative feelings, and finally, poor family and social connections. Implementing professional student success coaching based on Hope Theory is a focused intervention designed to support students as they struggle with the severe issues and symptoms inherent in Korean society generally and within the Korean education system specifically.

DESIGNING AN INTERVENTION USING PROFESSIONAL COACHING BASED ON THE HOPE MODEL TO OFFSET DEPRESSION AND TO IMPROVE STUDENT SUCCESS METRICS

Lack of hope and depression could be offset through professional student success coaching built around the Hope Model with its processes and procedures for developing and supporting goals and pathways to success. Snyder (1994) explains that it is necessary for the individual to believe in their ability to generate a roadmap or pathway to success. This belief in the ability to build a pathway is as important as believing in the ability to achieve one's goals. These pathways and agencies working together in a transactional way with pathway thinking enhances agency thinking and in turn agency thinking enhances pathway thinking (Compton & Hoffman, 2013; Irving et al., 2004; Rand & Cheavens, 2009; Snyder, 1994). The research of Feldman and Dreher (2012) examined a single intervention with college students to improve hope and increase goal directed behavior and found that, "a single-session intervention can increase hope in the short term as well as lead to greater levels of goal progress as much as a month later" (p.756). By providing professional student success coaching to junior and senior year undergraduate students over several semesters, hope, goal identification, and the development of incremental pathways to enable goal attainment will increase hope resulting in the slow diminishment of the depressive symptoms.

The Hope Model uses three phases to direct goal-oriented thoughts when a goal is pursued: an individual's learning history, the pre-event phase, and the event sequence phase (Rand & Cheavens, 2009; Snyder, 1994). These pathways and agencies are combined with an individual's emotional state. Individuals with a history of accomplishments and of successfully overcoming obstacles will have an emotional set that is more hopeful, contains more positive feelings, and exhibits higher levels of confidence that goals can be reached (Compton & Hoffman, 2013; Rand & Cheavens, 2009; Snyder, 1994)

When pursuing a goal, the individual will enter the pre-event phase where they weigh the outcome value of the goal. If the outcome value of the goal is too low, the goal pursuit will cease. If the goal is determined to be important, the individual will move into the event sequence phase. As the individual starts to pursue the goal, the individual will monitor and assess the outcome value of the goal and regularly determine the outcome value of the goal. This constantly evolving assessment of goal pursuit is necessary because the outcome value cannot be appropriately determined until pursuit of a goal has commenced (Compton & Hoffman, 2013; Rand & Cheavens, 2009; Snyder, 1994). Ultimately, the goal is either attained or not and the resulting positive or negative emotions become part of the individual's learning history resulting in building hope and confidence or hopelessness and passiveness (Compton & Hoffman, 2013; Rand & Cheavens, 2009; Snyder, 1994).

There are two external sources that affect the individual during the event sequence: stressors and surprise events. Stressors are any obstacles that put the goal pursuit at risk. Stressors influence goal pursuit by effecting pathways, agency, and outcome values as well as the learning history of the individual (Synder, 1994). During goal pursuit, individuals who have low hope will likely see stressors as insurmountable challenges; whereas, individuals who have high hope will likely see stressors as an obstacle that can be overcome. In addition to stressors, surprise events are incidents that occur outside of the context of the goal pursuit. They can be either positive or negative in nature and effect the individual's agency by either increasing or decreasing motivation depending on the incident that has occurred (Compton & Hoffman, 2013; Rand & Cheavens, 2009; Snyder, 1994).

By embedding professional student success coaching into the student's academic course of study as part of a junior or senior year experience, Korean post-secondary institutions can proactively and effectively create and support hope, goal setting, and goal attainment (Black, 2016). Professional coaches can help the students to identify goals, and then, provide steps to establish smaller, incremental and reachable goals as a pathway to goal attainment that will help students to improve their ability to cope with surprise events and stressors, as well as, cultural expectations and limitations of their role within Confucian societies (Black, 2016). The results of the coaching should be increased feelings of success, competence, and finally, future goal attainment.

COACHING

The terms "executive coaching" and "life coaching" are recent constructs that can be traced back to the late 1980s, when they first appeared as new and more personalized applications of general counseling techniques (Bosworth, 2006). Witherspoon, Goldsmith, Lyons, and Freas (2000) explain that a coach helps people see beyond what they are today to what they can become tomorrow. An effective coach provides a sturdy shoulder to stand on, so one can see further than they might see on their own (p. 12). Professional coaching practices concentrate on moving an individual forward through motivation and targeted support to enhance skills and perspectives, resources, and creativity in a highly-personalized format (International Coach Federation, 2015). The International Coach Federation defines coaching as "partnering with clients in a thought-provoking and creative process that inspires them to maximize their personal and professional potential" (International Coach Federation, 2015). The core competencies of coaching are goal setting and managing incremental progress towards goals, as well as, individual accountability regarding progress towards the established goal. When these core competencies are combined with the personalized nature of coaching, a strategy for intervention to address the issue of hopelessness and depression for Korean post-secondary students could be extremely effective.

EMERGENCE OF COAHCING AS A RETENTION AND STUDENT SUCCESS TOOL IN POST-SECONDARY EDUCATION

In 2001, Inside Track, a for-profit coaching company in San Francisco, California became the first corporate coaching organization to develop specific coaching programs and services targeted to higher education institutions. Over the next decade several other educational coaching and academic mentorships organizations such as Aviso and Education Sales Marketing Corporation (ESM) began to develop services for supporting and assisting university level students.

Given the ineffectiveness of universities to increase graduation rates in the three decades from 1980 to 2010, by 2010 many educators and university administrators began to conclude that retention solutions were more likely to come from comprehensive re-examination of the status quo and by looking outside the institution for new approaches (Bettinger & Baker, 2013; Bosworth, 2006; Kelly & Schneider, 2012). This led some institutions to consider and begin utilizing individualized professional coaching services for certain segments of students or to consider developing dedicated executive coaching departments to address specific issues and challenges within their institutions. In support of the trend towards utilizing coaching in academic institutions, the U. S. Department of Education published findings in 2014 that established academic coaching as a promising practice for improving academic persistence and for addressing certain issues and challenges among certain underrepresented and minority student groups (IES, 2014). In evaluating student response to coaching across several studies, the Institute of Education Sciences (IES) found that students assigned to receive coaching were significantly more likely to remain enrolled at their institutions than students who do not receive coaching (IES, 2014).

THE ROLE OF GOAL SETTING AND INDIVIDUAL FOCUS IN COACHING

In assessing the value of coaching in the areas of goal setting and measuring progress towards goals, the study by Black (2016) found that in describing their work coaches uniformly preferred to use the term "coaching plan" rather than goal setting. Several coaches commented that urging a student to directly work on goal setting and goal development was often counterproductive and could impede the development of a trusting and authentic relationship between student and coach. This perspective is instructive in the Korean context as students suffering from anxiety, depression, and or hopelessness will need to work up to goal identification and goal setting slowly as they develop trust and confidence in the coach and the coaching process. An example of a typical coach's response on the role of the coach in goal setting from the study was:

Most students understand that they need a plan. As we develop the plan together, the skills and resources that the student needs to work on will come into focus. Then we will set steps around the skills we need to develop (Black, 2016, p. 87).

Another key to coaching as a promising intervention for Korean post-secondary students is that coaching combines a highly individualized and targeted student support with the caring, empathetic and non-judgmental counseling of a "missional" professional who authentically desires to help the student. It is this intentional two-part approach that leads to improvement in the individual student's feelings and perspective about their goals and opportunities for future career success.

The Table below excerpts quotations from the coaching study on the importance of individual student support within successful coaching.

It is this intentional two-part approach of carefully developing a plan that will meet each individual student's needs and addressing individual challenges and obstacles that most likely will lead to improvement in the individual student's feelings and perspective about their goals and opportunities for future career success (Black, 2016).

THE IMPORTANCE OF SELF-EFFICACY, NON-COGNITIVE FACTORS AND GROWTH MINDSET TO COACHING

The importance of self-efficacy and non-cognitive and growth mindset skills for student success comprise another core competency of effective professional coaching in post-secondary education environments (Black, 2016). Good coaches know how to identify, assess, and then develop self-efficacy, non-cognitive and growth mindset skills for those they are coaching as well as themselves. "Developing non-cognitive skills and supporting growth mindset development is considered central to student success" (Black, 2016, p. 89). The coaching study explained the importance of these skills for creating an environment for student success. One interviewee captured the majority opinion of the coaches by stating that the "biggest challenges are grit, resilience, fear, confidence and motivation. So, my coaching is less about scheduling, time management and work-life balance and more about the growth mindset" (Black, 2016, p. 89) These comments and the comments included in Table Two illustrate the central role of self-efficacy and non-cognitive and growth mindset skills. The importance of coaching work to develop confidence and self-reported feelings of student success are also detailed by the comments in Table 2 (Black, 2016).

Table 1. Tailoring coaching to each student's specific needs

"As a coach, you have to understand how each person thinks about and articulates their obstacles."
"The coach has to find the motivation elements and context that are going to be most helpful to each student and then figure out an approach that will most help that individual student."
"I work to customize my approach based on the understanding that the student has of their life and their challenges."

Table 2. Non-cognitive skill development as crucial to student success

"The biggest obstacles to completion usually come down to non-cognitive elements like attitude, grit, confidence and having the ability to fail and get back up and try again."
"The non-cognitive skills of grit and determination play a major role in retention and student success."
"The biggest challenges are grit, resilience, fear, confidence and motivation. So, my coaching is less about scheduling, time management and work/life balance and more about the 'growth mindset'."
"Getting them to commit and believe in their ability to succeed is the key to building a pathway to student success."

The study by Black (2016) concluded that the principal issues affecting retention of online degree completion students were deficiencies in non-cognitive skills and lack of a growth mindset or belief in one's ability to succeed. This finding confirms that the intrinsic skills of commitment, determination, attitude, grit, confidence, perseverance, and "believing I can do it" are the keys to student success (Black, 2016). Further, key predictors of both retention and student success are non-cognitive skill levels and whether a student is willing to create a growth mindset (Clark & Estes, 2008; Dweck, 2006; Dembo & Seli, 2012; Duckworth, Peterson, Matthews, & Kelly, 2007; Rueda, 2011; Markle, Brenneman, Jackson, Burrus, & Robbins, 2013). As such it is reasonable to suggest that combining key elements of Hope Theory within a comprehensive and professional coaching program that is specifically designed to tackle defined issues can significantly and positively impact Korean post-secondary students who are dealing with depression and hopelessness as well as self-reported feelings of low student success and an inability to reach career and personal milestones.

EXEMPLARY COACHING IS FREE OF JUDGEMENT, CATEGORIZATION, STEREOTYPING AND ASSUMPTIONS

The data from the coaching study also illustrated the almost universal perspective of good coaches to consistently and intentionally remove as many aspects of judgment, categorization, stereotyping or assumptions as they can across all aspects of their coaching. The study explained that good coaches avoid stereotypes and are sensitive to student culture and background. It illustrated the core values and beliefs of good coaches around student success and each student's ability to overcome obstacles and issues. For example, the following is a comment from a coach:

When we pick up the phone, each student is completely unique and every single student is perfectly capable. They are creative, they are resourceful and they are whole. Our job is to give them a space so that they can grow bigger and greater and have happiness and success not just in this class or program but in life and career (Black, 2016, p. 115).

This core value and belief of a student's ability to succeed and overcome obstacles and challenges through coaching is important in the Korean student context and is another reason why embedding coaching as a core element of a junior or senior year experience is a meaningful construct. The depth of each individual coach's sensitivity to student culture, life events, experiences and background is instructive to the creation of a meaningful and successful Hope Theory focused coaching intervention for Korean post-secondary students. Given the lack of positive feelings and the high levels of depression and hopelessness, having a coach who is non-judgmental and sensitive to the cultural and educational pressures, believes in the student and their ability to overcome obstacles is critical.

IMPLEMENTING THE JUNIOR OR SENIOR YEAR HOPE THEORY FOCUSED COACHING EXPERIENCE WITHIN THE SOUTH KOREAN POST-SECONDARY UNIVERSITY SETTING

The proposed Hope Theory driven coaching intervention would assemble a team of highly trained professional coaches to work with the Korean students on a face-to-face, individual basis over the course of one or two terms. Ideally, the program would occur during either the junior or senior year of the student's undergraduate program. The coaching program would be delivered in the university's language of instruction with Korean being the predominate language medium. Instead of crafting the program as an intervention aimed at alleviating depression or hopelessness, the program should be articulated as an important method for developing career readiness skills and as a program that is designed to support students with the upcoming transition from college to the work place. Framing the coaching program as a career development and a career readiness program will partially offset the negative cultural perceptions of Koreans regarding therapy sessions and asking for help with individual problems.

The proposal for intervention would draw on the elements of Hope Theory. Through professional coaching built around Hope Theory, students will, over time, construct healthy goals and begin to create meaningful pathways to success while developing positive emotions and feelings around the process. The goal is to guide and support the students as they develop a history of accomplishments and of successfully overcoming obstacles, which could optimally lead to an emotional array that is more hopeful, contains more positive feelings, and exhibits higher levels of confidence that goals can be reached (Compton & Hoffman, 2013; Rand & Cheavens, 2009; Snyder, 1994).

FUTURE RESEARCH DIRECTIONS

A longitudinal study of an implemented coaching program at one or more South Korean post-secondary institutions should be undertaken to determine the impact and efficacy of the recommended coaching program in combating the student success issues that are detailed in this study. The study methods should utilize pre- and post-coaching surveys and assessments that yield numerical scores across the full range of issues that the students are experiencing so that progress can be measured and then assessed on an issue by issue basis. The study should be designed to measure and assess non-cognitive and soft skill development as well as motivation and resilience. Ideally, the study will follow the students for at least a year after graduation to determine if there is a lasting impact to the coaching program or if the proposed solution provides only a temporary impact that dissipates once the coaching program's support structures are withdrawn.

Further study of the cost of coaching programs designed to address the significant individual success and psychosocial issues that are effecting young South Koreans should also be undertaken. The cost of coaching programs versus other existing and proposed approaches to the issues discussed in this chapter would be a meaningful and instructive addition to the current literature on coaching and psychosocial therapy. The perceived high cost of coaching and other one on one student retention and student success supports is a consistent criticism within the literature. However, the comparative cost of alternative approaches or the cost of doing nothing may be higher. Data driven decision making by senior university leadership on coaching implementation would be significantly improved if there was reliable return on investment (ROI) data related to coaching and other individualized student support programs focused on increasing student success and retention.

Finally, a review of the literature yields a lack of coaching studies assessing coaching programs that are delivered in languages other than English. The proposed intervention to address the low satisfaction and low success issues exhibited among South Korea's college going population is to deliver a specialized coaching program at Korean post-secondary institutions in Korean. One or more studies assessing how the language medium impacts the efficacy and the outcomes of a coaching program would be helpful in assessing the merit of coaching programs in non-English speaking countries.

CONCLUSION

Coaching based on the Hope Theory should be embedded within the junior and senior year post-secondary curriculum at South Korean institutions. Hope Theory based coaching can reduce stress and depression while helping students become better learners by developing and then supporting hope concepts, goal creation, and the skills necessary to deal with stress because of obstacles and surprise events. In the recommended approach of this chapter, the coaches would initially assess the student's self-efficacy, non-cognitive and growth mindset skill, and then develop a coaching plan to move the students forward toward a positive and growth oriented mindset. Using the tenants of the Hope Theory, coaches can provide a means and methodology for improving goal identification and support, as well as, the development of pathways and incremental steps along a continuum to enable reaching these goals along with support structures during the pre-event and event sequence phases. The purpose of creating hopeful students by facilitating the pursuit of reachable goals will guide the coaches as they work with the undergraduate students. Helping them to craft realistic pathways that are supported by their own agency is important for the proposed program. The desired result, instead of feeling hopeless and defeated when a goal is not achieved, would be students with skills to create successful learning histories. This will better prepare them to achieve realistic goals and find appropriate pathways thereby, crafting solutions that will lead to desirable and reachable goals in the future as they prepare to enter the Korean workforce.

Although the recommendation and conclusion to provide specialized professional coaching for Korean higher education students is worthwhile, the proposal suggests just one step towards addressing the broad and multi-faceted social issues associated with the quickly developing industrial economy that South Korea has experienced over the last 30 years. It is essential for students to have hope for a successful adulthood but if jobs are not available after graduation, this will further support the cycle of despair that seems to be prevalent. The larger picture of social change in the workforce also needs to be addressed by those with authority in the government and the larger industrialized economy.

REFERENCES

Ai, A. L., Cascio, T., Santangelo, L. K., & Evans-Campbell, T. (2005). Hope, meaning, and growth following the September 11, 2001, terrorist attacks. *Journal of Interpersonal Violence*, *20*(5), 523–548. doi:10.1177/0886260504272896 PMID:15788553

American Psychological Association. (2016). *Depression*. Retrieved from http://www.apa.org/topics/depression/

Arnau, R. C., Rosen, D. H., Finch, J. F., Rhudy, J. L., & Fortunato, V. J. (2007). Longitudinal effects of hope on depression and anxiety: A latent variable analysis. *Journal of Personality*, *75*(1), 43–64. doi:10.1111/j.1467-6494.2006.00432.x PMID:17214591

Asia Society Center for Global Education. (2016). *South Korean education reforms*. Retrieved from http://asiasociety.org/global-cities-education-network/south-korean-education-reforms

Black, R. C. (2016). *Coaching to increase retention in online degree completion programs*. Retrieved from http://digitallibrary.usc.edu/cdm/compoundobject/collection/p15799coll40/id/298507/rec/2

Brandtstädter, J., & Renner, G. (1990). Tenacious goal pursuit and flexible goal adjustment: Explication and age-related analysis of assimilative and accommodative strategies of coping. *Psychology and Aging*, *5*(1), 58–67. doi:10.1037/0882-7974.5.1.58 PMID:2317302

Brandtstädter, J., & Rothermund, K. (2002). The life-course dynamics of goal pursuit and goal adjustment: A two-process framework. *Developmental Review*, *22*(1), 117–150. doi:10.1006/drev.2001.0539

Calhoun, S. L., & Mayes, S. D. (2005). Processing speeds in children with clinical disorders. *Psychology in the Schools*, *42*(4), 333–343. doi:10.1002/pits.20067

Cho, M. J., Nam, J. J., & Suh, G. H. (1998). Prevalence of symptoms of depression in a nationwide sample of Korean adults. *Psychiatry Research*, *81*(3), 341–352. doi:10.1016/S0165-1781(98)00122-X PMID:9925185

Compton, W. C., & Hoffman, E. (2013). *Positive psychology: The science of happiness and flourishing*. Belmont: Wadsworth Cengage Learning.

Diener, E. (2000). Subjective Well Being: The science of happiness and a proposal for a national index. *The American Psychologist*, *55*(1), 34–43. doi:10.1037/0003-066X.55.1.34 PMID:11392863

Diener, E., Ng, W., Harter, J., & Arora, R. (2010). Wealth and Happiness Across the World: Material Prosperity Predicts Life Evaluation, Whereas Psychosocial Prosperity Predicts Positive Feeling. *Journal of Personality and Social Psychology*, *99*(1), 52–61. doi:10.1037/a0018066 PMID:20565185

Egeland, J., Rund, B. R., Sundet, K., Landre, N. I., Asbjornsen, A., Lund, A., & Hugdahl, K. et al. (2003). Attention profile in schizophrenia compared with depression: Differential effects of processing speed, selective attention and vigilance. *Acta Psychiatrica Scandinavica*, *108*(4), 276–284. doi:10.1034/j.1600-0447.2003.00146.x PMID:12956828

Feldman, D. B., & Dreher, D. E. (2012). Can hope be changed in 90 minutes? Testing the efficacy of a single-session goal-pursuit intervention for college students. *Journal of Happiness Studies*, *13*(4), 745–759. doi:10.1007/s10902-011-9292-4

Hedayati, M., & Khazaei, M. (2014). An Investigation of the relationship between depression, meaning in life and adult hope. *Procedia: Social and Behavioral Sciences*, *114*(21), 598–601. doi:10.1016/j.sbspro.2013.12.753

Institute of Education Sciences (IES) Website. (2014). *What works clearinghouse.* Retrieved from http://ies.ed.gov/ncee/wwc/

Irving, L. M., Snyder, C. R., Cheavens, J., Gravel, L., Hanke, J., Hilberg, P., & Nelson, N. (2004). The Relationships Between Hope and Outcomes at the Pretreatment, Beginning, and Later Phases of Psychotherapy. *Journal of Psychotherapy Integration*, *14*(4), 419–443. doi:10.1037/1053-0479.14.4.419

Irving, L. M., Snyder, C. R., & Crowson, J. J. Jr. (1998). Hope and the negotiation of cancer facts by college women. *Journal of Personality*, *66*(2), 195–214. doi:10.1111/1467-6494.00009 PMID:9529663

Mascaro, N., & Rosen, D. (2005). Existential Meanings Role in the Enhancement of Hope and Prevention of Depressive Symptoms. *Journal of Personality*, *73*(4), 985–1013. doi:10.1111/j.1467-6494.2005.00336.x PMID:15958142

Monitor, I. C. E. F. (2014, January 23). *High performance, high pressure in South Korea's education system.* Retrieved from http://monitor.icef.com/2014/01/high-performance-high-pressure-in-south-koreas-education-system/

OECD. (2016). *Population with tertiary education* [Data file]. Retrieved from https://data.oecd.org/eduatt/population-with-tertiary-education.htm

OECD. (2016). *Suicide rates* [Data file]. Retrieved from https://data.oecd.org/healthstat/suicide-rates.htm

Ohayon, M. M., & Hong, S. C. (2006). Prevalence of major depressive disorder in the general population of South Korea. *Journal of Psychiatric Research*, *40*(1), 30–36. doi:10.1016/j.jpsychires.2005.02.003 PMID:15878179

Rand, K. L., & Cheavens, J. S. (2009). Hope theory. In S. J. Lopez & C. R. Snyder (Eds.), *Oxford handbook of positive psychology* (pp. 323–333). New York, NY: Oxford University Press.

Snyder, C. R. (1994). *The psychology of hope: You can get there from here.* New York: Free Press.

Tarbuck, A. F., & Paykel, E. S. (1995). Effects of major depression on the cognitive function of younger and older subjects. *Psychological Medicine*, *25*(2), 285–296. doi:10.1017/S0033291700036187 PMID:7675916

The Economist. (2011, May 12). *Depressed students in South Korea: We don't need quite so much education.* Retrieved from http://www.economist.com/blogs/banyan/2011/05/depressed_students_south_korea

Tsourtos, G., Thompson, J. C., & Stough, C. (2002). Evidence of an early information processing speed deficit in unipolar major depression. *Psychological Medicine*, *32*(2), 259–265. doi:10.1017/S0033291701005001 PMID:11866321

Wang, L. C. (2013). The effect of high-stakes testing on suicidal ideation of teenagers with reference dependent preferences. *Monash University*, *29*(2), 1–33.

KEY TERMS AND DEFINITIONS

Coaching: The chapter uses the definition of coaching from the International Coach Federation website. The International Coach Federation defines professional coaching as "partnering with clients in a thought-provoking and creative process that inspires them to maximize their personal and professional potential" (International Coach Federation, 2015). The International Coach Federation website lists the core competencies of professional coaching as goal setting and managing incremental progress towards goals, as well as, developing individual accountability measures regarding progress towards one or more collaboratively established goals or skill proficiencies.

Depression: The chapter utilizes the 2016 American Psychological Association (APA) website definition of depression. "People with depression may experience a lack of interest and pleasure in daily activities, significant weight loss or gain, insomnia or excessive sleeping, lack of energy, inability to concentrate, feelings of worthlessness or excessive guilt and recurrent thoughts of death or suicide."

Event Sequence Phase: If a goal is determined to be important, an individual will activate efforts to attain the goal by moving into the event sequence phase. During the event sequence phase, the individual will monitor and assess the outcome value of the desired goal. During the event sequence phase the individual's assessment is constantly evolving because the outcome value of the goal cannot be appropriately determined until pursuit of a goal has commenced (Compton & Hoffman, 2013; Rand & Cheavens, 2009; Snyder, 1994).

Growth Mindset: The chapter utilizes Carol Dweck's definition of growth mindset from a 2012 interview with OneDublin.org. "In a growth mindset, students understand that their talents and abilities can be developed through effort, good teaching and persistence."

Hope: Snyder (1994) adopts the most prevalent definition of hope which he expresses as "a construct of pathways and agencies, with pathways being the ability to find ways to reach one's goals, and agencies being the motivation to reach one's goals" (p. 355).

Hope Theory: Hope Theory is centered on the premise that well-being is enhanced when people have well-developed goals and believe or perceive that they have the capacity and resources to attain their stated goals (Rand & Cheavens, 2009).

Learning History: The learning history of the individual relates to positive and negative past outcomes that impact an individual's confidence and self-efficacy to attempt a future goal. It is the collective past experiences of an individual. If an individual's learning history is positive it can result in the development of hope and confidence that a goal can be attained. If the individual's learning history is negative it can reinforce hopelessness and passiveness (Compton & Hoffman, 2013; Rand & Cheavens, 2009; Snyder, 1994).

Non-Cognitive Skills: The chapter utilizes the Educational Testing Service (ETS) definition of non-cognitive skills. ETS and through its subsidiary, the ACT, defines noncognitive or "soft skills" as key elements and drivers of student motivation, integrity, and interpersonal interaction. Non-cognitive skills may also involve intellect, but they do so more indirectly and less consciously than cognitive skills. Non-cognitive and soft skills are often associated with an individual's personality, temperament, and attitude and they are a reliable marker for student success and retention in post-secondary institutions.

Pre-Event Phase: Before deciding to pursue a goal, an individual will enter the pre-event phase where they weigh the outcome value of the goal. If the outcome value of the goal is too low, the goal pursuit will cease. If the goal is determined to be important, the individual will move from the pre-event phase into the event sequence phase. (Compton & Hoffman, 2013; Rand & Cheavens, 2009; Snyder, 1994).

Self-Anchoring: The chapter uses the definition of self-anchoring from Compton and Hoffman (2013), which can be summarized as the level of satisfaction that an individual has with life.

Self-Efficacy: The chapter utilizes Albert Bandura's definition of self-efficacy from his seminal 1977 work, Self-Efficacy: Toward a Unifying Theory of Behavioral Change. Bandura's definition can be summarized as an individual's confidence in their ability to achieve desired goals and intended outcomes.

Social Psychological Prosperity: There are two separate types of prosperity, economic and social psychological. Social psychological prosperity is often used as a national ranking measure by the World Bank, the OECD and other global organizations. It is commonly defined as, an individual's feelings and perspectives around how well-being and how people's thoughts, feelings, and behaviors are influenced by the actual, imagined, or implied presence of others (Compton & Hoffman, 2013).

The Suneung: The Suneung is the name of the Korean national college entrance exam that is accepted by South Korean colleges and universities and that determines college placement. The test is offered once a year in November. The Suneung is the Korean equivalent of the US administered ACT and SAT exams.

Section 3

Assessing the Impact of Retention and Success Programs

Chapter 10
Boosting Retention Through a Foundation of Academic Advising

Sean Nemeth
Brandman University, USA

ABSTRACT

While there is no direct causal link between academic advising and increased student persistence, the role of the academic advisor can be key to an institution's success. This chapter examines one university's approach to redesigning the academic advising model from the ground up and committing to a philosophy of continuous improvement in academic advising, retention and student success. A decade in the making, the tools and approaches created through this process now play an important part in the institution's success and can be a road-map for other institutions to follow as they aspire to revise and improve their academic advising models and to improve student success.

INTRODUCTION

Mission: The mission of this chapter is to provide an institutional example of improved student persistence through systematic improvements in academic advising and focused assessment of academic advising programs.

Many researchers point to academic advising as having an integral impact on student retention (Anderson, 1997; Kuh, 1997; Gardner, 2003; Tinto, 1987, 1993, 2004). However, there is no direct causal link between academic advising and improved persistence outcomes. Yet, in absence of this direct link, few would argue that academic advising plays an important role in the student experience (Habley, 1987; Clark, 1989). This chapter will share the example of one institution's successful efforts to use academic advising as a foundation for improving student retention.

Nearly every college student can recall interactions with an academic advisor since that individual often helps the student select his or her first academic courses and sometimes helps in the selection of the major program. It is the idea of open, unfettered student access throughout the academic lifecycle

DOI: 10.4018/978-1-5225-2998-9.ch010

that puts academic advisors in such an influential position when it comes to retention. There are few opportunities for students to build an impactful and meaningful relationship with a staff member of the university. According to Habley (1994), "academic advising is the only structured service on the campus in which all students have the opportunity for on-going, one-to-one contact with a concerned representative of the institution" (p. 10). While faculty members will change from one semester to the next, an advisor can be a consistent thread that helps to tie all of the semesters together for a student. The Council for the Advancement of Standards (CAS) provides recognized guidelines for the field of academic advising. CAS explains the importance of academic advising as, "a crucial component of the college experience, academic advising encourages students to cultivate meaning in their lives, make significant decisions about their futures, and access institutional resources" (Council for the Advancement of Standards, 2013, p. 3).

BACKGROUND

Brandman University (BU) is a private, non-profit institution. It is a part of the Chapman University System and was established in 1958 at El Toro Marine Base in California. The university's roots were originally established by serving military students and over time program offerings expanded to locations on several military bases. Though BU continues to serve military students, the university has expanded to include additional campuses located in civilian areas. These military and civilian campuses serve working adult student populations who wish to pursue a college degree or certificate or to complete a degree program or course of study that was previously interrupted or put on hold. This expansion has produced a distributed university that has 26 physical campuses, throughout California and Washington, as well as fully online programs. The central office, located in Irvine, coordinates strategic efforts for the university while faculty and administrative staff at each location deliver course material and support the day-to-day operations of each local campus. The student body at BU is non-traditional and diverse. The average student age is over 30 years old and the ethnic background of students qualifies the university as a Hispanic Serving Institution (HSI). Nationally, adult learners and Hispanic students have traditionally retained and graduated at rates below the average student. This makes the success at Brandman worthwhile to examine at a deeper level.

TOTALLY STUDENT CENTERED

In a recent accreditation evaluation report (2012), the Western Association of Schools and Colleges (WASC) described BU as "totally student centered." The crux of their comments centered on the university's willingness to put the experience of the student at the center of all processes and decisions. From creating assurances of student learning in the classroom to creating student services that do not bounce students around to multiple offices to get answers, the university has committed to making sure students have an outstanding experience. One example of this commitment to students is the academic advising model that began to be developed in 2007. This chapter will focus on the role that academic advising played in helping the institution to achieve higher retention and graduation rates. Success in retention is the outcome of several key areas and it would be impossible to separate and quantify the impact of each of those areas. The strategies that follow are not solely responsible for the success at Brandman University, but they certainly played a role in that success and there is a good chance they could help other institutions as well.

THE FIJI CONCEPT

When the university began looking at improving its academic advising model, one of the first issues that the team encountered was creating a consistent framework for building this new model. As a distributed system, faculty and other resources are distributed as well. What was available at one campus often was not available at others. An idea that was mentioned by one location would be quickly shot down by another due to differences in resources and culture. The team needed a way to overcome this obstacle without losing sight of the real difficulties that plagued each location.

A breakthrough occurred when a humorous idea of opening a campus on the island of Fiji was suggested at a meeting in 2007. After the jokes subsided about who would be the first to volunteer for such an exotic campus location, a tool was born that helped the team to improve focus immensely. Fiji began to represent an idyllic campus with unlimited resources and no baggage regarding prior processes or misconceptions. Ideas were quickly given the litmus test of whether they would work at a utopian campus in Fiji. Negative discussions about why an idea could not work at one location quickly turned to more open conversations about whether this was the best way to address a particular task in Fiji. Once the team had this framework to handle differences across business segments, they were able to tackle key aspects of building the foundation for the academic advising model.

The trip to Fiji began by developing a mission statement for the academic advising department. Mission statements are an important step in ensuring that sub-units of the institution are aligned with the mission of the overall university (Bart 1997; Kirk and Nolan 2010; Swales and Rogers 1995). As part of the project, the cross-functional retention initiative reviewed the university mission statement. Several traits and concepts were identified that would be important for the foundation of the academic advising department. These traits and concepts were then refined and boiled down to a more concentrated list. After roughly a month of deliberations, the mission statement for advising at Brandman University was developed as follows:

The mission of academic advising... is to guide, support, mentor, and engage students effectively so they can accurately plan their educational experience and be inspired and self-empowered to achieve their academic, career and life goals.

Upon closer examination, the phrases and position of words take on a great deal of meaning for the institution. "Guide, support, mentor, and engage" calls out the intention to have a developmental advising model that is a partnership between advisor and student. "Effectively" and "accurately" speak to the importance of providing students with the right tools and guidance for decision making. This can be further extrapolated to show that these factors can be measured and potentially assessed. The phrase "so they can accurately plan" points to an expectation of student ownership of the planning process. "Inspired and self-empowered" shows the aspirational hope of the committee as to what an advisor could provide on a deeper level for students. "Academic, career, and life goals" is an indicator of the holistic view of students that advisors will take. The mission is a single short sentence that, when done correctly, can remain a guiding force for the organization for many years to come (Bart, 1988; Pearce and David, 1987). As the group travels towards Fiji, this is the Northern Star that points the direction.

If the mission statement is the Northern Star, the map for navigation would be the standards developed by The Council for the Advancement of Standards, also known as the CAS Standards. The Council for the Advancement of Standards in Higher Education has developed a series of guiding documents for higher education. The Standards for Academic Advising (2013) outline 12 general standards that "provide a framework for developing strong academic advising programs" (p. 3). These standards cover the foundational aspects of developing a mission and overall vision for the program. The CAS Standards for Academic Advising is an 18-page document, but the guidance that it contains can take an academic advising program years to develop, implement, and master. They address the fiscal aspects of an advising program, including financial resources, technology, and facilities. Finally, the Standards address the importance of assessment and ways to create a culture of continuous improvement. They offer guidance on structural aspects such as governance and human resources. Making the right choices on the human resource aspects of your advising department can be one of the more difficult parts of this process.

REVIEW OF THE LITERATURE ON ACADEMIC ADVISING

Student to advisor ratios can make or break a university's ability to have meaningful impact on key student outcomes. Before wading into the ratio conversation, there are several key questions that you will need to answer first. Who will be doing the advising at your institution? Faculty advisors? Professional advisors? Peer advisors? Perhaps, a combination of these options? Is advising provided at a department/school level or is it provided centrally? Habley (1983, 1997) and Habley and McCauley (1987) described the various forms of administrative and organization structure that support academic advising at numerous institutions. The original 7 structures that were defined included: faculty only, supplementary, split, dual, total intake, satellite, and self-contained.

Will your advising team be more prescriptive or developmental in their approach? Is advising mandatory or on request? How proactive, or intrusive, will your advising model be? Crookston (1972) and O'Banion (1972) created the developmental approach to advising and encouraged more open negotiation between student and advisor. Intrusive advising suggests key interventions initiated by the advisor (Frost, 1991; Garing, 1992). The staffing and philosophical choices that you make will also have a heavy impact on the cost and budget of the department itself. To give some context to the wide disparity in advising ratios, we can look to a recent survey by NACADA, the recognized professional community for academic advisors, on Advising Load Statistics (2011). The median undergraduate student load for advisors at a private bachelor's degree institution is 325 students. At a two-year community college, this number rises to 1500 students per advisor. At a public doctorate institution, the median is 3375 students per advisor. Accordingly, the approach of an advising department would be different across these groups of institutions, given the difference in student loads. Assessing the willingness of the department and the institution to consider smaller academic advising roster is important, especially if smaller rosters can be tied to increased retention and graduation rates. If that willingness exists, developing a strong approach to assessment will be a key part of developing those opportunities.

ASSESSING THE IMPACT

"What gets measured gets done." It's a phrase you have probably heard a variation of previously. The citation for this phrase is missing because, as with a number of great phrases of the world, the origin is in question. This particular quote is most often attributed to management guru Peter Drucker. There is just one problem. According to the Drucker Institute, he never said it (Zak, 2013). Variations of the quote have been attributed to Tom DeMarco (1982), but you can trace this train of thought back to Lord William Thompson Kelvin, a Scottish physicist and engineer, in the late 1800s.

When you can measure what you are speaking about, and express it in numbers, you know something about it; but when you cannot measure it, when you cannot express it in numbers, your knowledge is of a meager and unsatisfactory kind; it may be the beginning of knowledge, but you have scarcely in your thoughts advanced to the stage of science. (1883)

This linguistic sidetrack could go further down this rabbit hole and discuss a possible version from Georg Joachim Rheticus, a cartographer from the early 16th century, but the point has already been made. Measurement is an elusive topic that has been discussed for many centuries, yet it is necessary for success.

Within the realm of academic advising, Crockett (1988) was one of the first to recognize the importance of measuring the efficacy of advising. He detailed a four-step process for administering an Academic Advising Audit. This is considered one of the earliest tools for measuring advisor effectiveness. More recently, the CAS Standards (2013) indicate how important it is to gather data to assess academic advising effectiveness. As you build, renovate, or review your advising model, it is important that you build some level of assessment into the process. Without it, as Lord Kelvin would say, "your knowledge is of a meager and unsatisfactory kind" (Kelvin, 1883).

In order to show improvement, academic institutions will need to carefully measure where they are currently. To some degree historical information can be used. Census files maintained by institutional research provide the ability to look backwards and develop past metrics in some relevant areas: term-to-term retention, year-to-year retention, GPA performance, graduation rates, etc. While it can be tedious to wade through this historical information, it can be one of the most useful exercises for improving performance. At the very least, simply sharing current performance can have positive benefits on institutional performance.

This relates to an approach that can assist smokers in reducing their consumption. The approach is to simply start counting the number of cigarettes that they smoke in a week. The act of counting raised their awareness around the number that they were consuming and their usage typically went down (Chandra, 2007). A similarity could be drawn to retention data. Simply sharing the overall results could have a positive benefit because people will be more aware of the metric. Results can be segmented by key business units- perhaps schools or campuses within an institution. However, some of the business units may be frustrated with the information that is being shared… usually the units that do not look as strong in comparison to the other groups. Dissonance leads to discourse. Discourse leads to new thoughts, new ideas, and sharing of best practices. These are the activities that are important to spur, and it is why simply sharing the data will begin to have a positive impact on metrics even if there are a few bumps in the road initially. Sharing baseline data can only do so much. The next stage will be to collect information that can encourage specific action.

Most institutions survey student opinions on their overall satisfaction with their university experience. Several consulting firms offer ready-made or semi-customizable surveys geared towards evaluating student preferences, expectations, and satisfaction levels. These surveys typically measure whether students find a particular element or aspect of the program or the student service suite important and then separately assess whether the student is satisfied with this identified element. They allow institutions to prioritize resources to those areas that students have deemed important and where satisfaction does not meet those expectations. Academic advising is often one of the most important services to students outside of the classroom (Tinto, 1987; McGillin, 2003). This can provide important leverage in competitive budget discussions where resources are scarce. Determining that students find the area important and that there is room to improve the service can show a better chance for having a return on the investment. Taking a longitudinal view of these metrics over several years can show directional indicators on whether the situation is getting better or worse. While these "off the shelf" surveys can be useful and easy to deploy, there will be times where institutions will need to develop surveys to meet their specific needs.

In this example, there were measureable performance gains based on the development of the Fiji model of advising, the institutional sharing of retention and other performance metrics, and the utilization of longitudinal satisfaction survey data. However, the results began to plateau making it necessary to find other opportunities for continuous improvement. By this point, the institution was committed to a developmental advising model that provided holistic support for students. The mission that was initially developed by the steering committee was being fully actualized by the institution. As a result, the leadership team began to focus on assessing individual advisor-student relationships to create opportunities for training and development.

Understanding the key aspects of the student-advisor relationship would require a specialized survey and not one that was widely available. Wes Habley's (2008) presentation on assessment and academic advising became a foundation used to create a homegrown assessment tool called the Advising Assessment Survey (AAS). The presentation called out the differences between an evaluation and an assessment. Evaluations look at the "merit, worth, and significance of someone or something" (p. 2). Assessments are focused on fostering improvement rather than evaluating (Robbins, 2009, 2011). Other research (Aiken-Wisneiwski, 2010, Campbell, Nutt, Robbins, Kirk-Kuwaye, & Higa, 2005) also advocated for an assessment approach if improvement is truly the intended outcome. Most organizations use some sort of annual review process to discuss and benchmark employee performance. The role of these evaluations is more geared towards identifying examples of employee achievement or deficiency with little guidance towards employee improvement. An assessment approach will focus more attention on the tasks that define someone performing the job successfully and clear direction on where improvement is needed to improve performance.

The commitment towards creating an assessment for improvement was an important goal. In the AAS example, promises were made to advisors about how the survey data would and would not be used. The data would not be part of individual evaluations. While it's easy to say something like that, it can be difficult to make sure everyone is on board. Direct managers needed to buy into this concept and had to hear the intention multiple times to truly understand. Managers typically expected that any information they had on an employee's performance could be used to evaluate that performance. When a manager had an employee that was not successful in the role, using the assessment data would have been an easy way to point towards that employee's shortcomings. However, for advisors to be willing to engage in the assessment, they needed to feel safe about the information being collected. The implementation structure was developed in such a way that would support this. The survey would be conducted annually and implemented over a three-year period. In year one, results were shared widely at the institutional level. Each advisor received their individual results, but no one else received it. In year two, results were shared at both the institutional level and business unit level, with the advisor again being the only person to see his or her individual results. The business units for the Brandman example were the 27 individual campuses. Most campuses had multiple advisors, so it still allowed room for anonymity of individual results. In the third year, results were shared at the institutional, business unit, and individual advisor level. Those in charge of the data needed to be prepared to fend off requests for this information. After the year one assessment, the officer in charge of developing documents for an accreditation review requested the data at an individual advisor level. Though accreditation is always an important event at an institution, maintaining advisor commitment to the assessment process was even more important. The officer only received the data at the institutional level and an explanation as to why they could not receive it further disaggregated. The trust of academic advisors was incredibly important to the long-term use of the survey.

For those embarking on the journey of building and implementing an assessment like this, there are some key lessons from the process to be shared. First, involve the stakeholders from the very beginning. Having the advisors assist in creating the questions that would be used in evaluating their performance was incredibly valuable. It took the mission statement to a new level and allowed for strategic discussions about the direction of the advising department. It was also a tremendous help with credibility. Though, the questions were created by a small committee of advisors, all advisors felt as though the approach was collaborative rather than top down because peers were involved and sought broad and specific input along the way.

Second, pilot testing is important. Once the committee developed the initial survey, they deployed it to a small number of students across five campuses. After reviewing the results, the committee made some key adjustments. Initially, a 5-point scale was used. The analysis showed that a number of respondents chose the neutral answer- especially when asked for their feedback on statements like "my advisor refers me to outside resources when appropriate." Did "neutral" mean that the student was neutral about the referrals or had never experienced a referral? As a result, the committee changed the survey to a four-point scale, which forced the student to decide whether the experience was positive or negative. Additionally, a "not applicable" category was added so that students could indicate if they had no frame of reference for a question being asked. The pilot test also allowed for other adjustments as well as statistical analysis to identify repetitive questions. Finally, there were lessons about how to best deploy the survey. Determining which email language worked best at encouraging student participation in the survey was valuable.

Third, build in plenty of time to discuss any specific training around the information uncovered in the survey. Without training, the assessment becomes something that is deployed once per year and then forgotten. In year one of the survey, the weakest category was "monitors student progress." The questions in this category focused on the proactive aspect of the academic advising relationship. Was the advisor proactively contacting the student about their progress? Did the advisor alert the students about important changes to their program? The average score for this category was 3.14 in the first year (Table 1). This category was emphasized with all advisors. Advisors shared that time management and tracking was a key issue. Training was created around time management and options like group advising to allow advisors to support more than one student at a time. Reports were built to help advisors identify students that needed outreach. As a result, scores increased to 3.26 in the second year and 3.31 in the third.

Table 1. Results of academic advising survey over five-year period

	2011	2012	2013	2014	2015
Monitors Student Progress	3.14	3.26	3.31	3.42	3.41
Overall Satisfaction	3.22	3.35	3.33	3.44	3.43

Building an assessment from scratch can be a time-consuming process, but if the institution has already identified a mission statement, it can provide the ability to create growth towards that mission that goes far beyond a survey that is purchased off the shelf.

THE OUTCOME

So, what was the impact of these changes and new approaches? Across the board, key metrics showed improvement during the implementation of these academic advising initiatives. Graduation rates, term-to-term retention rates, and student satisfaction rates all showed marked improvements. Before these initiatives began, student satisfaction with academic advising was one of the categories with the lowest level of student satisfaction— only parking was rated lower. Over the course of the 10-year implementation, those scores rose by 16% and advising is now towards the top of services for student satisfaction (Table 2). Term to term retention for all students rose by 4.7%.

Retention for first time students into their second term rose by 5.4%. Six-year graduation rates rose by 6%. Results at other institutions may vary, but the point is that an organizational commitment to focusing on academic advising can play a role in driving up key aspects of student performance.

FUTURE RESEARCH DIRECTIONS

There is no research study that currently links academic advising to increases in student persistence and there may never be. The ability to create a pilot and control situation to specifically test the impact of academic advising is what makes the task so difficult. All academic institutions provide some level of academic advising support to their students. How could an ethical study be created that would essentially prevent a segment of students from receiving this support? One option would be to compare students that utilize academic advising to those that do not. Unfortunately, this approach introduces a level of

Table 2. Tracking graduation and retention

	Pre-Advising Intervention	Post-Advising Intervention	Change
Graduation Rate (6-Year Undergrad)	69%	73%	6%
Term Retention Rate (All programs)	74%	78%	5%
First Term Retention Rate (All Programs)	73%	77%	5%
Satisfaction	3.8	4.41	16%

selection bias, since the students that seek out academic advising may have a higher engagement in school in general. Another possibility would be to consider a pilot program at a large state institution. The sheer size of the population at these schools, coupled with the fact that state institutions are expected to keep costs low to improve access, lead to more reactive advising models. Student-to-Advisor ratios are much higher at public institutions (NACADA, 2011), which results in more prescriptive discussions and less relationship building. Piloting a developmental model with lower ratios in this environment would help to further the body of knowledge in this area. However, it still would not answer the question as to whether it was specifically academic advising that had an impact or if it was the result of the student developing a relationship with a key employee at the institution. To do that, a study would need to include an intervention with a staff member that had no knowledge or training in academic advising.

Advising does not happen in a vacuum and controlling for outside factors can be next to impossible. Yet no one has the opportunity to build an education focused relationship with a student at the same level and consistency as an academic advisor. Advisors are the touchstone for the student experience and can help students build a connection to the institution, establish educational goals, and direct them to appropriate resources. If a student feels connected to a university and sees a means to reach their long-term goals, they will persist at a much higher rate. Advisors are one of the ways to accomplish that connection.

CONCLUSION

As enrollment pressures intensify across all institutions types, student persistence becomes a key factor in the long-term success of these institutions. One approach to improve student persistence is a purposeful focus on development and improvement of academic advising programs.

Academic advising is the single most important relationship offered to students by an institution of higher education. It is through this relationship that students will engage in a critical narrative process that will give shape and meaning to their curricular and life choices and through which they come to understand the interconnections of knowledge and the curricula. (McGillin, 2003, p. 88)

Increased student engagement and a clear understanding of how the academic work being done in the classroom connects to the long-term goals of the student will improve student persistence (Gardner, 2003; Tinto, 2004).

To accomplish this, institutions must begin by identifying a key individual as the party responsible for student retention. While the old axiom is true that retention is everyone's responsibility at the university, having a single person's success at the university tied to this area will encourage that person to look at academic and administrative discussions through a different lens. Beyond identifying a retention officer, a campus-wide commitment to reviewing and assessing the academic advising model will have positive benefits for students. Academic advising is often thought of as a secondary aspect of the student experience. This is a mistake. To provide the best learning experience for students, advising must be thought of as a teaching relationship that can connect the learning in the classroom to the overall outcome that the student seeks for the degree itself.

REFERENCES

Aiken-Wisneiwski, S. (Ed.). (2010). *Guide to Assessment in Academic Advising* (2nd ed.). Manhattan, KS: The National Academic Advising Association.

Anderson, E. (1997). *Chip*. Iowa City, IA: Academic Advising for Student Success and Retention. Noel-Levitz.

Bart, C. (1998). Mission Matters. *The CPA Journal*, *68*(8), 56–57.

Bart, C. K. (1997). Sex, Lies, and Mission Statements. *Business Horizons*, *40*(6), 9–18. doi:10.1016/S0007-6813(97)90062-8

Campbell, S., Nutt, C., Robbins, R., Kirk-Kuwaye, M., & Higa, L. (2005). *NACADA guide to assessment in academic advising*. Manhattan, KS: National Academic Advising Association.

Chandra, S., Shiffman, S., Scharf, D. M., Dang, Q., & Shadel, W. G. (2007). Daily smoking patterns, their determinants, and implications for quitting. *Experimental and Clinical Psychopharmacology*, *15*(1), 67–80. doi:10.1037/1064-1297.15.1.67 PMID:17295586

Council for the Advancement of Standards. (2013). Academic Advising Programs: CAS Standards and Guidelines. Retrieved from http://standards.cas.edu/getpdf.cfm?PDF=E864D2C4-D655-8F74-2E647CDECD29B7D0

Crockett, D.S., & American Coll. *Testing Program*. (1988). Academic Advising Audit: An Institutional Evaluation and Analysis of the Organization and Delivery of Advising Services.

Crookston, B. B. (1972). A developmental view of academic advising as teaching. *Journal of College Student Personnel*, *13*, 12–17.

DeMarco, T. (1982). *Controlling software projects: management, measurement & estimation*. New York, NY: Yourdon Press, c.

Frost, S. H. (1989). Academic responsibility: Can it be taught? *NACADA Journal*, *9*(2), 17–24. doi:10.12930/0271-9517-9.2.17 PMID:2593029

Gardner, J. (2003). as quoted by Joe Cuseo on the First-Year Assessment List FYA-LIST@listserv.SC.EDU, 02/05/03.

Garing, M. T. (1992). Intrusive academic advising. *New Directions for Community Colleges*, *21*(2), 97–104.

Habley, W. R. (1983). Organizational structures for academic advising: Models and implications. *Journal of College Student Personnel*, 1983, 24535–24540.

Habley, W. R. (1994). Key Concepts in Academic Advising. In Summer Institute on Academic Advising Session Guide. National Academic Advising Association, Kansas State University, Manhattan, KS.

Habley, W. R. (1997). Organizational Models and Institutional Advising Practices. *NACADA Journal*, *17*(2), 39–44. doi:10.12930/0271-9517-17.2.39

Habley, W. R. (2008). Using assessment to improve academic advising. *Paper presented at the annual Noel Levitz National Conference on Student Recruitment, Marketing and Retention*, Chicago, IL.

Habley, W. R., & McCauley, M. E. (1987). The relationship between institutional characteristics and the organization of advising services. *NACADA Journal, 7*(1), 27–39. doi:10.12930/0271-9517-7.1.27 PMID:2951508

Kirk, G., & Nolan, S. (2010). Nonprofit Mission Statement Focus and Financial Performance. *Nonprofit Management & Leadership, 20*(4), 473–490. doi:10.1002/nml.20006

Kuh, G. (1997). The student learning agenda: Implications for academic advisors. *NACADA Journal, 17*(2), 7–12. doi:10.12930/0271-9517-17.2.7 PMID:10538182

McCollough, M. A., & Gremler, D. D. (1999). Guaranteeing Student Satisfaction: An Exercise in Treating Students as Customers. *Journal of Marketing Education, 21*(2), 118–130. doi:10.1177/0273475399212005

McGillin, V. A. (2003). The role of evaluation and reward in faculty advising. In G. L. Kramer & G. L. Kramer (Eds.), *Faculty advising examined: Enhancing the potential of college faculty as advisors* (pp. 88–124). Bolton, MA, US: Anker Publishing Company.

NACADA National Survey of Academic Advising. (2011). Advisor Load Statistics. Retrieved from http://www.nacada.ksu.edu/Portals/0/Clearinghouse/documents/Chapter%206%20-%20Professional%20Advisor%20Load%20-%20FINAL.pdf

Nemeth, S. (2013). Advising Assessment Implementation: Moving from Fearful to Formative. *Presented at the annual Noel Levitz National Conference on Student Recruitment, Marketing and Retention*, Chicago, IL.

O'Banion, T. (1972). An academic advising model. *Junior College Journal, 42*(6), 62–64.

Pearce, J. A. II, & David, F. (1987). Corporate Mission Statements: The Bottom Line. *The Academy of Management Executive, 1*(2), 109–116. doi:10.5465/AME.1987.4275821

Robbins, R. (2009). Evaluation and assessment of career advising. In K.B. Hughey, D. Nelson, J. Damminger et al. (Eds.), Handbook of Career Advising (Ch. 12). San Francisco: Jossey-Bass.

Robbins, R. (2011). Assessment and accountability. In J. Joslin & N. Markee (Eds.), *Academic Advising Administration: Essential Knowledge and Skills for the 21st Century*. Manhattan, KS: National Academic Advising Association.

Swales, J. M., & Rogers, P. S. (1995). Discourse and the Projection of Corporate Culture: The Mission Statement. *Discourse & Society, 6*(2), 223–242. doi:10.1177/0957926595006002005

Thomson, W. (1883). Electrical Units of Measurement: The popular lectures and addresses. Macmillan and Co.

Tinto, V. (1987). *Increasing student retention*. San Francisco: Jossey Bass.

Tinto, V. (1993). *Leaving college: Rethinking the causes and cures for student attrition* (2nd ed.). Chicago: University of Chicago Press.

Tinto, V. (2004). Student retention and graduation: facing the truth, living with the consequences. The Pell Institute. Retrieved 2/9/14 from http://www.pellinstitute.org/downloads/publications-Student_Retention_and_Graduation_July_2004.pdf

Zak, P. (2013). Measurement Myopia. Retrieved from http://www.druckerinstitute.com/2013/07/measurement-myopia/

KEY TERMS AND DEFINITIONS

Advising Assessment Survey (AAS): The Advising Assessment Survey is a homegrown assessment tool created and utilized at Brandman University.

Assessment Approach: An assessment approach will focus more attention on the tasks that define someone performing the job successfully and provide clear direction on where improvement is needed to improve performance.

CAS Standards: The Council for the Advancement of Standards, also known as the CAS Standards, outline 12 general standards that provide a framework for developing strong academic advising programs.

Evaluations: Evaluations look at the merit, worth, and significance of someone or something.

Mission Driven: A mission driven institution uses processes and reviews to ensure that sub-units of the institution are aligned with the mission of the overall university.

NACADA: The National Academic Advising Association is the recognized professional community for academic advisors, and provides best practices and preferred approaches to advising across a range of academic institutions.

Totally Student Centered: To be totally student centered requires that an institution put the experience of the student at the center of all processes and decisions.

Chapter 11
Holistic Perspective on Student Retention: Focus Master's Programmes

Neeta Baporikar
Namibia University of Science and Technology, Namibia & University of Pune, India

ABSTRACT

Student retention is becoming critical as the budgets for higher education are shrinking and the economies are becoming more and more knowledge driven. Student retention if not addressed in a holistic manner would mean loss not only to the institution but also to the development of human resources and equitable economic development. This chapter intends to review the causes of student dropout, the current approaches adopted for retaining students with focus on master level programmes in management discipline at a Namibian University. In doing so it also addresses the socio-cultural aspects of student retention and intends to provide a holistic framework for enhancing the rate of student retention.

INTRODUCTION

The current interest in student retention has occurred in a climate where higher education has moved to a massified system with fewer resources to support more students. Chen et al. (2008) argue that academic success which underpins student retention requires more than acquisition of knowledge, and that the classroom is an important introductory point for helping students to begin to master key disciplinary concepts. In support of this, Meyer and Land (2005) put forward the pivotal role of students' understandings of what they call disciplinary threshold concepts for academic survival and success. The implication here is that the classroom needs to include active and interactive learning as the basis for developing an understanding of core disciplinary concepts. These core disciplinary concepts underpin academic success and have strong implications for student retention. There is thus a 'dynamic interplay' (Brysen and Hand, 2007) between student engagement, the quality of student learning and the teaching and learning context. In support of this view, Chen et al. (2008) identify engagement as being composed of the two aspects of, the degree of time and effort that students put into their education, but also the

DOI: 10.4018/978-1-5225-2998-9.ch011

'way an institution organises learning opportunities and services' (Chen et al, 2008, p. 340) so as to encourage students to take part in and thus benefit from the institution's various learning activities. The curriculum in a broad sense, or the teaching and learning programme, provides an ideal forum for approaches and strategies that encourage students to engage, as it is experienced in one form or another by all students (Crosling, Thomas and Heagney, 2008). Tinto (2000) also points out that the classroom is often the only setting in which students meet other students and their teachers and expands on some conditions that underpin students' engagement and thus their persistence in their studies. These include the institution and teachers holding high expectations of students in their learning. They also recognise that many commencing students may not be adequately prepared for the rigours of academic study and so there is the simultaneous need for academic support, in disciplinary contexts, that help students to 'know the rules (Tinto, 2000, p. 91). Hence, feedback about academic performance is important for a student's academic success, and involvement with fellow students in learning in the classroom is also an important element of student success (Tinto, 2000).

Student discontinuing even in professional courses like management programmes has been an observed problem at many South African Universities. Reducing the number of students that discontinue their studies is the motivation behind undertaking this study. Hence, this chapter intends to review the core constructs of student retention, the key elements of student retention, and current approaches adopted for retaining students with a focus on master level management programmes at a Namibian University. In doing so the study also addresses the socio-cultural aspects of student retention and intends to provide a holistic framework for enhancing the rate of student retention.

This study was conducted for a master's programme in management at one campus. Therefore, results may not be generalizable to the broader university populations. Some would consider this a limitation. However, retention is a programme and campus based phenomenon, and different types of campuses tend to attract different types of students (Berger & Lyon, 2005). According to Astin (1990), retention rates vary by campus and due to the differences in the types of students attracted and recruited by certain schools, and it is imperative that institutions provide an environment and climate that fits well with their particular student populations. Therefore, it is not only the responsibility of institutions, but also individual programme heads to help students persist. Furthermore, "each institution must tailor retention to fit the specific needs of its students and the context of that particular institutional environment" (Berger & Lyon, 2005, p. 3). A second limitation is that students who do not return to complete their thesis in the registered semester were considered a dropout although they may have merely 'stopped out' for a while and they may return at a future date to conclude their studies.

FOCUS AND BACKGROUND

The focus of this study is retention of students for master level programmes in the management discipline at a Namibian University. Namibia has basically two universities where master level programmes in management are offered. There are one or two private players but as they are more into training and distance learning programmes, this study is restricted to the university offerings only. Retention is an oft-studied topic, particularly more so in recent years due to its linkage to increased tuition revenue and student loans. However, retention studies have focused specifically on undergraduate students until now. Though some studies have been undertaken recently on non-traditional/adult learners, but the overall emphasis has been on undergraduate retention. A study of the literature shows a paucity of retention studies applicable to graduate students. Further, the majority are concentrated either on doctoral programs or on specific populations of students based on a common trait such as gender or ethnicity among various graduate study options. The largest group of graduate students – those pursuing master's degrees – has received the least attention in terms of retention research. The question of what factors affects persistence and retention among master's students is a valid one. The problem to be studied is the effects that different factors have on student persistence and retention at the master's level. One particular factor of interest is "institutional commitment," both in the traditional sense of a student's level of commitment to an institution, but also in terms of an institution's level of commitment to students. Very little scholarship exists that explores the factors that influence student persistence at the master's level (Alexander, Kohnke, & Naginey, 2011; Cohen, 2012). Master's programs can serve as an area of growth for an institution that faces stagnant undergraduate recruitment issues, such as a declining number of high school graduates. As more students earn a baccalaureate degree, the value of the degree is somewhat diluted as it becomes ubiquitous. Master's degrees offer a way for individuals (such as job seekers) to differentiate themselves from others who do not possess a graduate degree (Borchert, 2005).

This study focuses on master's program students at public universities in Namibia. The data were gathered from both Namibian public universities. Those involved in graduate certificate, certification-only, or similar programs were excluded from the study. Institution type is significant in this discussion, as it examines institutions in which the graduate offerings are primarily at the master's level. Public universities enrol greater numbers of students at the master's level than other institutional types, as public universities usually are less costly than their private or for-profit counterparts. They are also particularly sensitive to tuition revenue, thus examining retention issues on public campuses is relevant due to the budgetary implications that accompany enrolment changes. The methodological approach to this study was phenomenological by design (Patton, 1990; Creswell, 2009). Narrative data were extracted by document analysis, which entailed reviewing and analyzing from program participants over a 4-yr period. With a qualitative methodology, document analysis enables the researcher to make meaning out of student experiences based on what is reported (Patton, 1990). A phenomenological approach, in particular, allows the researcher to uncover core meanings by investigating commonalities in experiences among study participants, in this case among management students. Student narratives bring to light the lived experiences of these students as they progress through their degree program. The idea is that students, speaking for themselves, are far more expert in their own cultural ways of being and thus are well suited to comment on a particular program and offer suggestions for change.

Glazer-Raymo (2005) noted there are three main types of master's degrees: academic master's programs; vocational programs and professional programs. Academic master's programs tend to be more traditional in nature, offering students a theoretical or research-based degree option. The primary purpose of this type of program is not specific job preparation or professional credentialing but, rather, a broad exposure to a field of study. As the most general type of program, academic master's degrees offer several different paths (Borchert 2005; Cohen 2012; Glazer-Raymo, 2005). Cassuto (2015) argues that the academic master's degree does not suffer from having no meaning but, rather, a multitude of often-contradictory meanings. Vocational programs typically are the most narrowly focused of the three degree types, as the goal is employability in a specific field. It is as focused on practicality as the academic master's is on scholarly inquiry. These programs also tend more toward homogeneity than the others, as external review or licensure boards often set standards that institutions use as program guidelines (Borchert, 2005). Professional programs are designed for students who desire to gain deeper, specific knowledge about a relatively narrow field. More focused than academic master's programs, professional programs often are designed for students with previous experience in the field of study, such as in business or education. These students seek an expansion of their knowledge of an area (Conrad, Duren, & Haworth, 1998; Glazer-Raymo, 2005). Professional programs benefit from "doing-centred learning" (Conrad et al., 1998, p. 69), defined as having opportunities both in and outside the classroom to practice the lessons learned. Borchert (2005) defined this type of program as an "applied master's" (p. 6) and offered several examples, including the Master of Business Administration (MBA), which is the focus of this study.

LITERATURE REVIEW

An issue of concern in higher education institutions across the world is the retention and success of students in their studies. This is a particularly pressing issue in the context of widening participation for under-represented student groups, increasing student diversity and educational quality assurance and accountability processes. As well as the personal impact and lifetime chances that students are missing out on, non-completion has significant financial implications for students (and their families), and for society and the economy through the loss of potential skills and knowledge. There are also financial and reputational implications for higher education institutions. Students who do not complete may benefit to some extent by gaining skills like self-confidence and social-cultural awareness. These skills are developed by entering into the portals of professional education, in the current competitive and globalised higher education market, the reputational fall-out of low student retention and high student attrition figures can be damaging for institutions (Yorke and Longden, 2004).

Moreover, a college degree has value. Research has suggested that college attendance improves verbal, quantitative, communication, critical thinking, and moral reasoning skills (Pascarella & Terenzini, 2005). Furthermore, education has been linked to lower unemployment rates, greater job satisfaction, decreased reliance on social support and public assistance programs, lower rates of obesity, and higher reported levels of voting and volunteerism (Baum, Ma, & Payea, 2010). Attainment of a college degree is an important factor in improving one's earnings and financial security. Further, with globalization and knowledge society waves there are increases in the opportunity to participate in higher education and the positive benefits mentioned above have led to an upsurge in enrolments. However, persistence to degree completion has been a consistent problem for both the individual and the institution (Lloyd, Tienda, & Zajacova, 2001). One path toward achieving this goal will be to improve college persistence and ultimately graduation rates (Noel, Levitz & Saluri, 1985).

Defining persistence is not a straightforward task partly because different perspectives yield varied definitions (Pascarella, 1982). For example, persistence can refer to a student who continues enrolment by participating in any form of higher education under the jurisdiction of a state or national system (Tinto, 1982). By the same token, from an institutional perspective, persistence can be more narrowly defined to refer to a student who continues enrolment at any single institution. From the individual student perspective, persistence can refer to the continuation of enrolment at any institution of higher education, local, national, or international. Underscoring the difficulty in defining persistence, Berger and Lyon (2005, p. 7) defined eight terms that describe the voluntary or involuntary decision to remain in school: attrition; dismissal; dropout; mortality, persistence; retention, stop-out and withdrawal–departure of a student from a college or university campus. Each of these terms is potentially important for our review. The balance of this paper will focus on the theories of persistence and their impact on and relevance to improved retention within master's level programmes.

Theories of Persistence

The persistence literature relies predominately on two frameworks for understanding student departure: Tinto's *theory of student departure* (1975, 1987) and Bean's *model of student attrition* (1980, 1983).

The theory of student departure (Tinto, 1975, 1982, 1987, 1993) emphasizes the role of post matriculation campus-based interactions and integration on persistence. The essence of the theory of student departure is that persistence is a function of a longitudinal process of interactions between students and faculty, staff, and peers in academic and social settings (Tinto, 1993). Positive interactions and involvement in academic and social settings provide students with the means to understand and assimilate to institutional norms (termed *integration*), leading to a heightened commitment to completing college and to the institution itself. Conversely, negative experiences and factors that limit campus involvement weaken intentions and commitments and increase the likelihood of departure. Simply put, "other things being equal, the higher the degree of integration of the individual into the college, the greater will be his/her commitment to the specific institution and the goal of college completion" (Tinto, 1975, p. 96). A representation of the theory of student departure model is given in Figure 1.

Bean's Model of Student Attrition

A competing perspective to the theory of student departure is Bean's (1980, 1983) *model of student attrition*. Unlike the theory of student departure, which was based on traditional college students, Bean's model was generated to account for external factors that affect the persistence of nontraditional students. These factors, many of which are beyond the control of an institution, affect students by putting pressure on their time, resources, and sense of well-being (Rovai, 2003). However, conceptually, Bean's model

Figure 1.

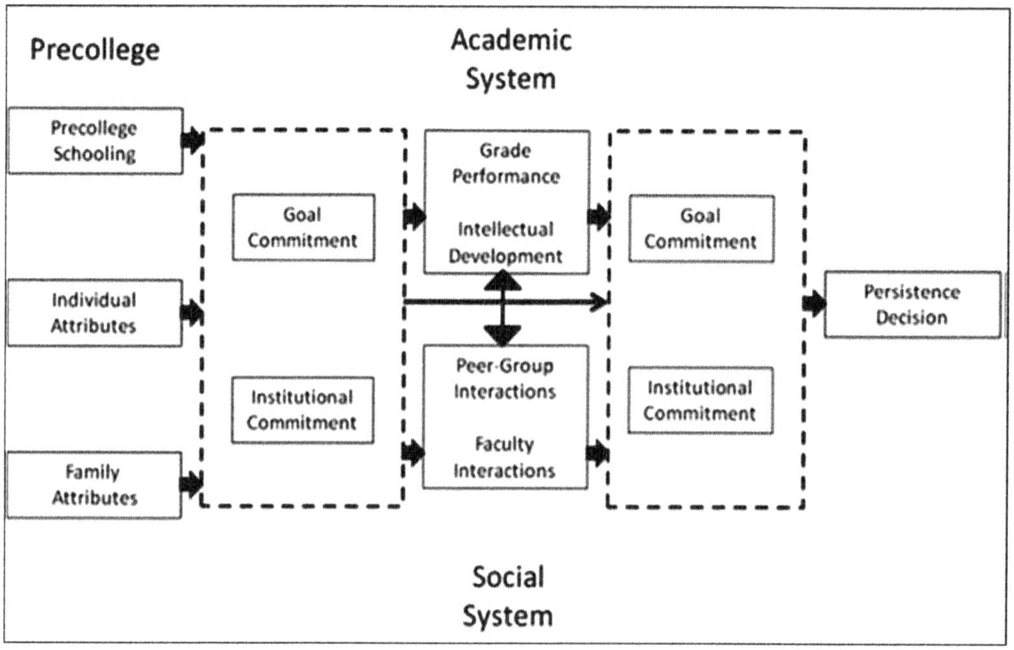

is very similar to the theory of student departure in that it emphasizes the ways in which background characteristics and interactions with an institution influence satisfaction, commitment to degree completion, and persistence (Bean, 1980, 1983). Bean's model stresses that student interactions and integration combine with subjective evaluations of the educational process, institution, and experience to directly influence satisfaction and indirectly influence intentions to persist (Himelhoch, Nichols, Ball, & Black, 1997). Simultaneously, external factors over which the institution has no control, such as opportunity to transfer, family commitments, and financial constraints, directly influence intentions to leave and drop out. Thus, external, attitudinal, and interaction factors collectively influence departure or persistence. A representation of Bean's model is given in Figure 2.

Various other scholars have also proposed ground breaking theories in this area, including the Psychological Model of College Student Retention (Bean & Eaton, 2000); College Readiness Model (Conley, 2007); Student engagement theory (Kuh, Kinsie, Buckley, Bridges & Hayek, 2007), Transition Theory (Schlossberg, Waters & Goodman, 1995); the interactionalist/integration theory (Tinto 1975); the Longitudinal Departure Model (Tinto, 1993); and psychological perspectives, including attribution theory and expectancy theory, highlighting a number of constructs related to student retention. Despite the increasing focus on student success and notable achievements therein, pervasive problems persist.

In an assessment of the status of first year experience initiatives in the United States at the beginning of the 21st Century, Barefoot (2000) identifies the central problem as a "continuous battle for status within the academy" (p. 17). She argues that most initiatives focus on retention rates only, rarely use discipline-based courses and are housed in marginal facilities, with limited budgets. She concludes that effectively, many "do not form part of a central and sustainable part of the institution's fabric" (p. 17). More recently, Kift, Nelson and Clarke (2010) cited similar concerns and specifically the absence of coherent frameworks. World-wide, student engagement is generally acknowledged as a key factor in student retention, and enhancing student engagement is a fundamental strategy for improving student retention, success and outcomes (McInnes and James, 1995; Horstmanshof and Zimitat, 2007; Chen, Lattica and Hamilton, 2008). Krause and Coates (2008) point out that in first year studies, it is crucial

Figure 2.

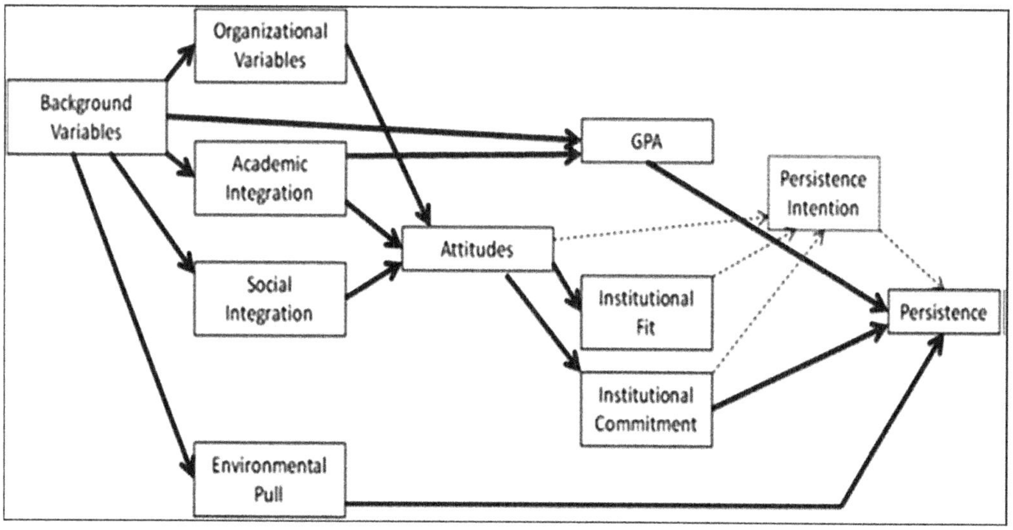

to encourage and assist student engagement as the foundation for successful study in later years. Student engagement is defined as a student's academic commitment and application (Astin, 1993) and shown in time and energy devoted to activities that are educationally purposeful. This also connotes the quality of student effort and students studying for meaning and understanding (Marton and Saljo, 1984) thus reflecting a constructivist approach to learning (Lawrence, 2005). However, engagement is not the sole responsibility of the student as it concerns students interacting with the learning environment (Bryson and Hand, 2007. Thus, managers and teachers have some responsibility to provide a setting that facilitates students' engagement and learning that 'gets students to participate in activities that lead to success' (Kuh, 2003; Kezar and Kinzie, 2006)

Student Retention Imperative for Management Programme

Davidson and Wilson (2013) observed that, despite Tinto's "paradigmatic" (p. 340) status in retention studies, applications of the model are limited when considering different student populations, and that institutional type and student type are variables that must be considered. This implies that graduate student retention issues are different from those of traditional undergraduates. Given the individuality of graduate students and programs, similar to those at the community college level, it is vital to "assess the needs of students…with the goal of understanding their unique needs" (Davidson & Wilson, 2013, p. 341). This was echoed by the Council of Graduate Studies (2013) in a global summit considering the use of big data in graduate education. Institutional experience tends to be markedly different at the graduate level than at the undergraduate level, especially in terms of institutional commitment to retention and persistence. The relative lack (or perceived lack) of institutional commitment to student success at the graduate level, as compared to a student's undergraduate experience, can negatively affect student persistence (Kern-Bowen & Gardner, 2010). Girves and Wemmerus (1988) presented a general model of graduate student degree progress, based on the academic integration identified by Tinto (1985). This model "replace[s] the idea of 'retention' or 'success' with the concept of 'degree progress'" (Girves & Wemmerus, 1988, p. 166). The study notes the difficulty in defining retention and success at the graduate level, unlike the undergraduate level, in which success was defined as earning a baccalaureate degree. Unlike most other models, Girves and Wemmerus delved further into the general model to create empirical models for degree levels and recommended further study using the empirical models, noting that the differences in students and programs should be considered. They also observed that "retention strategies may differ by group" (p. 187).

CONSTRUCTS INFLUENCING STUDENT RETENTION

Academic selectivity has a clear relationship with retention rates according to data from Ontario institutions (Wiggers and Arnold, 2011). More selective institutions typically have higher completion rates than those that engage a broader student body of traditionally underrepresented groups. Different program types, such as certificate programs versus applied degree programs, may have different student retention rates according to longitudinal studies. Institutional policies like course withdrawal deadlines and tuition refund policies and institution size can also influence student engagement and retention rates. These are all factors to consider when measuring student success metrics and benchmarking one institution against others. Based on the literature review, student retention in management programmes is based on the following constructs (Figure 3).

Figure 3. Constructs influencing student retention
Source: Self-Developed

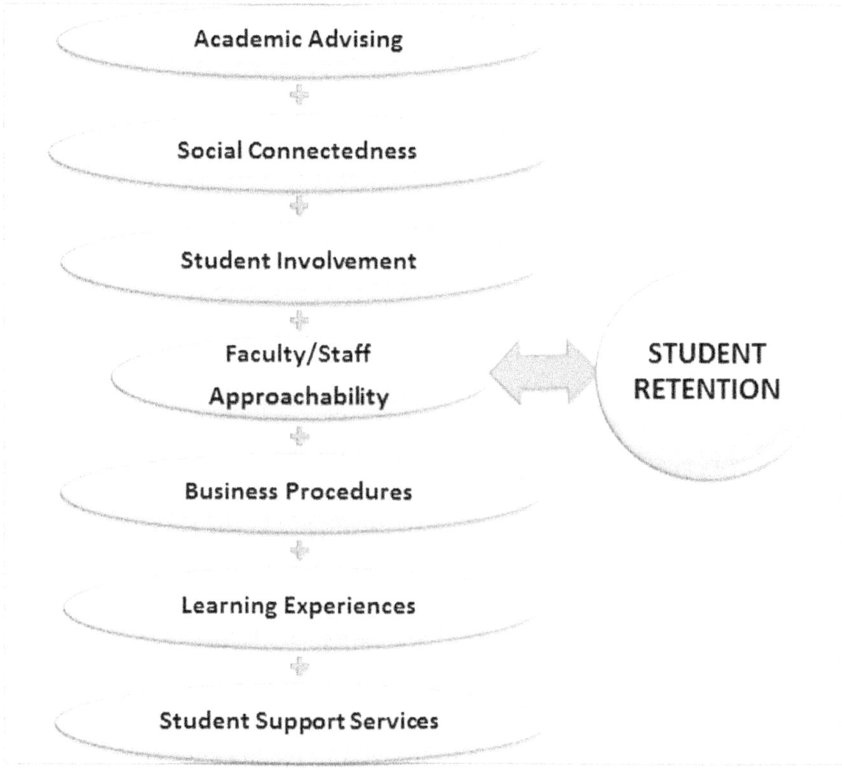

Academic advising: is considered "perhaps the most crucial aspect" of a student's interaction and engagement with an institution of higher learning. It is the relationship with his/her advisor that can play a major role in retention. Vincent Tinto, who has published extensively on student retention in the United States, suggests that social support for students through advising or mentoring can enhance retention and completion (Tinto, 1993). This is especially true for at-risk student populations like first generation and low-income students who may lack other resources for dealing with the challenges of completing a post-secondary degree program. Academic advising should be a process in which faculty and staff interacts with students as they develop, allowing and helping them realize what decisions should be made and subsequent actions taken to achieve their educational and career goals. Pascarella and Terenzini (2005) purport that academic advising plays a role in students' decisions to persist and also affects their chances of graduating. Many students who depart prematurely from college often report a poor academic advising experience, and one of the main components of any retention program is an excellent advising program (Tuttle, 2000).

Academic advising is much more than just scheduling courses and registering students for classes. Students take their relationships with their academic advisors seriously. Furthermore, academic advising might possibly be, as Hunter and White (2004) suggest, the only organized and structured attempt in which university faculty or staff have sustained interactions with students. When one considers the mentoring and counselling aspect of academic advising, it becomes obvious that helping students realize their purpose in higher education and why they are pursuing their current educational goals do not simply occur within one or two visits. Hence, academic advising is a process that occurs over time as students build relationships with their advisors. Williams, Glenn, and Wider (2008) elaborate on the benefits of these types of relationships stating, "[t]his relationship can improve the student matriculation processes and provides students with a sense of security. The relationship also provides a sense of connectedness where students feel that they belong to the school and that the school belongs to them" (p.1). Noel, Levitz, and Saluri (1985) state, "It is the people who come fact-to-face with students on a regular basis who provide the positive growth experiences for students that enable them to identify their goals and talents and learn how to put them to use. The caring attitude of college personnel is viewed as the most potent retention force on a campus" (p. 17). This is what academic advising entails as advisors must understand the importance of creating partnerships and building relationship with their students. The power of these relationships and the impact they have on student success cannot be overstated.

- **Social Connectedness and Student Involvement:** Also, important elements influencing student retention. Social connections allow students to "bond with other students to achieve a common goal," such as completing their degree program. Therefore, despite temptations to cut "soft activities" during budgetary cutbacks, it is important that students have a variety of opportunities to engage with peers through activities and organizations. Students who engage with faculty and staff in their programmes are more likely to persist to graduation. One of the important elements which affects college students' persistence is that of being socially integrated and connected with others, especially other students. For students pursuing professional graduate programmes such as management, the programme is a time for academic pursuit, personal growth and development. For many students, it is also an opportunity to explore or enhance themselves as social beings. Hence, there should be no barrier to this process. In fact, while some students desire to finish their programme, they do not consider themselves to be ultra-academic beings and instead want to partake in endeavours that develop them socially (Moxley, Najor-Durack, & Dumbrigue, 2001). While there is no doubt that the degree of social integration varies from student to student, Bean (2005) states, "Few would deny that the social lives of students in college and their exchanges with others inside and outside the institution are important in retention decisions" (p. 227). In fact, Kuh and Love (2000) claim that social integration consists of students' social and psychological comfort with their institutions' surroundings, associations with common groups of students, and a sense of belonging to the institution. These factors provide security which is needed to help in bonding with others to achieve common goals, one of the most important being to persist until completion of the programme.
- **Involvement and Engagement:** Students feel marginalized when they believe they do not fit it in. This leads to negative outcomes such as "self-consciousness, irritability, and depression" (Evans, Forney & Guido-DiBrito, 1998, p. 27). This feeling of marginalization causes students to wonder if they matter. Addressing this issue is important to student retention as it is an antecedent to student involvement in college activities and programs (Schlossberg, 1989). The most important step to becoming engaged and involved is for students to interact with their peers. According to Schlossberg, students' interacting with their peers is a requirement that must occur in order to make participation in campus activities and student organizations meaningful. However, the most important interactions with peers seem to reinforce the academic learning that takes place in the classrooms, and then the benefits of those interactions permeate into other areas of college life

(Pascarella & Terenzini, 2005). Whereas students must experience academic success to remain in college, it is also vital that they become involved and engaged in other areas of college life. Consequently, it is imperative for higher education administrators to work diligently to provide students with opportunities to get involved with campus organizations and activities (Tinto, 1993).

- **Faculty and Staff Approachability:** By providing adequate contact information, offices hours, and other opportunities for interaction, faculty members can greatly impact whether a student feels comfortable contacting a professor or advisor. It can be argued that management programmes can benefit from smaller class sizes and introductory courses taught by full-time faculty. This encourages, not only more faculty-student interaction, but it also has a positive effect to student success and retention. Further, scholars also agree that strong and engaged relationships between students and faculty are vital to student success in college (Kuh, Kinzie, Schuh, Whitt, & Associates, 2005). One of the principal aspects of facilitating these relationships includes faculty approachability. Approachability involves faculty being available and accessible both inside and outside class, especially at key junctures when students need them (Kuh et al., 2005). Many schools and their faculty members attain adequate levels of approachability by keeping regular office hours and promptly responding to student e-mails, texts and other social media discussion requests. However, approachability also means that faculty are easily reached outside of class. Faculty and student interaction can consist of many facets, including working on a research project or paper, participating in program activities, attending panel discussions and forums, discussing assignments and grades, and one on one faculty feedback on student performance. In short, the more contact a student has with a faculty member, the better chance he or she has in persisting until graduation (Pascarella and Terenzini, 2005).
- **Business Procedures:** Further burdensome procedures or institutional bureaucracy can cause students to develop a negative relationship. This can best be defined as the interaction that occurs between the student and the service providers at the institution (Bean, 2005). If students have unhelpful or unpleasant experiences with any administrative units this also has a negative impact on retention. Many students become discouraged when they perceive bureaucracy as more important to college staff than student service. As Bean (2005) asserts, "The bureaucratic aspects of the academy are soulless, deadening students whose spirits should be lifted by their academic experiences" (p. 230). Students can become equally disenfranchised with an institution when they feel they have been given the run-around or misled. All of these types of unhelpful experiences cause students to develop negative attitudes toward their institutions and, thus, they become less likely to graduate from those institutions (Bean, 2005).
- **Learning Experiences:** "The collective effort of faculty, staff and students." Tinto (1993) suggests that the classroom is the primary place for students to engage with their peers and with faculty. Hence, classrooms should be the starting point for retention initiatives: What would it mean for institutions to take student completion seriously? First and foremost, it would mean that institutions would stop tinkering at the margins of institutional educational life and make enhancing student classroom success the linchpin around which they organize their activities and efforts. Rather than beginning the conversation about improving student completion with the question "What programs should we employ?" institutions should first ask "What is the nature of the educational experience we want our students to have in attending our institution?" This is because one of the most important missions for institutions of higher learning is to provide meaningful learning experiences for their students. At the same time, students enter higher education with their

own expectations of learning experiences. These expectations impact how students respond to their environments and also act as precursors as students make academic decisions, such as choice of major (Pike, 2006). Expectations can also influence how students respond to their academic surroundings and impact their decisions of whether or not to remain in certain fields of study, or college in general (Bosch, Hester, MacEntee, MacKenzie, Morey & Nichols, 2008; Kuh, Gonyea & Williams, 2005; Pike, 2006). Meaningful learning experiences are an essential key to student retention and enriching learning experiences are also necessary to produce an economically independent and enlightened citizenry. Also, students with few chances to participate in meaningful learning experiences are denied the opportunity to integrate and apply the knowledge they have obtained in their classes (Kuh, Schuh, Whitt, Andreas, Lyons, Strange, 1991; Kuh et al., 2005; Moxley et al., 2001).

- **Student Support Services:** Such as academic support centers for academic writing, research methodology and data analysis, numerical analysis and case writing can help to ensure students succeed in management programmes. Adelman (1999) and Pascarella and Terenzini (2005) reported that academic resources such as these produced statistically significant positive impacts on student persistence. While academic support centres allow a university to accept students with a broader range of ability levels, such services can have a high cost as these students have an increased probability of taking longer to graduate and a higher risk of not completing their program. In order to promote student success and persistence through graduation, Tinto (1993), suggests a focus on improving the classroom and improving classroom practices in four ways: expectations, support, assessment and feedback, and involvement. Hence, a number of colleges and universities offer students a wide variety of services and resources intended to promote persistence by providing academic assistance (Pascarella & Terenzini, 2005). Both Miller (2005) and Seidman (2005b) contend that if students are admitted, then they should have expectations for being provided services that will help them succeed. Institutions that truly desire to increase student persistence should implement and advocate the usage of "responsive, learner-centered support services, such as peer tutoring and special labs for writing and mathematics" (Kuh et al., 2005). Thus, it is important for management schools to implement and maintain a range of academic resources and student support services that promote student success and increase retention.

STUDENT PERSISTENCE-ACHIEVEMENT-RETENTION LINKAGE

In terms of persistence and achievement, three particular forces account for 'student persistence-achievement -retention linkage' (Figure 4) leading to student outcomes. This entire spectrum consists of: cognitive, social, and institutional factors.

Cognitive Factors

Cognitive factors are the academic abilities, aptitude, attitude, intelligence, learnability, knowledge, rigor and desire to succeed that a student brings to the programme environment. These factors may be measured by such variables as course selection and completion in high school, aptitude, or extracurricular involvement in academic-related areas. Cognitive factors are important because they directly relate to the student's ability to comprehend and complete the academic portion of the curriculum. An important

Figure 4. Student persistence- achievement -retention linkage
Source: Self-Developed

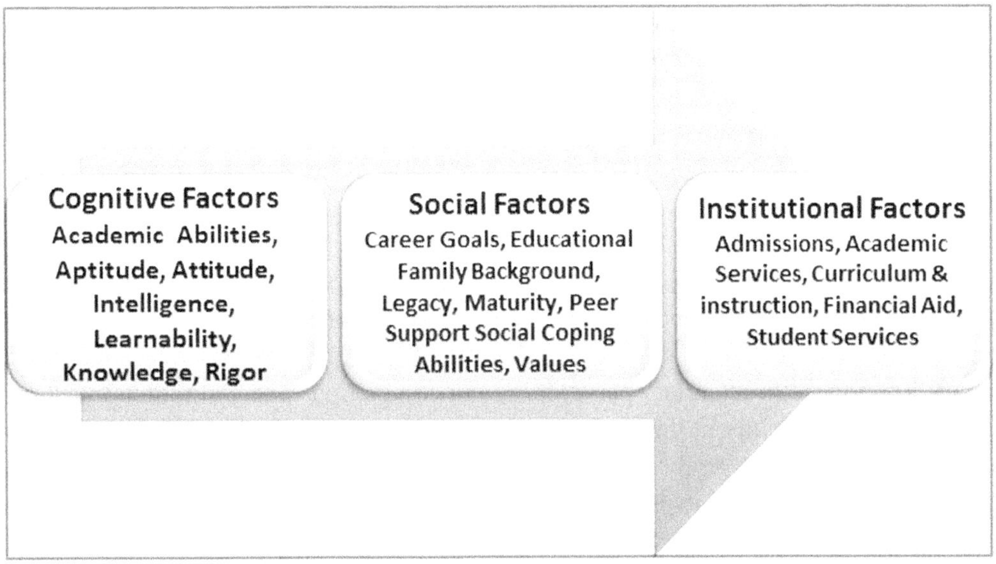

element of the cognitive factors relating to student persistence and achievement is the student's decision-making and problem-solving ability. The decision-making process is an important part of the models described earlier. Tinto (1975, 1993) describes the decision-making process regarding goal commitment and dropout. Bean (1982) describes intent to leave, and Anderson (1985) identifies value conflicts and career indecision among the important variables that a student controls through their social and cultural values. The student's decision-making process occurs within the confines of the geometric shape represented in the model presented above. It is here that the social and cognitive factors interconnect to form the decision-making process. Further, academic preparation is a contextual, school-level factor that considers the extent to which a student's educational environment has been academically rigorous. Just as with ability, rigor is a widely accepted and intuitive determinant of success. In two studies of national longitudinal databases, Adelman (1999, 2006) emphasized the importance of a rigorous high school curriculum. In addition, recent work found that those who took rigorous courses in high school were more likely to earn a bachelor's degree than those who took less rigorous high school courses (Long, Conger, & Iatarola, 2012).

Social Factors

The second factor related to student persistence and performance is the set of social factors impacting students. Such factors include career goals, educational family background, legacy, maturity, peer support social coping abilities, values. The social issues facing management students are of ever-increasing interest to higher education personnel generally, and to management educators and to industry and corporate sector leadership specifically. The research field generally agrees about the importance of social integration with regard to student retention and the fact that students have a difficult time persisting when they are not socially integrated into management learning environment. Thus, the identified social block factors, in the linkage are uniquely important to students' stability. A student's social underpinning and opportunities have crossover impact on his or her cognitive development. A student who is brought up in a culturally and educationally rich environment will develop skills critical to postsecondary, career, and personal success. Students hailing from less supportive environments may bring with them deficiencies in their self-esteem and efficacy, especially as they relate to academics when compared with students from more advantaged backgrounds.

Institutional Factors

This presents stresses, at some level, to all students. Substantial research exists on the stresses of freshman year, especially on minority and low-income students. Regardless of one's subscription to either social anthropology theory (Tinto, 1988) or biculturation theory, how the institution reacts to students is of primary importance to retention, persistence, and completion. The institutional factors block in the linkage relates to the ability of the institution to provide appropriate support to students during the programme years, both academically and socially. Issues related to course availability, content, and instruction affect a student's ability to persist, as do support mechanisms such as tutoring, mentoring, and career counselling. Although this block has a direct effect on a student's stability during the pursuit of the programme, it also can be seen as a flexible set of programs or conditions that the institution can mould to meet the diverse needs and attributes of individual students. The significance of setting institutional factors on equal ground with cognitive and social factors is to illustrate the importance of institutional participation and knowledge in students' social and academic development. This linkage places this set of factors in a linear way because it is the cognitive block that forms the foundation for programme and professional success. It is here that the institution can identify and match the needs of individual students, a student cohort group, or the student body as a whole. Briefly stated, the cognitive factors form the academic ability that is the strengths and weaknesses of the student. Social factors, such as the ability to interact effectively with other persons, personal attitudes, and cultural history, form a second set of external factors that characterize the individual. The third set is institutional factors. Institution factors refer to the practices, strategies, and culture of the university that, in either intended or unintended ways, impact student persistence, achievement and thereby retention.

Bamber & Tett (1999) found that non-traditional students, and particularly mature learners, benefited from formative feedback. For example, formative assessment can offer students:

- Space to explore, try out different approaches and develop their own ideas.
- An opportunity to become aware of their own progress and find out about themselves as learners.
- An opportunity to negotiate with tutors and/or peers on matters of assessment including the allocation of marks (Povey & Angier, 2004).

It can be argued that what goes on in the teaching and learning programme, that is the learning, teaching and assessment practices, play an even more important role in the retention and success of students from underrepresented groups (Yorke in Ferrier and Heagney (eds) 2008). For many of these students, time constraints mean the classroom is the only element of university life they experience. Hence, when faculty know something about the life experiences of their students, they can organise teaching programmes that facilitate the students' maximum participation.

- **Other Cultural Issues:** The classroom provides lecturers with opportunities to link into students' values such as the value of work, struggle, persistence, and resilience. Some students who are first in their family to access higher education also have to overcome barriers such as poverty and poor primary and secondary education experiences in order to get into university. Many of these students have extraordinary persistence and resilience skills which, if acknowledged by their teachers, can assist in further engaging them in their studies. Similarly, refugee students from war-torn countries may have exhibited great courage in re-locating to a new country. How often are their experiences outside the classroom acknowledged?
- **Practical Issues:** There are practical strategies that lecturers can employ in the classroom to assist students, particularly those from under-represented groups, to succeed and persist at university (Krause & Coates,2008). By talking about student support services in their first lectures for the year, teachers can play an important role in linking students to relevant supports such as counselling, disability services and career advice. While equity group students tend to need and use support services more than non-equity group students, they often lack the confidence to go and seek them out for themselves (Yorke, 2008).

IMPERATIVE COMPONENTS FOR RETENTION

Holistic perspective on student retention in master's programmes in management consists of imperative components (Figure 5).

SOLUTIONS AND RECOMMENDATIONS

Tinto (1993) offers a useful set of action principles for implementation of a retention program:

1. Institutions should provide resources for program development and incentives for program participation that reach out to faculty and staff alike.
2. Institutions should commit themselves to a long-term process of program development.
3. Institutions should place ownership for institutional change in the hands of those across the campus that have to implement that change.
4. Institutional actions should be coordinated in a collaborative fashion to insure a systematic, campus-wide approach to student retention.
5. Institutions should act to insure that faculty and staff possesses the skills needed to assist and educate their students.
6. Institutions should front-load their efforts on behalf of student retention.
7. Institutions and programs should continually assess their actions with an eye toward improvement.

Based on these action principles a holistic approach to student retention is suggested. Developing and implementing this holistic student retention program requires a commitment from leaders, faculty, and staff. Through our discussion with some of these individuals and based on the review of related research, a holistic student retention program must be based on:

Figure 5. Imperative components for holistic perspective on student retention

Sl No.	Imperative	Components
1	Mapping of term course schedules for every programme with the critical courses identified	Non-completion of a critical course in the term, it is recommended that a registration hold is made. Student to be provided with academic advising for understanding and analysing what occurred. Take appropriate action to ensure the student understands how to proceed. For example, if thesis student has not completed research methodology by the end of the first year, it is unlikely that the student can complete the degree in a reasonable amount of time.
2	Establish an automated contact system maybe through email, Twitter, Facebook, etc.	Let each student know that the institution is monitoring his or her progress. Send notes to students who do well as well as to those who need to improve.
3	Provide proactive academic advising	Move academic advisors to the students. Develop a structure that fosters professionalism and accountability among advisors. Drop-in advising should be made easy by having advisors after institution hours.
4	Create Interest Groups	Establish Interest Groups (known by several different names) Allow these small groups of similar interests to work together. Appropriate supervision will lead to high-achievers who in turn can be role models.
5	Attention to Students Residence.	Research shows that students who live on campus earn higher grades and are retained at higher levels than those who live off campus. Living-Learning Communities offer an even better opportunity to a smaller segment and enhances the retention rate.
6.	Scrutinize Course Grades	Courses with high percentages of Ds and Fs, particularly those with high enrolment, should be studied. Encouraging faculty to take attendance is a good idea for all courses but is particularly important in courses with high percentages of Ds and Fs.
7.	Exit Students Survey	Surveying students who choose to leave the university is important. Retention can be an issue for high-achieving as well as low-performing students and different strategies are necessary to engage and meet the needs of all student groups.
8	Attention to Full Range of Academic Support Services	Pay attention to full range of academic support services. Full range of these services if made available timely and properly help to engage and retain students. Involve academics in these support services so that teaching-learning gets better linked to student engagement thereby enhance retention rate.
9	Early Intervention and Action	Earn intervention followed by action is fruitful. Data shows that without intervention more 50% students who don't make the passing grade in first semester will drop out. Yet, an intervention and timely action like low cost mandatory course focusing on study habits, time management, note taking, etc. will significantly reduce attrition.
10	Deploy and Adopt New Approaches	Be willing to deploy new approaches. Examples include: Coaching, Wellness Programmes, Mentoring by Professionals, Soft Skills Development, Employability Programmes etc.

Adapted from: Practical Steps to Improving Retention and Graduation Rates, The Florida State University Student Success Team, U.S Department of Education, pp. 5 - 13.

1. **Proven Research:** Given the resources and effort that must go into a retention program, the final plan must be based on solid, proven evidence of success. It is a long way to travel with no idea of the outcome. If such an effort fails, the task of putting the pieces back together would be daunting, to say the least. Spending time looking at what works, and borrowing from the best is worth it.
2. **Suitability to the Particular Needs of the Programme:** Not all programmes are equal. That said, no "boxed" retention program works the same hence all efforts must be shaped to meet the specific needs of each programme.

3. **Institutionalization and Regularisation:** Any program at the beginning is usually a special project supported by outside funds. In the end, however, any successful effort must be institutionalized with respect to funding, policy, and practice. Outside funding does not last forever, and stated policy ensures that any interventions can become a mainstay in institutional-based practices.
4. **Involvement Across Departments and All Staff:** Everyone must be involved at some level. The most successful practices engage the entire staff, while the least successful strategies are very compartmentalized. There are many institution-wide programs that individuals in certain parts of programmes never knew about. Of course, they failed. But those institutions that had a broad outreach among faculty and staff, with clearly stated policy and practical objectives, tended to be successful.
5. **Consider Dynamics of the Change Process:** Change is difficult and uncomfortable. One should not underestimate the impact of change on one's ability to push through policy changes.
6. **Extensive and Appropriate Retraining:** Team members must be brought along and be given all possible opportunities to learn about the recommended interventions and to develop appropriate skills as necessary.
7. **Focus on Students:** Although this statement sounds like a given, many programs end up making the effort about themselves and not the clients. Everything should point to how it affects students. This mind-set is a good one for all institutional practices that often gets lost in the "career" mind-set of board members, administrators, faculty members, and staff. Simply out, students must be central to all operations.
8. **Viability:** Ensure that the approach is fiscally viable. Soft monies (grants, for example) provide a good foundation for start-up, but they are not a long-term solution to persistence at any institutions. An important component of a strategic plan for retention is to build in a long-term fiscal plan to ensure that the approach can operate without external support.
9. **Support Institutional Research:** In the monitoring of retention programs and students, data and analysis on all interventions, programs, and, ultimately, students are the saving grace. One must have the numbers to show whether movement has been made, either positive or negative.
10. **Patience:** All change takes time, and change theory tells us that change usually takes a negative tack before the eventual positive change occurs. It is important to understand that this trend is a normal mode and that some negative changes will happen before the positive yield will be seen. Therefore, leaders and other team members must be patient and understand that this long-term effort will have its rough spots.

Moreover, a holistic approach to student retention must be sensitive to students' needs and target the most needy student populations. All students can benefit from a retention effort, whether through improved tutoring programs or increased need-based aid. Hence, the development of student retention for management programmes requires supportive leadership, the willingness to evoke change and careful planning. Once this is understood, steps include: preplanning, planning, implementation, and student retention program monitoring. This strategic process can be developed in line with an institution's strategic planning schedule.

FUTURE RESEARCH DIRECTIONS

One of the shortcomings of the model is that it is still evolving. For example, more work is needed to refine the concept of "faculty readiness" as institutional and student readiness is well established in the literature. The faculty leadership's notion of academic freedom and how it may impact on uptake of the approaches discussed and the extent of involvement of faculty student leadership in student success matters also need further interrogation. There is also a need for cohort studies in which the impact of the initiatives aimed are tracked systematically. This researcher also believes that a focus on the faculty as an organisational unit will facilitate debates on further developments to support retention and student success. To date the literature review has not revealed a faculty-based model that integrates existing high impact student success initiatives and theories and at the same time guides institution-wide initiatives has not been reported in the literature. Monitoring, evaluating and refining practices, based on the model, and undertaking the outcome and impact studies that underlie the student academic development and excellence will provide vistas for continued research. Additionally, future research should include retention factors beyond the seven constructs and the scope should be broadened beyond traditional, face-to-face classes.

CONCLUSION

It is important for students to evaluate programs designed to benefit them (Oliver, Smith & Wilson, 1989). This research has utilized popular theoretical models developed by scholars such as Tinto (1975) and Bean (1980). South African universities have been investigating strategies for improving student success for a long time (Mabizela 1994; Potgieter, 2010; Rollnick, Mphahlele, & Ogude, 1997). However, key factors in the external environment, firstly a culmination of twelve years of poor schooling for the majority; an undifferentiated post-school system and the yet to be established predictive validity of the new National Senior Certificate examinations for university study (Collier- Reed, Wolmarans & Smit, 2010; Klingbeil, Mercer, Rattan, Raymer, & Reynolds, 2004; Mabizela 1994; Scott, Yeld & Hendry, 2007) severely impact on strategies that universities must adopt to bridge the school-university gap. Thus, in this chapter, student retention from a teaching and learning perspective, in terms of teaching and learning approaches that have an impact on students' decisions to continue with or withdraw from their studies in management programmes has been examined. The major need is to engage students in their studies, and hence adopting a holistic framework is one step in the right direction, not only from a student retention point of view, but also in creating effective managers through the master's programme in management. Given the dearth of graduate-specific study, it is useful to examine research on students that has transferability to graduate students. Graduate students tend to be a more heterogeneous group than undergraduates, and more so the management graduate students. This is another reason that less research has been conducted on graduate retention. Nevertheless, universities of the 21st century would be well served to examine their own policies and procedures to determine whether graduate and undergraduate students are treated equitably on their campuses.

REFERENCES

Adelman, C. (1999). *Answers in the toolbox: Academic intensity, attendance patterns, and bachelor's degree attainment (Document No. PLLI 1999-8021)*. Washington, DC: U.S. Department of Education, Office of Educational Research and Improvement.

Adelman, C. (2006). *The toolbox revisited: Paths to degree completion from high school through college*. Washington, DC: U.S. Department of Education.

Alexander, C., Kohnke, M., & Naginey, A. (2011) *Undergraduate and graduate retention: Two concepts, one outcome*. Retrieved 28 March 2017 from www.earth.callutheran.edu/fact_book/documents

Astin, A. W. (1990). *Assessment for excellence: The philosophy and practice of assessment and evaluation in higher education*. New York: Macmillan.

Astin, A. W. (1993). *What Matters in College?* San Francisco: Jossey-Bass.

Bamber, J., & Tett, L. (2001). Ensuring integrative learning experiences for non-traditional students in Higher education. *Journal of Widening Participation and Lifelong Learning*, *3*(1), 8–18.

Barefoot, B. (2000). The first year experience: Are we making it any better? *About Campus: Enriching the Student Learning Experience*, *4*(6), 12–18.

Baum, S., Ma, J., & Payea, K. (2010). *Education pays 2010: The benefits of higher education for individuals and society*. Retrieved March 28, 2017 from www.advocacy.collegeboard.org

Bean, J., & Eaton, S. (2000). A psychological model of college student retention. In J. Braxton (Ed.), *Reworking the departure puzzle: New theory and research on college student retention* (pp. 48–61). Nashville, TN: Vanderbilt Universities Press.

Bean, J. P. (1980). Dropouts and turnover: The synthesis and test of a causal model of student attrition. *Research in Higher Education*, *12*(2), 155–187. doi:10.1007/BF00976194

Bean, J. P. (1983). The application of a model of turnover in work organizations to the student attrition process. *Review of Higher Education*, *6*(2), 129–148. doi:10.1353/rhe.1983.0026

Bean, J. P. (2005). Nine themes of college student retention. In A. Seidman (Ed.), *College student retention* (pp. 215–243). Westport: Praeger Publishers.

Berger, J. B., & Lyon, S. C. (2005). Past to present: A historical look at retention. In A. Seidman (Ed.), *College student retention* (pp. 1–29). Westport, CT: Praeger Publishers.

Borchert, M. A. E. (2005). *Master's education: A guide for faculty and administrators. A policy statement* (2nd ed.). Washington, DC: Council of Graduate Schools.

Bosch, W. C., Hester, J. L., MacEntee, V. M., MacKenzie, J. A., Morey, T. M., Nichols, J. T., & Young, R. R. et al. (2008). Beyond lip service: An operational definition of learning-centered college. *Innovative Higher Education*, *33*(2), 83–98. doi:10.1007/s10755-008-9072-1

Bryson, C., & Hand, L. (2007). The Role of engagement in inspiring teaching and learning. *Innovations in Education and Teaching International*, *44*(4), 349–363. doi:10.1080/14703290701602748

Cassuto, L. (2015, June 1). The degree for quitters and losers. *The Chronicle of Higher Education*. Retrieved March 16, 2017 from www.chronicle.com/article/The-Degree-for-Quitters

Chen, H., & Lattica, L. (2008). Conceptualizing engagement: Contributions of faculty to Student Engagement in Engineering. *The Journal of Engineering Education*, *97*(3), 339–352. doi:10.1002/j.2168-9830.2008.tb00983.x

Cohen, K. E. (2012). *Persistence of master's students in the United States: Development and testing of a conceptual model* [Doctoral dissertation].

Collier-Reed, B., Wolmarans, N., & Smit, R. (2010, October). The impact of NSC mathematics on student performance in mathematics in first-year engineering programmes: Where does the gap lie? *Paper presented at the STEM, Mind the Gap Conference, Academy of Science of South Africa, Cape Town*. Retrieved January 19, 2017 from http://uct.academia.edu/BrandonCollierReed/Papers/362566/

Conley, D. (2007). *Towards a more comprehensive conception of College Readiness. A report prepared for the Bill and Melinda Gates Foundation*. Eugene, OR: Educational Policy Improvement Center. Retrieved January 1, 2017 from http://www.smallschoolsproject.org/PDFS

Conrad, C. F., Duren, K. M., & Haworth, J. G. (1998). Students perspectives on their master's degree experiences: Disturbing the conventional wisdom. *New Directions for Higher Education*, *26*(1), 65–76. doi:10.1002/he.10106

Council of Graduate Schools. (2013). *Completion and attrition in STEM master's programs: Pilot study findings*.

Creswell, J. W. (2009). *Research Design: Qualitative, Quantitative, and Mixed Methods Approaches* (3rd ed.). Thousand Oaks, CA: Sage.

Davidson, C., & Wilson, K. (2013). Reassessing Tinto's concepts of social and academic integration in student retention. *Journal of College Student Retention*, *15*(3), 329–346. doi:10.2190/CS.15.3.b

Evans, N. J., Forney, D. S., & Guido-DiBrito, F. (1998). *Student development in college: Theory, research, and practice*. San Francisco: John Wiley & Sons, Inc.

Girves, J. E., & Wemmerus, V. (1988). Developing models of graduate student degree progress. *The Journal of Higher Education*, *59*(2), 163–189. doi:10.2307/1981691

Glazer-Raymo, J. S. (2005). Professionalizing graduate education: The master's degree in the marketplace. *ASHE Higher Education Report*, *31*(4).

Himelhoch, C. R., Nichols, A., Ball, S. R., & Black, L. C. (1997, November). A comparative study of the factors which predict persistence for African American students at historically Black institutions and predominantly White institutions. *Paper presented at the annual meeting of the Association for the Study of Higher Education*, Albuquerque, NM.

Horstmanshof, L., & Zimitat, C. (2007). Future time orientation predicts academic engagement among first-year university students. *The British Journal of Educational Psychology*, *77*(3), 703–718. doi:10.1348/000709906X160778 PMID:17908382

Hunter, M. S., & White, E. R. (2004). Could fixing academic advising fix higher education? *About Campus: Enriching the Student Learning Experience*, *9*(1), 20–25.

Kern-Bowen, J., & Gardner, R. (2010). Creating campus community for graduate students through programs, services, and facilities. *The Bulletin, 78*(2). Retrieved March 28, 2017 from www.acui.org

Kezar, A., & Kinzie, J. (2006). Examining the ways Institutions create student engagement: The role of mission. *Journal of College Student Development*, *47*(2), 149–173. doi:10.1353/csd.2006.0018

Kift, N., Nelson, K., & Clarke, J. (2010). Transition Pedagogy: A third generation of FYE – A case study of policy and practice for the higher education sector. *International Journal of the First Year Experience in Higher Education*, *1*(1), 1–20. doi:10.5204/intjfyhe.v1i1.13

Klingbeil, N., Mercer, R., Rattan, K., Raymer, M., & Reynolds, D. (2004, June). Rethinking engineering mathematics education: A model for increased retention, motivation and success in engineering. *Paper presented at the American Society for Engineering Education Annual Conference & Exposition*, Salt Lake City, Utah.

Krause, K., & Coates, H. (2008). Students Engagement in first-year university. *Assessment & Evaluation in Higher Education*, *33*(5), 493–505. doi:10.1080/02602930701698892

Kuh, G. (2003). How are we doing at engaging students? *About Campus: Enriching the Student Learning Experience*, *8*(1), 9–16.

Kuh, G., Kinzie, J., Buckley, J., Bridges, B., & Hayek, J. (2007). Piecing together the student success puzzle: Research, propositions and recommendations. *ASHE Higher Education Report*, *32*(5), 1–182.

Kuh, G. D., Gonyea, R. M., & Williams, J. M. (2005). What students expect from college and what they get. In T. Miller, B. Bender, & J. Schuh et al. (Eds.), *Promoting reasonable expectations: Aligning student and institutional views of the college experience* (pp. 34–64). San Francisco: Jossey-Bass.

Kuh, G. D., Kinzie, J., Schuh, J. H., & Whitt, E. J. et al. (2005). *Student success in college: Creating conditions that matter*. San Francisco: Jossey-Bass.

Kuh, G. D., & Love, P. G. (2000). A cultural perspective on student departure. In J. M. Braxton (Ed.), *Reworking the student departure puzzle* (pp. 196–212). Nashville: Vanderbilt University Press.

Kuh, G. D., Schuh, J., Whitt, E., Andreas, R., Lyons, J., & Strange, C. et al.. (1991). *Involving colleges: Successful approaches to fostering student learning and personal development outside the classroom*. San Francisco: Jossey-Bass.

Lawrence, J. (2005). Re-conceptualising attrition and retention: Integrating theoretical, research and student perspectives', *Studies in Learning. Evaluation Innovation and Development*, *2*(3), 16–33.

Lloyd, K. M., Tienda, M., & Zajacova, A. (2001). Trends in educational achievement of minority students since Brown v. Board of Education. In C. Snow (Ed.), *Achieving high educational standards for all: Conference summary* (pp. 147–182). Washington, DC: National Academy Press.

Long, M. C., Conger, D., & Iatarola, P. (2012). Effects of high school course-taking on secondary and postsecondary success. *American Educational Research Journal*, *49*(2), 285–322. doi:10.3102/0002831211431952

Mabizela, M. (1994). Voices from first year students at UWC. In B. Leibowitz & M. Walker (Eds.), *AD Dialogues 3* (pp. 23–46). Bellville, South Africa: University of the Western Cape.

Marton, F., & Saljo, R. (1976). On qualitative differences in learning: Outcome and process. *The British Journal of Educational Psychology*, *46*(1), 4–11. doi:10.1111/j.2044-8279.1976.tb02980.x

McInnis, C., & James, R. (1995). *First Year on Campus: Diversity in the Initial Experiences of Australian Undergraduates*. Canberra: AGPS.

Miller, T. E. (2005). Student persistence and degree attainment. In T. Miller, B. Bender, & J. Schuh et al. (Eds.), *Promoting reasonable expectations: Aligning student and institutional views of the college experience* (pp. 122–139). San Francisco: Jossey-Bass.

Moxley, D., Najor-Durack, A., & Dumbrigue, C. (2001). *Keeping students in higher education: Successful practices and strategies for retention*. London: Kogan Page Limited.

Noel, L., Levitz, R., & Saluri, D. (Eds.) (1985). Increasing student retention. San Francisco, CA: Jossey-Bass.

Oliver, M. L., Smith, A. W., & Wilson, K. R. (1989). Supporting successful black students: Personal, organizational, and institutional factors. *Nat. J. Soc.*, *3*, 199–221.

Pascarella, E. T. (1982). Concluding thoughts. *New Directions for Institutional Research*, *36*(36), 89–92. doi:10.1002/ir.37019823608

Pascarella, E. T., & Terenzini, P. T. (2005). How college affects students (Vol. 2). San Francisco: John Wiley & Sons, Inc.

Patton, M. Q. (1990). *Qualitative Evaluation and Research Methods*. Newbury Park, CA: Sage.

Pike, G. R. (2006). Students personality types, intended majors, and college expectations: Further evidence concerning psychological and sociological interpretations of Hollands theory. *Research in Higher Education*, *47*(7), 801–822. doi:10.1007/s11162-006-9016-5

Potgieter, M. (2010, October). Conceptual gain in chemistry: Is the gap addressed effectively? *Paper presented at the STEM, Mind the Gap Conference, Academy of Science of South Africa, Cape Town.* Retrieved January 10, 2017 from http://www.google.com.au/

Povey, H., & Angier, C. (2004). *'I can do it, but it'll be a battle': finding her place as an undergraduate mathematician*. Sheffield Hallam University.

Rollnick, M., Mphahlele, M., Ogude, N., Green, G., Huddle, P., Mahooana, P., & White, M. (1997). A team work approach to workshop delivery for tutor development. *Academic Development*, *3*(2), 91–116.

Rovai, A. P. (2003). In search of higher persistence rates in distance education online programs. *The Internet and Higher Education*, *6*(1), 1–16. doi:10.1016/S1096-7516(02)00158-6

Schlossberg, N., Waters, E., & Goodman, J. (1995). *Counselling adults in transition: Linking practice with theory* (2nd ed.). New York: Spring Publishing.

Schlossberg, N. K. (1989). Marginality and mattering: Key issues in building community. *New Directions for Student Services*, *48*(48), 5–15. doi:10.1002/ss.37119894803

Scott, I., Yeld, N., & Hendry, J. (2007). *A case for improving teaching and learning in South African higher education. Higher Education Monitor No. 6*. Pretoria, South Africa: Council on Higher Education.

Seidman, A. (2005b). Where we go from here: A retention formula for student success. In A. Seidman (Ed.), *College student retention* (pp. 295–316). Westport: Praeger Publishers.

Tinto, V. (1975). Dropout from higher education: A theoretical synthesis of recent research. *Review of Educational Research*, *45*(1), 89–125. doi:10.3102/00346543045001089

Tinto, V. (1982). Defining dropout: A matter of perspective. *New Directions for Institutional Research*, *36*(36), 3–15. doi:10.1002/ir.37019823603

Tinto, V. (1987). *Leaving college: Rethinking the causes and cures of student attrition*. Chicago, IL: University of Chicago Press.

Tinto, V. (1993). *Leaving college: Rethinking the causes and cures of student attrition*. San Francisco: Jossey Bass.

Tinto, V. (1998). Learning Communities and the Reconstruction of Remedial Education. In *Higher Education, Replacing Remediation in Higher Education Conference*, Stamford University, January 26-27.

Tuttle, K. N. (2000). Academic advising. *New Directions for Higher Education*, *111*(111), 15–24. doi:10.1002/he.11102

Wiggers, R., & Arnold, C. (2011). Defining, Measuring and Achieving 'Student Success' in Ontario Colleges and Universities. The Higher Education Quality Council of Ontario. Retrieved January 19, 2017 from http://www.heqco.ca/

Williams, I. L., Glenn, P. W., & Wider, F. (2008). Nurtured advising: An essential approach to advising students at historically black colleges and universities. *Academic Advising Today*, *31*(1). Retrieved from January 10, 2017 from http://www.nacada.ksu.edu/

Yorke, M. (2008). Is the first year experience different for disadvantaged students? In F. Ferrier & M. Heagney (Eds.), *Higher Education in Diverse Communities: Global Perspectives Local Initiatives* (pp. 113–120). Dublin: Higher Education Authority.

KEY TERMS AND DEFINITIONS

Academics: The scholarly activities of a school or university, of or relating to a college, academy, school or higher education institution, especially one for higher education.

Faculty: The entire teaching and administrative force of a university, college, or school.

Higher Education: The act or process of imparting and acquiring knowledge, developing the powers of reasoning and judgment, the act and practice of imparting knowledge, especially at college, or university and generally refers to graduates and or master's programmes.

Individual Competence: Refers to the capacity of individuals/employees to act in a wide variety of situations. It includes education, skills, experience, energy and their attitudes.

Knowledge: Knowledge is acquaintance with facts, truths, or principles, as from study or investigation; general erudition, familiarity or conversance, as with a particular subject or branch of learning including acquaintance or familiarity gained by sight, experience, research or report.

Management Education: The act or process of imparting or acquiring knowledge to develop the members of the executive or administration of an organization or business.

Quality in Education: Definitions of quality in education follow the general definitions of quality. Quality education actually means excellence in education. Quality education can be defined as conformance of education output to planned goals, specifications and necessities. Quality of education is also considered as non-defective education procedure. In fact, quality education is a type of education which meets fineness in education.

Student: Pupil, a person formally engaged in learning, especially one enrolled in a school or college; any person who studies, investigates, or examines thoughtfully.

Teaching: To impart knowledge or skill; give instruction, inform, enlighten, discipline, drill, school, indoctrinate; coach to help to learn.

Chapter 12
Retention, Retention, Retention: Exploring Retention Solutions at a Private London University

Kate Armstrong
Regent's University London, UK

ABSTRACT

This case study is the result of a reconnoitre style approach to observation, analysis and subsequent planning. This approach comprised several (on-going) extensive informal interviews with staff across departments responsible for quality, administration, teaching and management. This has been supported with the analysis and observations of the University's staff and student culture, coupled with benchmarking these insights with other HEIs and current retention academic literature. Before presenting the case, it is necessary to ground the reader in current understanding of student retention in HEI's. The literature reviewed is drawn from both UK and international scholarship and following the presentation of current literature on student retention and the University's case study, further empirical review is integrated within the case study where appropriate, to contextualise, corroborate or reflect.

INTRODUCTION

The British Higher Education Institute (HEI) sector introduced a fee-paying structure in 2010. Since then the British education marketplace has experienced a stronger consumer mindset and predictions are that student expectations of education as a service will increase. Students are now squarely focused on a 'value for money' analysis that includes the number of contact hours with tutors (Docherty, 2011). The institution that conducted this retention analysis is considered an Alternative Provider University (APU) and as such, the students have always been fee-paying. This enables the institution to offer both a case study and a best practice analysis of the expectations and demands that fee-paying students throughout the British education system might experience.

This change of focus to a service industry analysis and consumer mindset has put the focus on a host of student success measures at a broad range of British post-secondary institutions. However, the area getting the most attention is program and institution retention rates. For the institution at which this study

DOI: 10.4018/978-1-5225-2998-9.ch012

was conducted, retention was a key area of concern because institutional attrition rates were increasing and because prospective students are increasingly focusing on this metric and measure of an institution as they compare and contrast institutions throughout the admissions process.

This chapter endeavours to outline the reasons for retention and attrition using the APU as a single case study, framed with current academic understanding that is brought in from the retention and student success literature. The study aims to share best practices, strategies and tactics that target enhanced retention rates through policies, mandates and an overall improvement to the student experience across the institution. The goal of the chapter is to explain the comprehensive assessment of retention and student success practices and policies at a single private institution so as to inform the development of retention solutions and recommendations at a wide range of academic institutions and organizations that are facing retention and student success issues and challenges.

REVIEW OF THE LITERATURE EXPLORING RETENTION

Siedman (2012) asks why we should be concerned with retention and outlines several reasons why institutions need to care about student completion of academic and personal goals. The crux of Siedman's reasons are paraphrased here to include: student development and financial reasons for both the University and the student; the development of critical thinking and ability to analyse skills; to engage in reflective and critical discourse; and ultimately to develop students who will be able to contribute to society. Retention moves beyond the acquisition of and matriculation of students and instead focuses on the preservation of each individual student's retention from commencement through to completion and graduation. This chapter uses a London, UK, Alternative Provider University (APU) as a single institution case study. The study also uses existing literature on retention as a structure to frame and support the discussion and to outline and give context to the recommendations and retention interventions, strategies and tactics that proved successful in the case study.

Thomas, Hill, O'Mahony, and Yorke, (2017) wrote a report titled 'What Works?', which was developed as a means to understand the best strategies and approaches to helping students remain committed to their studies and to ultimately be successful graduates. Their report provides specific suggestions for institutions concerning the types of interventions and best practices that were most effective at the APU that was the subject of this case study and that might prove effective at other institutions and in other educational contexts. This study follows the same 'what works' approach and focuses on practical strategies, approaches and best practices to inform and then improve retention and student success across a range of academic institutions focused on making meaningful and thoughtful improvements to their retention rates. The review of the literature also gives insights and practical examples of a wide range of institutional approaches and specific interventions that have made a genuine difference to students' retention and success. These external studies are integrated and reflected on throughout this chapter within the context of the case study.

Alternative Provider University (APU)

This paper uses as its case study a UK private University, that is defined as an alternative provider university or an APU. According to the Higher Education Funding Council for England (HEFCE, 2017) an alternative provider (AP) institution is defined as any provider of higher education courses which does not directly receive annual funding from HEFCE or its equivalent bodies in the devolved administrations, does not receive direct annual public funding, and is not a further education college. As of March 2017, the HEFCE announced that there are 115 AP's in the UK, seven of which have degree-awarding powers, and only one of which has permanent degree granting authority. The distribution of students across the AP providers are small as 33 institutes have fewer than 100 students and only 11 AP's have over 1,000 students.

The APU used as a case study in this chapter attracts cosmopolitan students from 140 countries with no skewed preference to one country of origin. The institute's retention strategy discussed in this chapter is in its first year of implementation. The institution recently identified attrition as an inherent problem that the APU wanted to address and reduce in order to facilitate the success, progression and gradation of its students. The APU strategy and the ensuing chapter presented here, is the result of a reconnoitre style approach to observation, analysis and subsequent planning. The study included extensive, but informal, interviews with faculty and staff members across the institution and its various academic departments including quality and assessment, administration, teaching and management. The findings and recommended interventions have been supported with the analysis and observations of the University's staff and the institution's student culture. Further, the study tackles benchmarking of these insights with other HEIs and with the current academic literature on retention and student success.

From the initial investigation, it was deduced that attrition issues at the APU fall into one or a blend of the following three broad retention areas or categories:

1. Fails, Resits and Retakes
2. Withdrawal and Break in Studies
3. Student Engagement

This is not to over simplify the problem of retention and attrition rates at this particular institution; rather, it aims to give structure and focus to the analysis and the approach to strengthening the student experience across all sectors of the institution. The author acknowledges that there are a plethora of issues and challenges that cause or contribute to retention issues at different institutions and that the range of possible strategies to address and improve them are vast, multi-layered, interlinked and have the potential to be confusing. Focussing on three key retention areas which overlap and are interconnected simplifies an otherwise potentially multifarious process and allows for meaningful examination into the detail associated with each of the three study areas. This three-pronged assessment and evaluation approach and process has facilitated a more comprehensive understanding of the connected nuances associated with student retention and accordingly, has contributed to the development of the institution's Retention Strategy Action Plan (RSAP).

To address the three overarching retention issues of: Fails, resits and retakes, withdrawals and breaks in study and student engagement the improvement design focused on three broad strategic planning areas:

1. Improving Quality Measure and Quality Management
2. Fostering a Sense of Belonging with Students
3. Ameliorate Communications

As identified, retention identification and solutions are complex and multi-layered; therefore, these improvement areas are embedded across all of the tactics and solutions offered throughout this chapter.

The chapter is structured and organized according to the three retention areas and draws upon additional intelligence, literature and data driven metrics that has been gathered in association to the issue. Underpinning these major retention focus areas are the 'Improvement Strategic Planning Areas' (ISPA), which are issue specific and reinforce the improved retention analysis and outcomes as well as contribute to the action items that are laid out in the RSAP.

Table 1 summarises the retentions areas, goals and tactics and then aggregates the main retention action plan points of this report.

The summary in Table 1 is not meant to be an exhaustive list and it presents only a summary of some of the key retention factors explored in this chapter.

Retention Area One: Fails, Resits, and Retakes

Fails, resits and retakes lead the discussion, as it is the key area where the majority of retention leakages at the APU were found to occur. As previously mentioned there are issues at each level of the undergraduate progression at the APU. However, attrition occurs mainly at level 3 or at the foundation level and at level 4 which is the first year of undergraduate study. As such the RSAP focused on these levels first, with the goal of increasing retention by 10 percentage points at both level 3 and level 4 by 2018.

Table 1. Summary of retention tactics

Issue	Goal	Tactic
Retention Issue 1: High Fail Rate	Reduce level of resits and retakes	Implement and communicate team teaching policy
		Marking efficiency plan
		Clear assessment planning
		Improved resit experience and support
		Improved and efficient exam review process
		Academic support for struggling students
		Enhance quality of staff assessment implementation
Retention Issue 2: Break in Studies and Withdrawal	Reduce number of withdrawals/ break in studies and log reasons through better database management	Nurturing a sense of belonging and empowerment through PTs
		Student communication plan
Retention Issue 3: Student Engagement	To enhance engagement of students and provide touchpoints for them to engage with our brand DNA to encourage a sense of belonging	Attendance management policy
		Pre-diagnostic testing [Maths and Language] and internal observation upon arrival
		Create more cross-faculty touchpoints for student to engage with staff + guest speakers across programmes showcased through Student Experience Roadmap (SER)
		Enhanced personal tutor scheme that provides the sense of belonging and a 'bijou' experience that a small, private London University can provide.
		Embed a Strength's Based Education

To provide some context to the issues and challenges inherent in the level 3 and level 4 retention work, a huge volume of the resit and retakes falls under the undergraduate area. Fail rates among the Business faculty were considerably high with forty students receiving failing grades. Of this number, it is estimated that approximately 10-15% of the students who take a resit will fail again and or are unlikely to return to the university. The remaining 25-30% may complete their studies, but probably not within the normal registration period.

The principle retention goal in this area is to reduce the number of resits, retakes and ultimately fails, before they occur, with the aspiration that this in turn will enhance retention through a comprehensive student engagement strategy. This will require a number of tactics to enable enhanced engagement and to reduce the number of students receiving failing grades. There is no one tactic-solution to any of the identified retention issues and all of the core retention issues identified in the study were subject to multiple interventions and approaches. To facilitate these multi-pronged strategies, analysis of the failure rates across programmes and modules is still underway as this chapter goes to publication. Therefore, the following sections concentrate on the early data and response to the study and describe the early stages of the retention interventions across the three core retention issue and challenge areas.

Module Landscape

To give context, the Business faculty currently has a large library of modules (n= 129), with a high volume of students (n = 7278) and a high level of resits following assessment (n = 760). Once the Business courses and programmes were identified as a key retention focal point, the next step in the retention strategy plan was to identify the specific modules of the Business curriculum that were responsible for the high numbers of students having to resist for a course. A rating system was introduced, modules with fail rates under 10% were rated Green (not at risk) modules with failure rates between 10% -15% were categorised Amber (at risk) and modules with failure rates in excess of 15% were rated Red (Critical). These data points are based on a limited data sample as the retention plan is in its initial year. However, the APU will continue to collect data over several years and these future data points will be integrated with future analysis. Thus, the APU is taking a longitudinal approach to its retention study.

The process of rating and coding the modules Green, Amber and Red revealed the nuances the team required to help focus their attention on the critical retention points in the curriculum. This review revealed where the highest number of failing grades were given. The next steps were for the team to analyse why these modules have such high failure rates. The goal was to understand the risks to retention in the individual modules and then to develop ways to mitigate the difficulties and to reduce the fail rates. The aspiration was to enhance student engagement through specific, informed and meaningful interventions. From the investigation, the reasons for the high fail rates on the modules coded Red fell into one or more of the following camps:

- Differences in standards of delivery
- Teaching quality
- Difference in academic experience + application
- Managing large groups of staff
- Differences in assessment and feedback
- Lack of engagement from student
- Poor attendance of students

Retention Tactic: Curriculum and Assessment Improvement Plan

One of the first tactics to begin to address the issues with the modules coded Red or Critical was to ask the Head of Programmes (HOP) to review all modules that had resit rates above 10%. After reviewing the module, the HOP would then meet with the module leaders. The purpose of this meeting is to ensure that formative assessments are linked to summative assessments in these modules and to illustrate that the APU's senior leadership are focused on key retention issues and challenges.

Additional strategies in this area include:

- HOP to review assessment strategy of modules with low pass rate and low overall average grades.
- Module leaders to communicate and provide explicit formative feedback for summative assessment improvement.
- Module leaders to provide overall assessment feedback upon the release of individual feedback, followed by suggestions for improvement in the future. The feedback should be ready 2-3 weeks after the submission.
- Module leaders to check the re-assessment list for their modules, and ensure re-assessment instructions have been provided on Blackboard (Bb).
- Module leaders to check if re-assessment links on Bb for students are current and live and then adjust and update the links as necessary.
- Deputy HOP to remind students about resits via the new specially designed email alert system.
- Module leaders to run resit debriefs or revision sessions either face to face or virtually using a flipped classroom model as appropriate.
- The HOP to organise meetings with students whose attendance puts them in the serious risk category using the specially designed email alert system for general risk categories.

Critical Communication and Team Teaching Policy

The Team Teaching Policy for the Business faculty explains how team teaching should be managed to create a common student experience and maintain academic standards. The rationale behind this is the notion that the institution must attain consistency in the delivery of 'mega modules.' Mega modules are defined as courses that have over 350 students and that demand large lecture capacity, and that have multiple numbers of faculty and staff members working on them. To ensure consistency of experience for students in these mega modules and to enhance their engagement the following interventions were put in place to improve communications and to increase engagement for both parties. A further goal is to outline a framework for dialogue and structure for planning for the Module Leader (ML) and teaching team. The structure for this policy includes: administrative management, peer observation, weekly bulletins, and assessment communications.

Administrative Management

- The faculty will arrange [before induction] scheduled time to talk to their teams.
- The Module Leader will use the allocated time to communicate course content, changes, important facts, assessment times common approaches, key dates in the calendar and so on. These meet-

ings will be repeated every two to three weeks and the Module Leader will discuss with the team detailed class plans for the ensuing two to three weeks.
- Discuss assessment in detail and communicate common approaches in terms of style of delivery, focus on theory.
- It is critical that all of the team need to uphold the same view so that students should appreciate the team teaching effort and the diversity in expertise and teaching nuances this offers.
- The common goal is to ensure that classroom experiences are as identical as possible (notwithstanding the distinction of individual teaching styles).

Peer Observation

- Peer Observation should be mandatory across the team internally and the ML should ensure that they observe all team members at least once an academic year and feedback on improvements/ successes.
- Share best practice and the team voice: ML's need to create the feeling of a collegiate group team, where ideas are shared, issues expressed and 'one module voice' on approaches and interpretations of module content are clear to students.
- ML to create an improvement plan for any members of staff who seem to be struggling and or have a high level of student complaints around performance.
- In extreme cases, where staff are not demonstrating 'team spirit' the ML should escalate to the appropriate line manger to discuss.

Weekly Bulletins

- Through weekly bulletins students will be reminded of upcoming classes and the dissemination of lecture slides. The weekly bulletin can also remind students to log on and read information on the Blackboard system. Further this approach will create a platform for team dialogue around sharing ideas, experiences and best practices.

Assessment Communications

- Faculty should provide reminders for any upcoming formative and or summative assessments and provide an indication of the assessment strategy, what the faculty member is looking for, what will be rewarded, penalised and so on.
- Use the common assessment framework for grading and ensure that all team members are using the common assessment framework properly.
- Agree on a common feedback strategy. A common assessment rubric attached to assignments is a good place to start to enhance a common approach to marking. Ensure that all of the team use the rubric and agree that in text comments will also be used, in addition to a short reflective paragraph of commentary in the 'comments' to rationalise the grade. All team members to agree the common approach.
- Assessment Benchmarking Meeting: A day or two after summative assessment has been submitted the ML will organise an 'assessment meeting' where assessment can be marked, feedback compared and academically discussed so that a common approach and benchmarking of grades

can be agreed. Students will also know that this approach is taken to manage expectations and continue creating the 'common team approach'.

- Post marking benchmarking: Each tutor should to submit to the ML an empirical analysis of the marks, mean and standard deviation for the given assessment. The ML is responsible for detecting discrepancies and then acting upon them.

Retention Tactic: Student Communication Resit Plan and Analytics

One of the barriers facing the APU is a lack of student engagement with resits and retakes processes. The resist rates at the APU are disproportionately high in comparison to peer institutions and the rest of the UK based APU sector. Therefore, this issue appears to be unique to this APU. At this APU there are a high volume of students who are High Net Worth Individuals (HNWI's) and they do not wish to return to the APU (often their country of origin is not the UK) for a resit or retake and are simply happy to either trail credits or pay to take a retake in a following term. This presents unique challenges, as those that fail do not seem to be motivated to timely address the issue or to engage with faculty or staff members in a timely manner. As such, the following strategy was derived to improve the resit experience from a broad perspective, with communication and a personal touch inherent to the plan:

- Skills based content will be taught in large groups.
- Break out groups into subject specific area.
    - **Rationale:** This will reduce the number of staff and resource required, it will also improve the student camaraderie and likelihood of attending.
- Face to face or virtual revision sessions (to ensure that even those who do not return have access to content).
- Periodical communication to students and module leaders of re-assessment instructions, by using the personalized email alert system that communicates directly to students that they have a resit.
- Module leaders will be reminded to provide clear re-assessment instructions on the Blackboard area and this will be supported by communicating with the student's Personal Tutor (PT).
- Resit debriefs or revision sessions will be organised by Module leaders for both Spring and Summer resit period.

With regards to the revision sessions, they can be organized locally but delivered virtually, as the students at this APU are typically cosmopolitan and return home as soon as the term has finished. This appears to be the case regardless if they have revision session to attend or resits to take. It is recognised that attrition rates are 20% to 25% higher for online courses than they are for traditional, face-to-face courses (Nistor & Neubauer, 2010; Street, 2010). While fully online courses are not directly comparable to delivering revision and review sessions virtually in a flipped classroom model, it is something to be mindful of if retention is the primary focus. Given the improved retention outcomes of the blended learning approach it is recommended that revision and review sessions are offered in multiple delivery modalities to ensure multiple opportunities for engagement and so that students can take the modules in their preferred learning format.

The implementation of the engagement alert system for absence-heavy students will, apart from improving communication, also provide a useful source of data analytics. It is hoped that at-risk students could be identified through patterns of attendance, and, in the next phase, this data can be linked to assessment data to provide a form of ongoing, reflective, learning analytics for the programme. Clarifying ideas about student risk or propensity therein is a fundamental starting point in designing analytics-based strategies that aim to prevent or reduce academic failure. Supporting this Thomas et al., (2017) maintain that evidence should be drawn from existing reports and literature on student retentions as well as data, evaluation and feedback from the institution.

Retention Tactic: Improving Student Engagement

The aim is to create workshops for the APU that improve standards according to internal requirements for student engagement. These workshops would improve staff quality, expertise and credibility. They will also support the wider work of the University which includes the Postgraduate Certificate in Higher Education (PGCHE), while engaging the faculty and enhancing the overall quality of teaching. A possible option is to create groups for those with an education certificate or training and those without, in order to benchmark and compare best practices. Some of the workshops that might be included are listed below and could be created and supported with Academic Practice or staff attendance at these workshops can be emphasised as crucial to the overall retention goals of the institution. Line managers should impress upon their staff in appraisals who needs to attend.

- Exam writing best practices
- Grading efficiently
- Writing pedagogically
- Linking formative and summative assessments

A series of workshops could provide legitimate time and space for these pedagogic issues to be explored along with best practices of what tools and skills are required to enhance these skill sets.

Retention Issue Two: Students Withdrawing and Breaks in Studies

One of the key problems to addressing and then improving retention resides in not knowing the full or principle reasons for withdrawal or break in studies of the students at the APU. This is in part due to the lack of a coherent data system that captures knowledge, insight and crucial data that leads to meaningful conclusions around a student's departure and likelihood to return. Building knowledge in this area is a continuous mandate of the Retention Working Group (RWG) as accurate data points and analytics are key to any future strategy and planning in this area. The problem is that the institution does not currently have a central database that denotes 'leaver codes' to track why students leave. However, the primarily reasons cited for withdrawal or a break in studies include:

- Students will not return after the end of semester with no reason given
- Financial Constraints
- Decide to work and move out of education

- Health issues
- Non-arrival (deposit given, but no show)
- Some students remain 'on the books' but have not been seen for years
- Some students change to other HEI's (we know this as they ask us to write references).
- Foundation can often be used as a 'gateway' to facilitate this switch

These primary reasons for non-retention can grouped into four key themes: 1) financial, 2) illness, 3) personal and 4) visa unauthorised.

In sum, the reasons for leaving are mostly anecdotal. Thomas et al., (2017) support the collection of institutional data and qualitative research, which can be used to understand the local context (such as the characteristics of students who withdraw or have other success issues), before selecting interventions. The hardest challenge is knowing whether a student has actually withdrawn or left the APU. Attempts to code the leaver are currently challenging, as many students do not communicate once they have departed the APU. If we can understand reasons for withdrawal or break in studies with more clarity, we can identify where best to focus resources on improvements through clear analytics. As such, a central database is required for managing these data points. Leaver codes should be created along with administration resources specifically designated to identifying reasons for leaving or non-returning through a clear workflow with managerial 'touchpoints' which should include heads of programmes, administration and staff members and personal tutors. This would ensure that engagement with the student happens at multiple points in the 'departure journey' to ensure that any opportunity to engage with the student and perhaps help to mitigate them leaving is exploited. If these data are available, then a clear retention plan may be created to address the issues. It is important however to distinguish between withdrawal issues that can be managed (e.g. attendance, engagement, student experience) versus those that are out of our control (financial, personal loss and so on).

Retention Area Three: Engagement, Nurturing Belonging, and Empowerment

Research from other UK and European based higher education institutions combined with the APU's internal measures and assessments of student satisfaction and success suggest that precursors to disengagement or withdrawal are not attending class, disinterest, inability to keep up, language barriers, lack of motivation, and immaturity. As such, the retention tactics in this area centre on the implementation of the personal tutor scheme, attendance management policies and creating a sense of belonging and enhancing the personal nature of the small APU experience. As a small APU in the heart of London, the 'bijou' education experience is often a precursor to students choosing to enroll at the APU. As metrics, KPIs, and strategy have dominated the educational experience and the retention strategy, Nichols (2010) reminds us that interviews and conversations with students themselves can lead to important retention and student success insights. Nichols (2010) explored the success of retention tactics pre and post implementation. He identified that after the intervention was in place that contact with the course tutor, the course set-up or course design and factors of self-motivation all contributed to enhanced experiences and consequently course continuation and retention. The inherent importance of the student's self-determination, even if they are not self-aware of this internal trait is essential; but when combined with their experiences with engaging and well-prepared tutors, the educational provider, and student support services, their success becomes even more likely. The next section outlines the retention tactic of personal tutors, which aims to enhance a sense of belonging and is particularly important at a small APU.

Retention Tactic: Personal Tutors

Black (2015) argues that coaching can facilitate non-cognitive skills and a positive growth mindset in students. This philosophy is in accordance with the retention tactic of personal tutors, which aims to foster a sense of belonging and nurture students at the APU. Researchers Yeager, Walton, and Cohen (2013) support the notion of a growth mindset and the need to nurture students to enhance retention.

This APU launched a personal tutor (PT) initiative and it has been established for one year. It was created in accordance with the Learning, Teaching and Assessment Strategy of the APU and aims to offer students a personalised learning experience, close contact with academic staff, and support to identify their learning needs and address them. Another goal of the assigned personal tutor initiative is to nurture a culture of belonging within an academic and social community which will enhance retention through a personalised learning experience. The personal tutor scheme is in its early stages but once entrenched within the APU culture, its stated goal is to be the funnel through which a plethora of retention tactics may be processed and then delivered to the student in an individualized approach.

Research into engagement and retention by the HEA (2012) states that students who think about leaving are less satisfied with their University experience and appear to be less engaged with their peers and their institution. Students who did not think about leaving appeared to have a better understanding of the university processes and were more likely to report a positive relationship with staff and students. Students are particularly likely to consider leaving after Christmas and during the first semester, which supports the wider evidence that the majority of students who leave do so during their first year. Students who withdrew had the lowest rates of satisfaction with their higher education experience in general, and their academic experience in particular. The aim through the PT scheme along with the other retention interventions, is to foster a connection with the student and the institution, particularly at levels 3 and 4, where PT meetings are more frequent.

Retention Tactic: Attendance Management Policy

The chapter proposes multi-methods to tackle the issue of attendance with a strong drive to reduce and reverse the culture that missing class is acceptable. Leadership on the attendance issue must come from top down and the heads of programmes need to communicate and emphasize the attendance policy to students regularly. The goal behind this is to nurture students through a sense of belonging but also to provide them with a clear directive on attendance and to create culpability for missed classes. The proposal is that all HOPs will take ownership of communicating directly with students that are at risk due to continued non-attendance. Attendance flags and warnings are currently sent to HOPs on a weekly basis by members of the administration team. These flags and warning codes are identified by the coloured traffic light system, to indicate students that are most at risk.

The Assistant Dean of the Business Faculty initiated an 'engagement prompt database' so that the students at risk may be communicated with before they become suspended. The engagement drive for retention purposes is to prevent the 'blue' students from becoming 'red' as red is a critical state and one stage before the student is suspended. The heads of programmes have implemented this task and are trialling a scheme whereby students coded blue now receive a standard letter template, that is personalized to them by the HOP, that warns them that they are close to being suspended. In addition, the student's personal tutor is also sent a personalised email that encourages the tutor to talk to the student in their next meeting and via email.

Retention Tactic: Reduce Suspensions

Progress on reduction in the number of suspensions is largely anecdotal at this stage, due to the lack of data issues that permeate the problems outlined in this chapter. However, the administrative team have remarked that the current registration data indicates 'at least a 20% drop in suspensions', which would constitute a significant improvement. Students flagged red are invited in for a one-to-one with the HOP. Feedback from students who have been labelled as red and have had a one-to-one with their HOP or personal tutor have been positive, as the student has appreciated the personal approach and has assured the HOP of their determination to be more responsible for their learning, attendance and engagement in the future.

The retention aspiration is to embed this culture of communication between the registrar, the Heads of Programmes and Personal Tutors to ensure a strong nurture track of student communication, culpability and an action plan to get students back on track before further barriers to engagement occur. This strong communication culture needs to be supported by a systemic engagement focus and a deep understanding of the importance of personalized student focus.

SUMMARY OF KEY RETENTION INTERVENTIONS

This chapter has explored and presented a retention strategy for an APU in the UK. The scope and strategy are unique to the student cohort and the nuanced issues that they bring, along with the distinct problems and engagement issues the APU faces. However, the strategies, approaches and processes have been framed around a review of relevant retention literature in the field. In spite of the individualisation of this APU's retention approach there are meaningful retention interventions that other higher education providers can glean from this study. The Retention Tactics of this APU are summarised below:

- Curriculum and Assessment Improvement Plan
- Critical Communication and Team Teaching Policy
- Student Communication Resit Plan and Analytics
- Improving Student Engagement
- Identifying reasons for withdrawal and Break in Studies
- Attendance Management Policy
- Creating a sense of belonging and nurturing

FUTURE RESEARCH DIRECTIONS

The future ambitions of the retention strategy are to enhance the retention elements that have not been explored fully yet. Strengths based education and coaching is one retention project that is supported by the research. The ambition is to include coaching training into the Personal Tutor (PT) experience. In addition, the implementation of psychometric testing to match students and personal tutors in a more meaningful way and to identify student's strengths in order to enhance student and staff engagement and interaction is planned. The goal is for the personal tutor experience to become more significant in the development of the student's skills and progression. The future conversations and studies in this area will be focused on nurturing talent, and developing growth mindset.

To support the coaching element of the enhanced PT experience, a Regent's OAR scheme may be considered that aims to encourage students to attend and be engaged with the PT experience. Further aims of this approach are to considerably augment the student's University experience through the achievement of a specific range of skills that are level and programme specific that are documented and captured as ongoing achievement records either in an e-portfolio or similar system so that they may be conveyed to future employers. Research also indicates that this tactic to enhance engagement and a sense of purpose is a method that ensures higher levels of retention.

A retention pedagogy toolkit could be implemented to help with transition and progression, diversity, curriculum design, engagement, assessment and evaluation and captures the inherent tactics of the retention strategy, so that they may be communicated in a meaningful way. A longitudinal study of this approach would be helpful to understand the impact of this approach.

Another future direction is the development of a set of retention awards to recognise and reward staff engagement with the overall retention strategy and with various specific elements of the strategy. Finally, evidence of nurturing belonging or direct correlation with retention and success would also be rewarded. A study of the impact of this positive reward strategy would be helpful in determining if such programmes should be expanded or modified.

CONCLUSION AND RECOMMENDATIONS

This chapter has attempted to outline the retention strategy in response to the issues and challenges associated with the UK APU. The chapter has explored and presented the factors contributing to these outcomes and strategy, framed within the context of previously established literature. From the case and the research reviewed, it is evident that there is no single set of variables that are able to predict success factors, progression or retention. The themes that emerge from this chapter centre around what the APU can 'do' for the student, but we must remember that in education, the learner needs to take responsibility for their actions, commitment and motivation.

Because analytical data are crucial as a foundation upon which to plan strategy and then to operationalize the strategy, one of the inherent limitations of the APU's retention strategy is the lack of useful analytics. From the review of literature in this chapter, it became apparent that this is essential for improvement; therefore, the APU aspires to improve the collection of meaningful analytics during the next iteration of the planning and operation. The appropriate analytics can then be used to decipher whether the changes were effective and what additional improvements need to be made in the strategy and operations. It is also important to obtain a global perspective of retention statistics. This should also be part of the initiatives around retention for the APU.

Through this chapter, the aspiration is that it has made some small but meaningful contribution to expanding understanding of retention and student success initiatives at a British APU. As such, the hope is that educational leaders will be able to positively effect and inform meaningful exchange across a wide variety of retention projects.

REFERENCES

Attract Project. (2012). Retrieved from http://attractproject.org/

Ball, S. (1997). Policy sociology and critical social research: A personal review of recent education policy and policy research. *British Educational Research Journal*, *23*(3), 257–274. doi:10.1080/0141192970230302

Ball, S., Maguire, M., & Macrae, S. (1997). The Post-16 Education Market: Ethics, interests and survival. *Paper presented at BERA Conference*, York, UK. September 11-14.

Black, R. (2016). Coaching to Increase Retention in Online Degree Completion Programs, a Promising Practice Study. Retrieved from https://rossier.usc.edu/files/2015/06/Ruth_Claire_Black.pdf

Davismoon, S. (2016). Supporting Transition, Transforming Culture, Improving Retention and Student Success Through Personal Tutoring. *Paper presented at the HEA Conference 'Ready for Retention: Effective Systems for Transition and Student Success' at University of South Wales.*

Kuh, G. D., Cruce, T. M., Shoup, R., Kinzie, J., & Gonyea, R. M. (2008). Unmasking the effects of student engagement on first-year college grades and persistence.

Nichols, M. (2010). Student perceptions of support services and the influence of targeted interventions on retention in distance education. *Distance Education*, *31*(1), 93–113. doi:10.1080/01587911003725048

Nistor, N., & Neubauer, K. (2010). From participation to dropout: Quantitative participation patterns in online University courses. *Computers & Education*, *55*(2), 663–672. doi:10.1016/j.compedu.2010.02.026

Rosen, A. (2011). *CHANGE.EDU: Rebooting for the New Talent Economy*. New York: Kaplan Publishing.

Seidman, A. (2012). College Student Retention Formula for Student Success (2nd ed.). Alan College Student Retention: Formula for Student Success (The ACE Series on Higher Education). Plymouth, Rowman & Littlefield Publishers.

Thomas, L., Hill, M., O'Mahony, J., & Yorke, M. (2017) Supporting student success: strategies for institutional change What Works? Student Retention & Success programme. Retrieved from https://www.phf.org.uk/wp-content/uploads/2017/04/Full-report-Final.pdf

Tresman, S. (2002). *Towards a Strategy for Improved Student Retention in Programmes of Open, Distance Education: A Case Study From the Open University UK.*

Yeager, D., Walton, G., & Cohen, G. L. (2013). R&D: Addressing achievement gaps with psychological interventions. *Phi Delta Kappan*, *94*(5), 62–65. doi:10.1177/003172171309400514

KEY TERMS AND DEFINITIONS

APU: Alternative Provider University. According to the Higher Education Funding Council for England (HEFCE, 2017) an alternative provider (AP) is defined as any provider of higher education courses which: Does not directly receive annual funding from HEFCE or its equivalent bodies in the devolved administrations, does not receive direct annual public funding and is not a further education college. As of March 2017, HEFCE announced that there are 115 AP's in the UK, seven of which have degree-awarding powers (and only one is permanent). The distribution of students across providers are small as 33 institutes have fewer than 100 students and only 11 have over 1,000 students.

Bb: Blackboard Learning Management System.

HEFCE: The Higher Education Funding Council for England. The HEFCE is charged with creating and sustaining supportive conditions for the UK system of higher education.

HEI: A British Higher Education Institute is subject to the regulatory authority of the HEFCE.

HOP: The Head of Programmes.

ISPA: Improvement Strategic Planning Areas are specific approaches to improving retention in focused subject matter and administrative areas of the institution.

Mega Modules: Mega Modules are defined as courses that have over 350 students and that demand large lecture capacity, and that have multiple numbers of faculty and staff members working on them.

RSAP: Retention Strategy Acton Plan of the APU. The institution's guiding retention plan and strategy document.

RWG: Retention Working Group. The cross discipline and cross functional working group at the APU charged with developing and implementing the comprehensive retention plan.

Section 4

Pedagogy and Program Design to Support Retention and Student Success

Chapter 13
Maximizing Retention and Progression to Graduation in Online Programs:
A Case Study in "Designing With the End in Mind"

Mary E. Mancini
University of Texas at Arlington, USA

Daisha Jane Cipher
University of Texas at Arlington, USA

Darab Ganji
Academic Partnerships, USA

ABSTRACT

This chapter uses the literature on student retention in online programs as a foundation to present a case study illustrating how a well-designed, affordable and high-quality online program substantially increased access, retention rates, and progression to graduation. The case study covers how, by using the principles of designing with the end in mind. The University of Texas at Arlington College of Nursing and Health Innovation (CONHI) developed and implemented a highly successful, award-winning online Bachelor of Science in Nursing (BSN) completion program for registered nurses (RN) who had entered the profession through an associate degree program. This RN-to-BSN program was specifically designed to overcome issues known to impact enrollment and completion in nursing programs by leveraging technology and the best practices for online education. The case study also explores how best practices in online education were incorporated into the development of the RN-to-BSN program.

DOI: 10.4018/978-1-5225-2998-9.ch013

INTRODUCTION

This chapter uses the literature on student retention in online programs as a foundation to present a case study illustrating how a well-designed, affordable and high-quality online program substantially increased access, retention rates, and progression to graduation. The University of Texas at Arlington College of Nursing and Health Innovation (CONHI) developed and implemented a highly successful, award-winning online Bachelor of Science in Nursing (BSN) completion program for registered nurses (RN) who had entered the profession through an associate degree program. This RN-to-BSN program was specifically designed to overcome issues known to impact enrollment and completion in nursing programs by leveraging technology and best practices for online education. In 2008, responding to the national demand for a more highly educated nursing workforce, the CONHI initiated its accelerated online RN to BSN program and has since enrolled over 28,000 students, with progression to graduation rates of 80 to 90 percent. The case study explores how best practices in online education were incorporated into the development of the RN to BSN program.

REVIEW OF THE LITERATURE

The number of college students enrolled in at least one online course increased for the ninth straight year, according to the Babson Survey Research Group's annual survey of more than 2,500 colleges and universities—including both nonprofit and for-profit institutions (Allen & Seaman, 2011). Over 6.7 million students were taking at least one online course during the fall 2011 term, an increase of over 6.1 million in 2010.

With increased enrollment in online programs comes a greater need for strategies to retain and progress successful students. Student retention is a complex and multifaceted problem with no single issue as a key component (Jeffreys, 2007). Retention strategies in online education have been attempted in many formats, but there are few empirical studies of effective retention methods. Despite the lack of empirical studies of retention strategies in online education, there is general consensus on effective practices. Hill (2017) outlined a set of commonly implemented strategies:

1. **An "Early Alert" Program:** A student receives an email from the instructor if s/he has not responded during the first few weeks of the course.
2. **An Online Tutoring Program:** Providing tutors, both phone and online, is presumed to increase student success.
3. **More Tutoring:** Provide tutoring above and beyond online tutoring, such as offering a distance learning success course.
4. **Student Success Course:** Provide a student orientation or a student success course that teaches learning skills to help students understand how to learn online.
5. **Learning Communities:** Have a cohort-type system whereby a group of cohorts move through a program together.
6. **Peer Tutoring:** Implement a system whereby students who are ahead of new students and who are successful in their programs tutor new students.
7. **Introduction to Distance Learning:** Provide a tutorial on online communication, such as how to use a discussion board.

8. **Metrics:** Measure course completion rates and persistence rates several times per year.
9. **Focus on Individual Courses:** Identify the courses that are most and least successful.
10. **Read the Research:** Keep apprised of empirical support for online retention.
11. **Involve Faculty:** As the key service providers, involve the faculty in assessment and planning.

Leeds and colleagues conducted one of the few experimental studies to improve retention (Leeds, Campbell, Baker, Ali, Brawley, & Crisp, 2013). The intervention group received video orientations, welcome e-mails, personal phone calls from professors, course contracts, and quizzes on the syllabus content. Results from 162 students divided between the experimental group and the control group found that there were no statistically significant differences in the retention rates between the groups (70% versus 69%). The authors noted that among those students who withdrew, regardless of assigned group, the most common reason reported was the intense course workload. Those students appeared to underestimate the extent of the work required for course completion.

In summary, the development of, and enrollment in, online programs has increased dramatically in the past decade. Attrition rates in online programs are higher than in-seat programs, and this results in a significant loss for both the institution and the student. An argument can be made that marginal increases in retention can have favorable impacts on the individual, the institution, and even society. There is a need for scientific studies of retention strategies in online education that move beyond anecdotes and case studies. Identifying methods that impact outcomes, as well as those that do not, is the next step in the empirical exploration of online retention.

THE DEMAND FOR HIGH-QUALITY ONLINE EDUCATION IN NURSING

In 2010, the Institute of Medicine recommended that in order to meet the nation's healthcare, 80 percent of registered nurses should hold a minimum of a BSN degree. This was an ambitious goal given that at the time approximately 50% of new RNs were graduating with an associate's degree and approximately the same percentage of the nation's 3.1 million RNs held a BSN degree (U.S. Department of Health and Human Services, 2013). As a strategy for meeting the needs of the community, addressing capacity constraints, and meeting the needs of students who had to work while attending school, schools of nursing developed online programs (AACN, 2007; AACN, 2015).

To date, many colleges have created new RN-to-BSN programs or have expanded their existing ones. According to the American Associations of Colleges of Nursing (AACN), there are currently 679 RN-to-BSN programs available nationwide and 28 new RN-to-BSN programs are currently in the pipeline. Over 400 of these programs are offered at least partially online. The online programs vary in length between 12 months to 24 months depending upon the school's curriculum, program requisites, and the student's previous academic accomplishments. The Fall 2014 survey of AACN revealed that the enrollments in RN-to-BSN programs increased by 10.4 percent between 2013 and 2014, which accounts for the 12th consecutive year of increases in RN-to-BSN enrollments (AACN, 2015).

Approximately 79% of RNs with an associate degree do not continue on to a BSN degree. Major reasons for not returning to school include: lack of access to BSN-completion programs, cost of attendance, and competing work and family demands (Dacanay, Vaughn, Orr, Andre, & Mort, 2015; Robertson, Canary, Orr, Herberg, & Rutledge, 2010). When RNs do enroll in RN-to-BSN programs, the attrition rate has been reported to be anywhere between 10 to 50% (Mancini, Ashwill, & Cipher, 2015; Seldomridge

& DiBartolo, 2005; Rosenberg, Perraud, & Willis, 2007). Factors noted to be associated with attrition include demographic variables such as age, prior academic performance, part-time student status, as well environmental factors such as family responsibilities and financial demands (Robertson, Canary, Herberg & Rutledge, 2010) may explain the wide variation in reported attrition rates.

As with all online programs, RN-to-BSN programs also struggle with effective ways to retain and progress students toward program completion. Little has been written about the optimal design characteristics of online nursing programs and little is known about the retention and attrition of nurses enrolled in RN-to-BSN programs. RN-to-BSN students are generally non-traditional learners who tend to be older, married, working, and going to school part-time. Thus, their attrition in BSN programs needs to be understood in the context of individual, family, and academic responsibilities (Dacanay, Vaughn, Orr, Andre, & Mort, 2015; McEnroe-Petitte, 2011; Robertson, Canary, Orr, Herberg, & Rutledge, 2010).

Most existing research on retention focuses on undergraduate pre-licensure BSN programs. The collective findings are that factors contributing to retention can be differentiated into academic and non-academic categories. Academically, there is an association between grade point average (GPA) of pre-requisite courses and retention. Students who have higher GPAs in earlier semesters tend to remain in the program, while those with poorer grades often withdraw before program completion (Blair, 2015). Nursing schools require high levels of critical thinking and academic rigor, which can pose a challenge for many students. Students who have failed more than one nursing course during their program may be less likely to complete the program. Abele, Penprase, and Ternes (2013) found that each course failure in a BSN program lead to a less than 36% chance of completing a BSN program.

Non-academic factors found in the literature to affect retention were lack of time, financial issues, lack of family support, poor stress management skills, and low self-esteem or self-confidence (Megginson, 2008). The lack of time due to family and work responsibility has been reported to be the primary reason among RNs with associate degrees for not completing the BSN programs (McEnroe-Petitte, 2011). Lack of family and peer support has been noted to be a significant stressor for returning students and financial issues can adversely affect their academic achievements (Porter, 2008). There is a likely association between students' high levels of self-esteem or self-confidence and their ability to find academic success in BSN programs (Kenny, Kidd, Nankervis, & Connell, 2011). Students who have low self-esteem or low self-confidence often do not do learn skills and concepts well and thus, they are unable to manage a demanding course load (Carr, Walker, Carr, & Fulwood, 2012). Students with low self-esteem probably lack the emotional skills necessary to persist in nursing programs (Shelton, 2012). Other personality traits found to be associated with retention in nursing programs include high levels of self-motivation, enthusiasm, and conscientiousness (Baldacchino & Galea, 2012).

In summary, registered nurses motivated by the prospect of enhancing skills and advancing careers nevertheless face time, financial, and personal constraints and need a flexible course of study to successfully acquire additional educational degrees. Accordingly, there was a need for the design of a highly scalable online RN-to-BSN program that was "fit for purpose," reflecting the best practices in online education.

DEVELOPMENT OF THE PROGRAM

In 2008, the University of Texas at Arlington College of Nursing and Health Innovation responded to community needs by developing a highly scalable, accessible, and affordable online option to complement its on-campus degree completion program. To achieve its goal, UTACONHI teamed with Academic

Partnerships, a global, Dallas-based higher education service provider that assists public universities in converting their traditional degree programs to online formats. Additionally, the provider has a multi-faceted approach to the recruitment and retention of qualified students. Together, they studied the practical barriers that prevent associate-degree holding registered nurses from continuing their education and designed strategies for recruiting and retaining students through graduation.

The online RN-to-BSN program (termed the Accelerated Online program or "AO") was intentionally designed with the "end in mind" – that is, producing an online program that is engaging, student-centric, focused on student success, allows for maximum flexibility, and leverages technology for recruitment and retention. During program and course development, faculty used evidence-based best practices for online education, focusing specifically on ways to meet the needs of the intended student population -- working RNs. This program incorporated recommendations promulgated by the Sloan Consortium (is now the Online Learning Consortium OLC) - an organization of individuals, institutions, and organizations committed to quality online education (http://sloanconsortium.org; Moore, 2008). Orienting faculty to these recommendations before they developed courses for the AO program kept their focus on the best practices in online education in order to maximize program effectiveness.

By design, both UTACONHI and Academic Partnerships exercised their core competencies in developing the AO program. Continuing today, UTACONHI ensures academic quality by controlling curriculum, qualifications for faculty and coaches, admissions, grading, student advisement, tuition, and student aid. Academic Partnerships brings expertise in online instructional design, assists faculty in transforming and updating their courses for online success, and provides the university with process mapping, integrated marketing, and recruitment and retention services – with specific focus on course-to-course persistence through to program completion.

Professional instructional designers with expertise in online pedagogy were brought in to work with faculty members to assure that the overall look and feel of the courses would meet student expectations. Specific design characteristics were discussed and were used consistently in all courses. For example, all core courses used the same placement for navigation icons, course syllabi, and assignments.

To maximize flexibility and accommodate the professional and personal demands on working RNs, the AO RN-to-BSN program was designed to offer ten start dates a year. This compares to the typical two to three start dates per year in a traditional semester-bound system (fall, spring, and summer). Moreover, courses are delivered in five-week blocks, allowing students to complete the ten required nursing courses in approximately 1 year. With courses being offered 10 times a year, students are able to "step out" for a five-week term without losing significant progress toward degree attainment. In a traditional semester-based system, when students need to take time off for a personal or work reason, they typically have to take a full semester off resulting in a delay of five to six months. Stepping-out for a semester often results in students failing to return and complete the program. It was reasoned that designing the program in a manner that recognizes the demands placed on the learner would increase student satisfaction and ultimately progression to graduation rates.

Academic researchers have reported that students in online programs often struggle with staying on time and on task with their schoolwork (Dacanay et al., 2015; Robertson, Canary et al., 2010). This struggle reflects two common characteristics of online programs. First, students often report a sense of isolation when they take online courses as they do not have the opportunity to interact in robust ways with other students in the class. Second, online courses can have high faculty-to-student ratios fostering the students' sense of isolation in the course. To address these issue, the program was specifically designed to include Master's or doctorally-prepared academic coaches (functioning as teaching assistants). These individuals are assigned to support small groups of approximately 25 to 30 students. Working under the direction of UTACONHI faculty, academic coaches are an additional source of social presence in the course and assist students to stay on time and on task.

BUILDING FOR SUCCESS: KEY CHARACTERISTICS FOR SUCCESSFUL ONLINE PROGRAMS

Since its inception in 2008, UTA's online RN-to-BSN program became the largest in the U.S. among public universities, growing to 13,046 students online nationally in 2016 (see Figure 1). Enrollment in the program now represents 6% of the national RN-to-BSN market and 65% of the Texas RN-to-BSN market (Academic Partnerships, March 26, 2017). To date, UTACONHI has enrolled over 28,000 online RN-to-BSN students since the inception of the program.

Several retention strategies were built into the partnership's RN-to-BSN model, which emphasizes flexibility and accessibility for its non-traditional students. Prerequisite courses are offered online and multiple start dates are offered through the year. Delivering curriculum in five-week sessions instead of the traditional 15-week sessions allows students flexibility in starting the program and progressing at a pace that works for them. Students can begin the program 10 times a year rather than wait to begin classes once or twice a year, as in typical semester-designed programs. With classes running in five-week modules, the program employs a "carousel model" that lets students step off the carousel for a five-week term if needed to accommodate their schedules or disruptions that emerge in their personal lives. The convenience and flexibility of the model has been a significant factor in positive student persistence and retention.

Figure 1. UTA CONHI RN-to-BSN AO and on-campus enrollment trend (2006-2016)

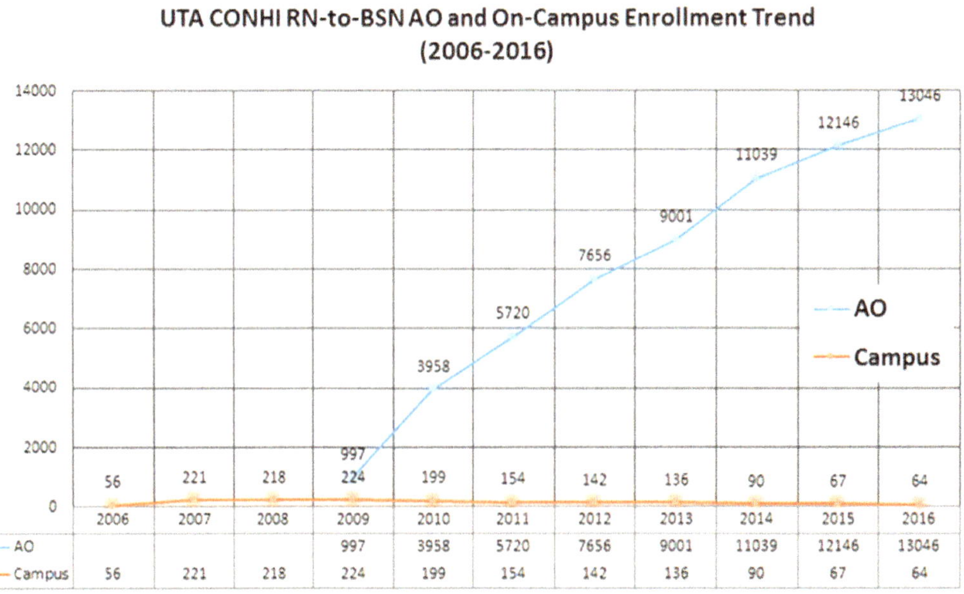

	2006	2007	2008	2009	2010	2011	2012	2013	2014	2015	2016
AO				997	3958	5720	7656	9001	11039	12146	13046
Campus	56	221	218	224	199	154	142	136	90	67	64

A course-by-course payment structure has also aided student retention. The UT Arlington-Academic Partnerships model reduces monetary outlays and offers a competitive tuition. Many healthcare facilities offer their nurses tuition reimbursement that covers some portion or all tuition costs. They typically pay after a course is successfully completed. Thus, students in online RN-to-BSN program can begin their studies for the cost of a single course, with reimbursement for their first course covering their second course and so on to the end of the program.

Student retention and progression to graduation are supported by the expertise of both UTACONHI faculty and Academic Partnerships. Coaches aid students in learning and meeting course deadlines and provide early recognition and remediation of at-risk students. The coaching model allows students to feel as if they are working in small classes rather than large sections. The model augments faculty engagement with students while maximizing the number of students a faculty member can handle. Assisting students to stay on time and enrolled is essential, particularly with five-week online courses and when students "step off the carousel." It is also vital to communicate with students on a routine basis to assure they successfully progress from course to course and don't miss registration and payment deadlines.

The following table lists factors that were integral to the success of the UTACONHI RN-to-BSN program.

As shown in Table 1, the program was developed with instructional designers who work with faculty to assure that the online courses are of high quality and "fit for purpose". They assist faculty with program and course level planning, instructional design and quality reviews. In addition, the program provides appropriately qualified Academic Coaches (instructional assistants) who support the faculty-of-record and provide students with a small group experience. Students are sectioned into cohorts of approximately 25 individuals in undergraduate and 20 in graduate programs. Coaches answer questions and provide timely feedback on student work while facilitating the student experience such that small group interactions are maximized.

The program optimized course sequencing and planning that includes many flexible options. For example, there are multiple opportunities for students to start in their program of choice (there are 10 starts per year). Courses are offered at a frequency that allows the student to progress rapidly through courses to graduation. In addition, the appropriate program and course duration is determined by detailed blueprinting of the program and each course.

Another key characteristic of the program is the course carousel, which was designed to allow students to "step out" for short periods of time in response to work and personal needs (see Figure 2).

Table 1. Key characteristics of the UTACONHI Online RN-to-BSN Program

1. Competitive time to completion of the program
2. Designed with instructional designers with expertise in online pedagogy
3. Five-week nursing courses that let them step off the carousel
4. Ten starts a year
5. Online prerequisite courses
6. Coaches
7. Retention specialists
8. Flexible payment options

This recognizes the unique needs of the working adult and maximizes the opportunity for students to progress to graduation. The carousel paradigm is critical to successful program operations and student persistence. An optimized course sequence benefits the university by helping students complete the program without step-outs. In addition to valuing student time-to-complete, UTACONHI seeks a course schedule arrangement that can best accommodate faculty availability.

Another key characteristic of the program is the collaboration with a partner who has expertise in student recruitment, student retention, online instructional design, faculty development, and process improvement. UTACONHI works with Academic Partnerships (AP), whose mission is to help public and private not-for-profit universities increase enrollment and revenue through online delivery of high-quality instruction. AP helps faculty members convert their on-campus courses and programs into an online format, recruits qualified students for those programs and helps retain enrolled students through graduation. As illustrated in Figure 3, the program incorporates a multi-pronged approach to retention. Advisors engage with individual students at important milestones throughout their program of study. They prepare students with the knowledge, resources, and confidence to thrive as online learners from enrollment to graduation. Academic Partnerships' student retention specialists – in collaboration with UTACONHI advisors – provide comprehensive support and retention services to the online RN-BSN students from the first day through successful program completion. They regularly engage with students through various outreach campaigns, payment reminders, check-ins on their well-being, and proactively intervene with all "at-risk" students.

In 2012, in recognition of the successful public-private partnership between UT Arlington and Academic Partnerships in relation to UTACONHI's online RN-to-BSN program, the Texas Higher Education Coordinating Board awarded UTA and Academic Partnerships the Texas Higher Education Star Award for their exceptional contributions toward increasing student participation and success in higher education.

Figure 2. UTACONHI RN-to-BSN program course carousel

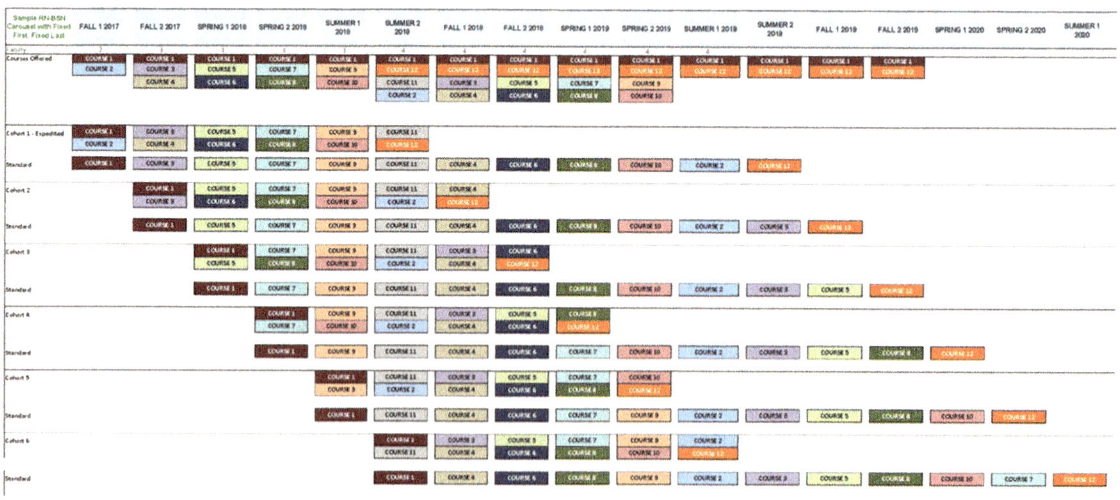

Figure 3. Dedicated retention specialists

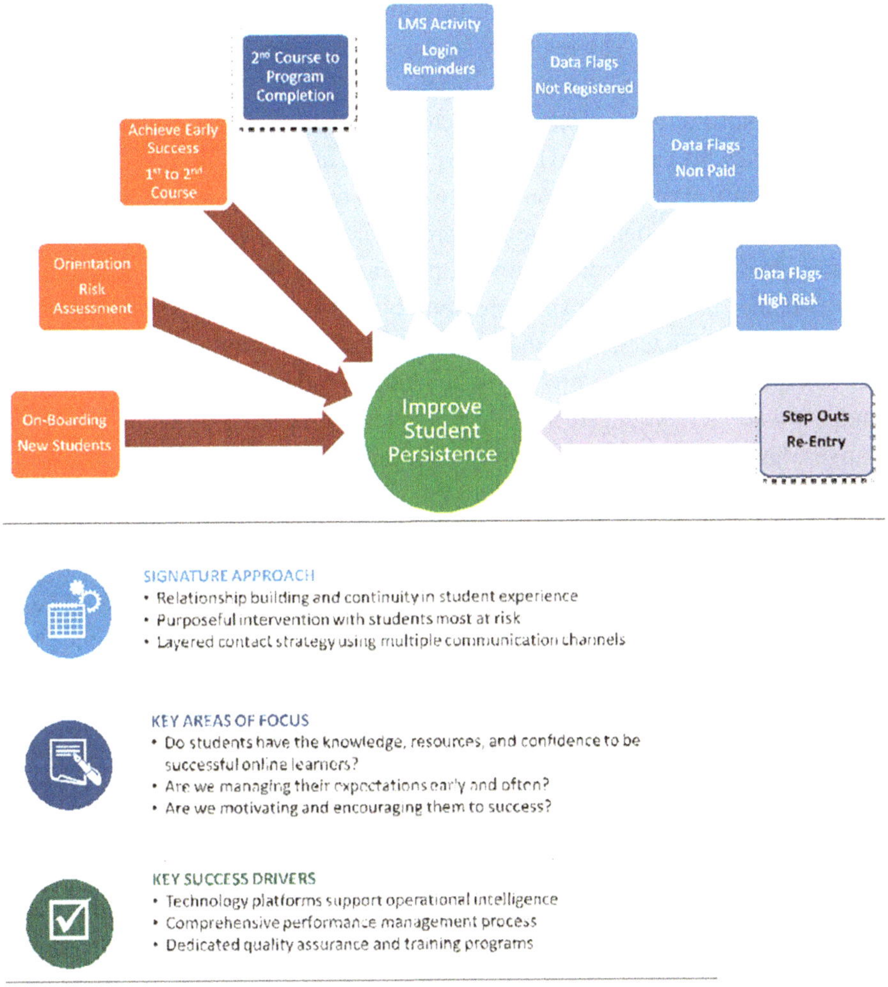

EMPIRICAL FINDINGS

A recent quantitative study by Cipher, Mancini, and Shrestha (in press) reported a retrospective analysis of 9,567 students enrolled in the UTACONHI AO RN-to-BSN program from 2008-2016. Student demographic characteristics (gender, race, age, and prior degree status) academic variables (financial aid status, number of dropped courses, number of withdrawals, and number of failed courses) and outcome variables (progression to graduation, timely graduation, and retention) were collected for AO RN-to-BSN student cohorts who enrolled during eight academic years. The sample included all students were marked as "inactive", meaning that they had graduated, discontinued, or failed out of the program. The final sample included data from a total of 9,567 students, all of whom started and ended the program in the AO RN-to-BSN program.

Most of the students were female (87.4%), and the mean age was 37.7 ± 9.4 years. The ethnic breakdown of the sample was primarily white (59.2%), followed by Black/African-American (15.7%), Hispanic/Latino (12.6%), and Asian (5.9%). The majority of the sample enrolled in the program with an associate's degree as their highest level of education (75.4%), and 15.9% had a previous bachelor's degree in another field. Over one-third of the sample (41.7%) received financial aid for all or part of the program.

Most of the students (79.8%) graduated from the program, and 73.3% graduated on time (6 semesters or less). The remaining students either discontinued (19.8%) or failed out of the program (0.4%). Regression analyses were performed to identify potential predictors of the likelihood of graduation, likelihood of timely graduation, and risk of discontinuing the program. Predictors tested were age at enrollment, gender, ethnicity, whether the student had a previous baccalaureate degree in another field of study, whether the student received financial aid, the number of dropped courses, the number of withdrawals, and the number of failed courses. The predictive model of graduation revealed many predictors to be significant. Students who were younger, receiving financial aid, had a previous baccalaureate degree, fewer failed courses, fewer withdrawals, and more dropped courses (a counterintuitive finding which is discussed below) were significantly more likely to graduate, even after controlling for all other predictors in the model. Students with a previous baccalaureate degree were 59.7% more likely to graduate, and students with financial aid were 19.9% more likely to graduate. With every additional failed course, students were 24.4% less likely to graduate, and with every withdrawn course, students were 22.8% less likely to graduate.

The predictive model of on-time graduation, defined as six semesters or less, found significant predictors; younger students, those from an ethnic minority background, those receiving financial aid, had a previous baccalaureate degree, fewer failed courses, fewer dropped courses, and fewer withdrawals were more likely to graduate on time. Likewise, the predictive model of program retention found similar predictive associations. Students who were younger, receiving financial aid, had a previous baccalaureate degree, fewer failed courses, fewer withdrawals, and more dropped courses were significantly more likely to remain in the program.

Higher numbers of dropped courses were associated with a higher likelihood of graduation and retention, but a lower likelihood of timely graduation (as would be expected due to the delay caused by dropping a five-week course). Dropping courses was common for students in the study, as 69% of the sample dropped at least one course. This is likely due to the flexibility that the AO program offers to accommodate the professional and personal demands of working RNs. With courses starting every 6 weeks, the AO RN-to-BSN students are able to take short breaks from classes to meet work or family demands and be able to resume classes quickly without waiting for up to eight months for the next semester to begin. Many students reported taking advantage of this option and taking two or three "breaks" from course work, or needing to drop courses and resume the program fairly quickly. This may explain why dropped courses were associated with a higher likelihood of graduation and retention.

The graduation rate was 80%, which is consistent with previously published national graduation rates of pre-licensure baccalaureate nursing programs as well as BSN completion programs (NLN, 2016). This particular analysis reveals that the discontinuation rate (19.8%) was consistent with the national average, and the percentage of students who had failed out of the program was extremely low (0.4%). Retention has historically been recognized as an important issue in nursing programs and proves to require further study. Reasons for discontinuation are currently not noted in the university data warehouses, and the CONHI is seeking ways to create a vehicle for following up with those students who did not progress.

LESSONS LEARNED

In order to maximize student retention and progression to graduation, online programs need to be designed with the end in mind so they are fit for purpose. To create effective and efficient online programs requires attention to key factors for success. One of the most important factors is the provision of faculty development that assures that instructors know and apply the best practices associated with this transformative educational technology (see Table 2). For this reason, working with instructional designers familiar with the format is essential. Another valuable lesson learned as the portfolio of nursing programs moved toward the accelerated online format was the importance of having faculty within a degree program work together to translate a successful on-campus program to the online format. Doing so provides opportunities for peer-to-peer learning while maximizing course alignment and achievement of program objectives. This approach can be contrasted to instructional designers working one-on-one with faculty to design courses – a less efficient and more expensive undertaking.

As student enrollment grew, specific strategies to maximize student engagement were implemented. Strategies included the use of engaging videos with high production value, flexible course sequencing (through the design of the carousel), and ongoing analysis of data through analytic software such as Blackboard Analytics or Civitas (see Table 3). The continuous data analysis allows instructors and coaches to identify early warning signs that a student is at risk for failure and needs intervention, such as when there is a delay in opening courseware or submitting an assignment. In addition, it is important to remember that in order to meet accreditation standards (regional and/or discipline specific), ongoing comparison of student, course, and program outcomes between online and on-campus programs is required. These comparisons provide program faculty and administration with robust information to support continuous course and program improvements.

Table 2. Elements of program design involving faculty

1. Faculty development in best practices for online education
2. Blueprinting at the program and course level
3. Faculty working with instructional designers to develop courses
4. Faculty working in teams to design courses

Table 3. Strategies to maximize student retention

1. Use of engaging videos with high production value
2. Flexible course sequencing
3. Ongoing data analyses of student, course, and program outcomes

FUTURE RESEARCH DIRECTIONS

There is a need for scientific studies of retention and student success strategies in online education that move beyond anecdotes and case studies generally and this need is intensified for science intensive programs such as nursing. Furthermore, additional studies designed to identify retention and student success methods that positively impact outcomes, as well as those that do not, is the next step in the empirical exploration of online retention.

CONCLUSION

This UTACONHI case study demonstrates that it is possible to create online programs that are both high quality and highly scalable. Student retention and progression in accelerated online courses require intentional focus on appropriate design, understanding of student needs, and a commitment to continuous program evaluation and improvement. It is inadequate to simply load classroom instruction on to an online learning management system and assume this will produce the desired results. Leveraging technology, expert partners and best practices in online education can be transformative to higher education.

REFERENCES

Abele, C., Penprase, B., & Ternes, R. (2013). A closer look at academic probation and attrition: What courses are predictive of nursing student success? *Nurse Education Today*, *33*(3), 258–261. doi:10.1016/j.nedt.2011.11.017 PMID:22177474

Allen, I., & Seaman, J. (2011). *Going the Distance: Online Education in the USA 2011*. Wellesley, MA: Babson Survey Research Group.

American Association of Colleges of Nursing (AACN). (2007). Alliance for Nursing Accreditation Statement on Distance Education Policies.

American Association of Colleges of Nursing (AACN). (2015). New AACN Data Confirm Enrollment Surge in Schools of Nursing.

Baldacchino, D. R., & Galea, P. (2012). Student nurses' personality traits and the nursing profession: Part 2. *British Journal of Nursing (Mark Allen Publishing)*, *21*(9), 530–535. doi:10.12968/bjon.2012.21.9.530 PMID:22585266

Blair, J. (2015). Predicting success in nursing programs. *Research in Higher Education Journal*, *28*, 1.

Carr, J., Walker, W., Carr, M., & Fulwood, D. (2012, May). Reflect for success: Use of mentor recommendations to help failing students. *British Journal of Community Nursing*, *17*(5), 226–228. doi:10.12968/bjcn.2012.17.5.226 PMID:22584401

Dacanay, A. P., Vaughn, S., Orr, M., Andre, J., & Mort, K. (2015). Factors contributing to nursing distance education student success. *Journal of Nursing Education and Practice*, *5*(2), 34.

Hill, C. (2017). 11 tips for improving retention of distance learning students. Retrieved from http://cdn.facultyfocus.com/wp-content/uploads/2015/02/Strategies-for-Increasing-Online-Student-Retention.pdf

Hooper, J., McEwen, M., & Mancini, M. E. (2013). A Regulatory Challenge: Creating a Metric for Quality RN-to-BSN Programs. *Journal of Nursing Regulation*, *4*(2), 34–38. doi:10.1016/S2155-8256(15)30156-3

Institute of Medicine. (2011). *Future of Nursing: Leading Change, Advancing Health*. Washington, DC: The National Academies Press.

Jeffreys, M. R. (2007). Nontraditional students' perceptions of variables influencing retention: A multisite study. *Nurse Educator*, *32*(4), 161–167. doi:10.1097/01.NNE.0000281086.35464.ed PMID:17627208

Kenny, A., Kidd, T., Nankervis, K., & Connell, S. (2011). Mature age students access, entry, and success in nurse education: An action research study. *Contemporary Nurse*, *38*(1-2), 106–118. doi:10.5172/conu.2011.38.1-2.106 PMID:21854242

Leeds, E., Campbell, S., Baker, H., Ali, R., Brawley, D., & Crisp, J. (2013). The impact of student retention strategies: An empirical study. *International Journal of Management in Education*, *7*(1/2), 22–43. doi:10.1504/IJMIE.2013.050812

Lindley, M. K., Ashwill, R., Cipher, D. J., & Mancini, M. E. (2017). Expanding capacity with an accelerated online BSN program. *Journal of Professional Nursing*, *33*(1), 5–10. doi:10.1016/j.profnurs.2016.03.010 PMID:28131147

Loftin, C., Newman, S., Dumas, B. P., Gilden, G., & Bond, M. L. (2012). Perceived barriers to success for minority nursing students: an integrative review. International Scholarly Research Network Nursing.

Mancini, M. E., Ashwill, J., & Cipher, D. J. (2015). A comparative analysis of demographic and academic success characteristics of on-line and on-campus RN-to-BSN students. *Journal of Professional Nursing*, *31*(1), 71–76. doi:10.1016/j.profnurs.2014.05.008 PMID:25601248

McEnroe-Petitte, D. M. (2011, April). Impact of faculty caring on student retention and success. *Teaching and Learning in Nursing*, *6*(2), 80–83. doi:10.1016/j.teln.2010.12.005

Megginson, L. A. (2008). RN-BSN education: 21st century barriers and incentives. *Journal of Nursing Management*, *16*(1), 47–55. PMID:18211335

National Council of State Boards of Nursing. (2009). Innovations in education regulation report: Background and literature review. Retrieved May 15, 2016 from https://www.ncsbn.org/Innovations_Report.pdf

National League of Nursing. (2016). Retention rates in RN programs. Retrieved May 15, 2016 from http://www.nln.org/newsroom/nursing-education-statistics/retention-rates-in-rn-programs

Porter, K. B. (2008). Current trends in student retention: A literature review. *Teaching and Learning in Nursing*, *3*(1), 3–5. doi:10.1016/j.teln.2007.09.001

Poster, E. C., Mancini, M. E., & Ganji, D. MOOCS and More: Expanding Online Access for Nurses. *Educause Review Online*. Retrieved August 5, 2013 from http://www.educause.edu/ero/article/moocs-and-more-expanding-online-access-nurses

Robertson, S., Canary, C. W., Orr, M., Herberg, P., & Rutledge, D. N. (2010). Factors related to progression and graduation rates for RN-to Bachelor of Science in nursing programs: Searching for realistic benchmarks. *Journal of Professional Nursing*, *6*(2), 9–107. doi:10.1016/j.profnurs.2009.09.003 PMID:20304377

Rosenberg, L., Perraud, S., & Willis, L. (2007). The value of admission interviews in selecting accelerated second-degree baccalaureate nursing students. *The Journal of Nursing Education*, *46*(9). PMID:17912993

Seldomridge, L. A., & DiBartolo, M. C. (2005). A profile of accelerated second bachelors degree nursing students. *Nurse Educator*, *30*(2), 65–68. doi:10.1097/00006223-200503000-00007 PMID:15785342

Shelton, E. N. (2012). A model of nursing student retention. International Journal of Nursing Education Scholarship, 9(1). *Article*, *6*, 1–16. doi:10.1515/1548-923X.2334

U.S. Department of Health and Human Services, Health Resources and Services Administration (2010). The Registered Nurse Population: Findings from the 2008 National Sample Survey of Registered Nurses.

U.S. Department of Health and Human Services, Health Resources and Services Administration (2013). The U.S. Nursing Workforce: Trends in Supply and Education.

KEY TERMS AND DEFINITIONS

AACN: The American Association of Colleges of Nursing is the national society in the United States for baccalaureate and graduate nursing education.

Academic Coaches: Academic Partnerships (AP) uses Academic Coaches who are instructional assistants that support the faculty and provide students with a small group experience. Students are sectioned into groups of 20 and they work directly with their coach. The coach answers questions and provides timely feedback on student work while facilitating the student experience.

Academic Partnerships (AP): A global, Dallas-based higher education service provider that assists public universities in converting their traditional degree programs to online formats and recruits and retains qualified students through graduation.

Course Carousel: The Course Carousel is designed to allow students to step out of the program for short periods of time in response to work and personal issues and challenges.

End in Mind: Designing a course or degree program with the "end in mind" means producing an online program that is engaging, student-centric, focused on student success, allows for maximum flexibility, and leverages technology for recruitment and retention.

Non-Traditional Learners: RN-to-BSN students are usually non-traditional learners. Non-traditional learners are usually, older, married, working full time and going to school part time.

Pre-Licensure BSN Programs: Pre-Licensure BSN degree programs are for students who do not already have an associate's degree (ADN) or a diploma in nursing, and do not already have an RN license.

RN-to-BSN: The RN to BSN programs are for students who do have an associate's degree (ADN) or a diploma in nursing, an RN license, and who would like to complete their BSN degree.

Sloan Consortium: The Sloan Consortium is now called the Online Learning Consortium (OLC). It is an organization of individuals, institutions and organizations committed to quality online education.

UTA: The University of Texas at Arlington.

UTACONHI: The University of Texas at Arlington College of Nursing and Health Innovation.

Chapter 14
From Cramming and Convenience to Engagement and Retention:
Four Pillars of Online Course Design for Student Retention

Laura Michelle Galloway
Brandman University, USA

ABSTRACT

Online learning has shown persistent and unrelenting growth over the past few years and serves a wide variety of modern educational needs. However, attrition can be a troublesome phenomenon unless course designers and instructors develop methods of engaging students through use of the Four Pillars discussed in this chapter. The strategies and tactics associated with and springing from these Pillars will go a long way in retaining students in online programs. Face-to-face instructional methods simply will not work in the online environment, and the innovative methods discussed in this chapter will yield student engagement and completion of their educational goals.

INTRODUCTION

The needs of the 21st Century learner are unique. Therefore, 21st Century course design must evolve to serve and meet these unique needs. Students in higher education systems today require learning that is meaningful, applicable and engagement-centered. The more college students experience learning that is relevant, captivating and transformative, the more likely they will stay at it and graduate. Higher retention rates and deep student learning experiences are inextricably linked, you simply cannot have one without the other. Unfortunately, the converse is also true. Students are likely to drop-out of a course or school altogether if they do not think their education experiences are important, relevant or interesting. With the rise of online learning in higher education, this delicate relationship between effective pedagogy and student retention is of even greater concern. This chapter will provide practical tips and applicable

DOI: 10.4018/978-1-5225-2998-9.ch014

practices on how to repurpose traditional notions of pedagogy in order to design and deliver online courses that meet the unique needs of 21st Century students while improving retention and increasing student success.

REVIEW OF THE LITERATURE

The introduction and projected use of technology in Higher Education challenges the traditional notions and practices of education at the University level. "Distance education enabled by e-learning is at the forefront of university participation in an increasingly connected world. Physical, temporal, cultural and educational borders are becoming both less rigid and less predictable than before" (Hedge & Hayward, 2004, p. 1). Mainstream pedagogy is and will continue to shift toward integrating online technologies and online facilitation methods (Hicks, Reid & George, 2001).

The following illustrates the trends in higher education toward the growing use of technology in higher education.

Online learning systems are here to stay. The challenge is to meet the needs of students through well designed and facilitated courses. The 2015 Survey of Online Learning, *Online Report Card: Tracking Online Education in the United States*, conducted by the Babson Survey Research Group, reports that the number of higher education students taking at least one online course in 2015 was up by 3.9 percent from 2014. This report also shows that 2015 was the 13th consecutive year of growth in the number of students taking courses at a distance (Babson, 2015). Assessments of online education outcomes have shown slow but steady improvements over time (Babson, 2015). In the 2014 report, over half of the academic leaders surveyed said that online and face-to-face student learning outcomes were the same (Babson, 2015). As of 2017, it is undisputed that online learning has become a permanent part of most university's strategy, practice and purpose.

The challenge this trend presents to educators is how to ensure that online learning experiences are robust, relevant, interesting, engaging and challenging. Unfortunately, and far too often, it is assumed that online course design and facilitation merely follow the same logic and practice as the established face-to-face pedagogy. Too frequently the online course environment is crammed and cluttered with resources, and even used as a mere repository to house videos, documents and other class and program resources. Offering courses online has changed how content is delivered, redefined traditional educational practices and roles, and created new opportunities for learning. Further, instructors must be trained to use technology and shown how to change the ways that course content is delivered. (Palloff and Pratt,

Table 1. Trends in higher education for online learning

An Increase in Online Course Offerings	A 3.9% increase in 2015 of the number of distance education students up from the 3.7% rate recorded in 2014. (Babson, 2015)
	More than one in four students (28%) now take at least one distance education course (a total of 5,828,826 students, a year-to-year increase of 217,275). (Babson, 2015)
	The percent of academic leaders rating the learning outcomes in online education as the same or superior to those in face-to-face instruction is now at 71.4%. (Babson, 2015)
Strategic Plans in Higher Ed	The portion of chief academic leaders reporting that online learning is critical to their long-term strategy has grown from 48% in 2002 to 70.8% in 2014 (Allen & Seaman, 2013).

Figure 1. Relationship trends in higher education

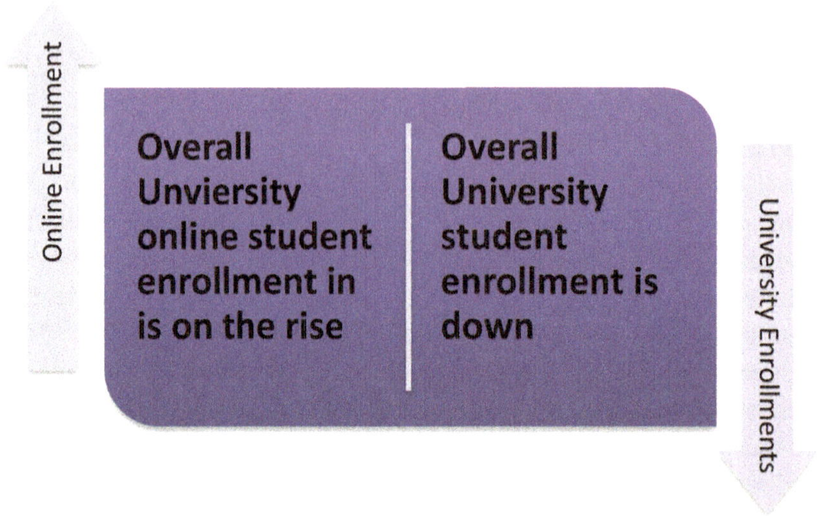

2013). Doing so, Palloff and Pratt (2013) posit, "can increase the potential for learners to take charge of their own learning process and facilitate the development of a sense of community among them" (p. 30).

WHAT THE MOOC EXPERIENCE TEACHES

The educational attempt known as "Massive Open Online Courses" ("MOOC") at first had spectacular growth. Harvard and MIT launched the first MOOC platform in 2010, called "edX." It was a joint effort of Harvard and MIT to bring elite virtual education to anyone in the world with an Internet connection. It was an immediate hit as roughly 155,000 students enrolled in this first MOOC (Allen, E., Seaman, J., 2013). The main idea was that a few star professors would record online lectures to thousands, thereby enabling sharing of their knowledge and enthusiasm for the subject matter with a global audience.

Soon thereafter, Sebastian Thrun, a Stanford professor, co-taught a MOOC in 2011 on artificial intelligence for 160,000 students from 190 countries. Almost as quickly as MOOC's grew in popularity criticism followed (Allen, E., Seaman, J., 2013). First, an experiment at San Jose State University revealed that students who took two online pilot courses created with MOOC course content did worse than those who took the same classes on campus. Additionally, the University of Pennsylvania Graduate School of Education released a study showing that only about half of students who registered for a course ever viewed even one recorded lecture. Even more concerning and more importantly, only about 4 percent of those registered for MOOCs completed the courses (Babson, 2015).

The 2015 Survey of Online Learning conducted by the Babson Survey Research Group revealed MOOCs are a rarity: only 8% of higher education institutions offered one and another 5.6% reported MOOCs are still in the planning stages (Babson 2015). Furthermore, the percentage of academic leaders who believe that MOOCs represent a sustainable method for offering online courses dropped to 16.3% (Allen, E., Seaman, J., 2013). Simply put, the "cram and clutter" and "sage on the stage" recorded lecture-style method of course design, on which the MOOC is based, has proven to be an ineffective method for online learning and has resulted in high student attrition rates.

A WORD ABOUT ADULT LEARNERS

Online learning has enabled countless students, most with families and daytime work obligations, to nevertheless further their education, be it a Bachelor's undertaking or graduate work. Adult learners, beset as many of them are by work, family and financial concerns, face constant temptation to simply drop out of higher education. Thus, course design and course delivery must be sufficient to overcome such temptations and retain the adult learners in higher education. In a sense, the adult learner environment is a crucible for course and program designs, including relevancy to their long-term goals.

Extensive research (Cranton, 2006, Mezirow, 1981) shows adults learn in dramatically different ways than children. Children learn by lecture and repetition and adults learn through application. Knowles (1975, 1980, 1984) noted that adult students are especially motivated to learn because "they experience a need to learn to cope more satisfyingly with real-life tasks or problems" (Knowles, 1975, p. 14). Adults learn best when they are actively engaged and take responsibility for their own learning. This kind of adult learning results in the synthesis of ideas or new knowledge, the ability to reflect, the development of problem-solving skills, achieving effective recall retention and application, and even transformed ideologies (Russell, 2006). Both learning and retention are improved when adults can apply what they have learned to information they have already gathered through personal experience. This means the students in college-level online courses will learn not by listening to long lectures or by watching endless PowerPoints, or by reading lengthy papers, but through dynamic engagement with the material, the instructor and with one another.

From lessons learned in working in the adult learning environment, the Four Pillars of online course design emerged and will be discussed next.

Figure 2. Four Pillars of online course design for student retention

RETENTION USING THE FOUR PILLARS OF ONLINE COURSE DESIGN

Pillar One: Building Learning

From 1996 to 2008 the U.S. Department of Education studied online learning in schools and findings were released in 2010. That report noted "students in online conditions performed modestly better, on average, than those learning the same material through traditional face-to-face instruction" (p. xvi). Although these results suggested that online learning can be more effective, it is important to note the research concluded "online learning is superior as a *medium*." (p. xviii, original emphasis). Also mentioned were additional learning time and materials and opportunities for collaboration. These aspects can serve as building blocks for effective online learning.

Backward Design

The most common approaches to designing courses typically begins with deciding the most appropriate methodologies for teaching content. In other words, traditional pedagogy focuses first on *how* the content will be taught and considers what is to be taught secondary. Wiggins and McTighe (2005) revolutionized these notions of traditional pedagogy and proposed to flip this process. They called it "Backward Design". They posited that designing a course "backwards" is the best way to ensure authentic learning. Because it flips the classic approach, backward design begins with the end in mind. It considers what is to be taught first, and how it should be delivered second. In other words, how the content is delivered should be adapted to the overall purpose or intent of the learning outcomes. This approach challenges the traditional "sage on the stage" model of pedagogy which emphasized the instructor's personality and presence as the gateway to student learning (Garrison & Kanuka, 2004, p. 96) and instead puts the learner in the center. An effective student-centered online instructor is someone "who is open to giving up control of the learning process" by making the learning drive the design of the course and its process (Palloff and Pratt, 2013, p. 24). Thus, once the course designer determines exactly what students need to learn and knows the focus, the task becomes designing methods for teaching the content and meeting learning goals.

The importance of designing online courses with the end in mind is imperative to student learning. Wiggins & McTighe (2005) recommend beginning the design process with questions such as the following:

- What relevant goals (e.g. Content Standards, Course or Program Objectives, Learning Outcomes) will this design address?
- What are the "big ideas"?
  - What specific understandings about them are desired?
  - What misunderstandings might occur?
- What key concepts and skills do students need to master by the end of the course?
- What key knowledge and skills will students acquire because of this unit?
- What should students eventually be able to do because of such knowledge and skill?

Figure 3. Wiggins & McTighe backward design (2005)

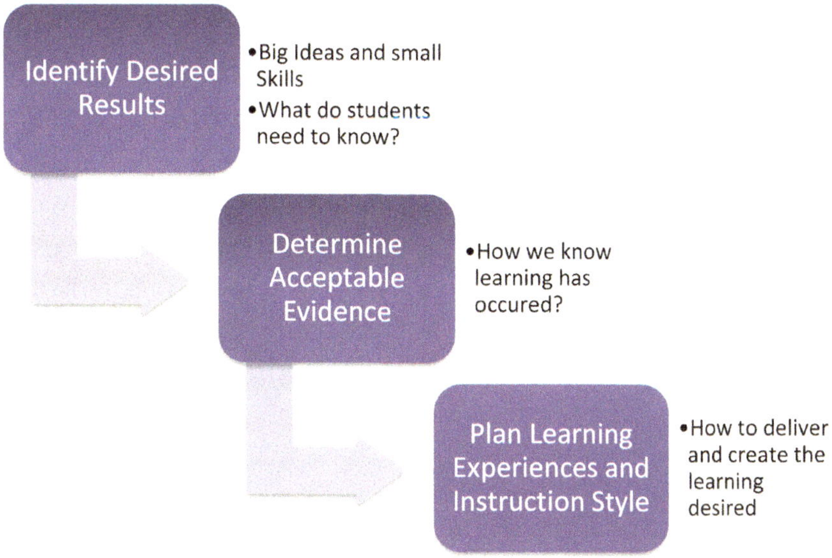

In "Backward Design" defining the ultimate learning goals of the course is the starting point of online course design for every course. Answers to the above questions will assist in developing an effective course design; its content and pace, assignments, activities, assessments, as well as choosing the appropriate technology (Caulfield, 2011).

Assignment Progression

Assignment progression means a design of assignments which will have multiple deliverables which evolve as the course progresses. After the course designer organizes the content into modules or units, the task turns to building course assignments which can be repurposed and improved or added to as the students gain understanding and proficiency. This allows students to think about the new knowledge they are presented with, reflect and evaluate it, discuss their ideas and hear others', and then apply or integrate that new information into their 'new normal'. The goal is to design an assignment that can cycle through this process more than once during the duration of a course.

Designing a few assignments in a course which can go through this progression cycle more than once also allows the students to immerse themselves in the content in such a way that it requires their attention and leads to depth of understanding. Assignment progression creates learning experiences where students must demonstrate their understanding by making meaning and by connecting and reconnecting their knowledge over time.

Active Engagement

Confucius over 2000 years ago knew active engagement was the key to deep learning and analytical ability. He said, "Tell me and I will forget, show me and I will remember, involve me and I will understand." Vygotsky (1978) argued learning occurs when students play an active role in learning. In this

Figure 4. Assignment progression cycle

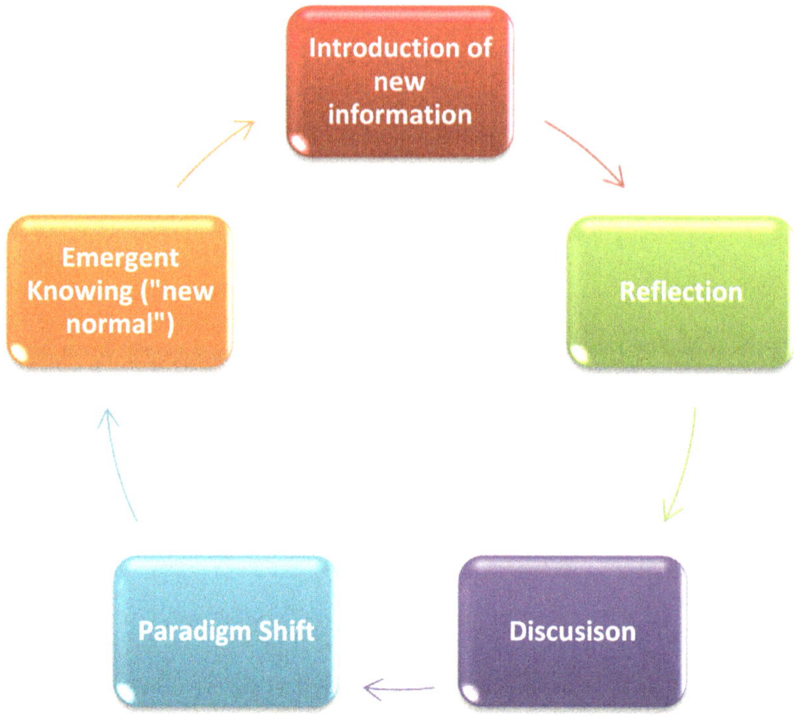

way, the traditional roles of student and teacher may be reversed: the teacher collaborates with students to bolster and to support their engagement with and integration of the course content. Learning therefore becomes a reciprocal and interactive process between the students and the students and the teacher. Actively involving students in their own learning, means creating learning environments that requires "…students to draw from their experiences and other disciplines, where lateral thinking is rewarded, where collaboration rather than competition is celebrated, and where failure is understood to be part of the journey to success…" (Helfand, 2015). Students learn best when they are actively engaged and given the opportunity to take responsibility for their own learning. "Learning is not a spectator sport… [Students] must talk about what they are learning, write reflectively about it, relate it to past experiences, and apply it to their daily lives. They must make what they learn part of themselves…" (Chickering and Ehrmann, 1996, p. 5).

Pillar Two: Creating Communities

Engaged Learning Communities

A recent popular topic looks to the benefits of learning communities in online learning environments. Several studies have found a significant correlation between student collaboration and true learning (Palloff & Pratt, 2013). The online environment adapts to the specific learning styles and schedule constraints of learners which yields more effective learning opportunities than face-to-face courses. Moreover, online discussions can encourage whole-class participation by creating a "level playing field" for the students who typically might not speak up (or have the chance to do so) in face-to-face settings. For example, in the online setting, questions or assignments are provided in advance so students have equal opportunities to compose thoughtful responses and have their voices heard. Additionally, the online classroom provides ample time for careful and thoughtful exchanges which have different effects than a face-to-face session where time for discussion is minimal or, depending on the instruction style, not at all.

This is borne out in studies which attempt to explain and predict the power of learning communities. Bandura's (1977) classic theory of social learning emphasized the correlation between the environment and individual behavior. He wrote, "most human behavior is learned observationally through modeling; from observing others one forms an idea of how new behaviors are performed and on later occasions this coded information serves as a guide for action" (p. 67). Palloff and Pratt (2013), in a recent study described the benefits of learning communities as not only assisting collaborative learning but also the additional benefit of reducing any feelings of isolation when working at a distance.

In another recent study at Columbia University, Wong and Faller (2014) discovered that students experienced deep learning as a direct result of the online learning community. They noted "the cohort structure contributed and impacted the sense of belonging and group identity building, providing the context for critical conversations" (p. 246). Moreover, they found "students looked at the cohort as space for collaboration, sharing, supporting one another, building personal and professional relationships as well as the space to be open to new ideas and challenge assumptions" (p. 246). This phenomenon of learning in the online community can be explained due to the high quantity and quality of interaction, trust, openness and support that leads to new meaning, understanding and awareness (Wickersham & Dooley, 2006). Developing a learning community and community of care is noted as effective for higher order learning (Garrison & Archer (2000); Southern, 2007). When individual learners engage in an online learning community with a high amount of interaction, developing new meaning or critical reflection and understanding through collaborative discussion is more likely (Wickersham & Dooley, 2006). Similar conclusions were reached by Lee & Nicolaides (2014),

> *...conscious and continuous dialogue enables the group to explore underlying assumptions of their meaning making, connect new ideas and furthermore build new ideas collaboratively... Indeed, more importantly, it was observed that the collaborative reflexive inquiry made space for learning that guided students to extend their own knowledge and understanding (p. 276).*

Provocative Material

A mistake which often leads to student attrition in online courses or programs is simply converting print materials for an online course or posting videos of recorded lectures. The online course becomes a mere repository for documents which students rifle through, read, understand and learn from in isolation. This makes learning difficult and can even discourage learning. The content that makes its debut in an online course shell should come alive to the students, have movement, provoke reactions, inspire thoughtful dialogue and cause the students to see and understand their world a little differently. This can be done in two ways. First, instructors and designers should consider integrating a variety of content and media sources that inspire deep thought and engage all learning styles and learning preferences. The Internet provides an abundance of interactive resources that can be used to increase engagement and appeal to each student's learning style or learning preference. Interactive resources include videos, podcasts, screencasts, video conferencing, assessments, articles, blogs, and presentation software.

The second method of introducing provocative material is for instructors and course designers to allow students the opportunity to respond and reflect on one another's work. This provides students opportunities to understand the same content from another perspective. One recent study (Galloway, 2016) examined the factors that supported transformation of students in an online social justice course. "All of the student-participants of this study shared how they grew because of their experience with the content and their experience of listening to and reading about their classmate's experience with the content. This perspective-shift allowed students to see the world through a new set of eyes, exposing new and unknown dimensions of the content (social justice) and themselves." (Galloway, 2016, p. 58)

Pillar Three: Supporting Authenticity

Authentic Conversations

Validating the "real" experiences of students' lives is a key part of the puzzle of finding ways of keeping students engaged and enrolled. Allowing students to report out on the content and explain how it applies (or does not) to their past or current life experiences makes the learning process relevant and personal. This is important, because of what we know of how and why students attend and finish college; it will help them address personal and professional opportunities. Thus, creating learning activities that are complex, yet also related to real-life experiences that can also be applied to future activities should be one of the main driving forces for design and facilitation of the online classroom (Knowles, 1975)

Studies support the importance of this Pillar. Southern (2007), described the connection between authentic dialogue and deep learning. Additionally, Taylor (2009) noted that the presence of authenticity in both the faculty and learner fosters trusting relationships which provides the conditions and safe environments for deep learning to take place. Similarly, Cranton (2006) suggested the notion of authenticity as a key component of the transformative relationship. These studies have in common the notion that students remain connected to their learning when it is "real". In other words, the quality of relatedness to the content and authenticity with self and others creates an internal motivation and sense of profound responsibility to their own learning. The following excerpt from a study also shows the importance of the authentic learning experience to students' success (Galloway, 2016, p.70).

Interactions between students that were rooted in authenticity, trust, risk, and acceptance were important to each participant in this study. This also allowed students to see oppression in a new light, through the eyes and experiences of other students. Several students expressed how their transformations began because of the vivid and real experiences with oppression. For example, when one student would explain his or her experience with being the oppressed, another student would begin to realize his or her own participation in the oppressive system as the oppressor. The authentic nature of these interactions pulled students into a shared experience from opposing viewpoints. This created a new level of commitment to the learning process as students began to understand how their contributions were vital for others' development as well.

Additionally, when students are given the opportunity to examine, reflect, share, and shift their ideologies, they learn new skills, new aspects of themselves, new abilities and new ways of being. To them, this kind of "real" learning, engages and creates the kind of internal motivation which keeps them enrolled.

Open-Ended Options vs. Restrictions and Uniformity

"I think there is a direct correlation between the amount of restrictions we place on students and their lack of interest in what we are doing" Sean Michael Morris (Stachowiak, 2016). There is no such thing as a neutral education process. The field of critical pedagogy is built on the premise that learning is never neutral and it is based on the notion that learning -- true learning -- should find the edge of a student's knowledge and nudge that boundary. Recent research proves that the mature learner is both capable and inspired by this type of learning design. Teaching which includes a focus on critical consciousness or what Freire (1998) refers to as "emancipatory" facilitation, focuses on students achieving an in-depth understanding of their current situation in the context of the world and others around them. While some college subjects are easier to achieve this level of learning with than others, there is always some element of the critical conscious pedagogy that can be integrated. This is where the content of an online course can break away from the flat "cram and clutter" and "sage on the stage" recorded lecture format. Course developers and online instructors must consider how to evoke a deeper sense of critical consciousness in and throughout their courses. The precursors for transformative learning can serve to be an excellent guide to presenting and designing online content. While the following are considered 'precursors' or evidence of deep learning, course designers and instructors could use the following as a list of possible assignment design ideas:

- Disorientating dilemma
- Critical reflection on assumptions
- Recognized discontent shared through dialogue,
- Exploration of new roles
- Self-examination of assumptions (usually expressing some level of shame or guilt)
- Tried a new role
- Planned and/or implemented new courses of action
- Acquired knowledge/skills to implement/plan
- Built competence/confidence and
- Reintegration to life (Brock, 2010; Mezirow, 2000, 2009; Snyder, 2008)

Course developers and designers can review the above list to help inform the development of assignments, or use the list as a checklist for an online course. For example, an undergraduate business course could ask the student to write an analysis on a situation from the perspectives of a CEO, the Human Resource Director and a front-line employee. This 'exploration of new roles' would require the student to move outside their unique perspective and begin to understand the larger and varying dilemmas from different perspectives on a single issue.

Pillar Four: Facilitating Thinking

Higher education should be in the business of opening and preparing young minds for the adaptive challenges (socially, economically, financially, politically, environmentally, etc.) they will face in their lifetime. Creating critical thinkers is the common purpose that connects every course.

The Art of Powerful Questions

Thoughtful and provocative questions will foster inquiry, understanding, and transfer or application of learning. Thoughtful questions push students to think beyond their present knowledge or understanding. Powerful questions can come from the instructor or from other students in the course. Often the powerful questions and the initiation of a transformative learning event come from the students themselves and from rich and engaging student to student interactions within the course. Below is an excerpt from a recent study on student transformative learning in a blended learning environment where a student described the impact faculty questions had on their learning experience (Galloway, 2016, p. 65),

I want to give a nod to the powerful questioning of the faculty. There were some real pivotal moments and the faculty did not let me off the hook...early on there were a couple of moments and I could have slid by but they kept probing. The questions forced me to go deeper and it was not comfortable, but it forced me to go places I was anxious to go. And it was really, important to go there and I gained some incredible insights from that. So, courageous faculty and supportive, creating a supportive environment for all of us to go to some dicey places together.

The use of powerful questions in an online course can help students connect theory and abstract concepts to their own lived. Such questions are helpful for focusing attention, connecting ideas, discovering deeper meaning and creating forward movement.

The art of crafting powerful questions can create deep learning experiences for students and can make a critical difference in whether students' minds are opened to or narrowed to possible considerations. Consider the following powerful questions continuum to understand the difference between less and more powerful questions (Vogt, 2003):

With movement from left to right, from simple yes or no questions to how, what and what if, the questions stimulate more critical and creative thinking and produces more authentic responses. A caution about "why" questions: unless a why question is carefully constructed, it can evoke a defensive reaction and response from the receiver. A "why" question can often be interpreted as a request to justify one's position or opinion as opposed to fostering a spirit of curiosity and exploration. Below is a sample list of powerful questions which could be effective in provoking deep and critical thinking.

Asking powerful questions in an online course is the most effective method for creating a dynamic, engaging and personalized learning experience. These types of stimulating questions inspire students to think about and feel their learning, which is exactly the kind of intrinsic motivation necessary for long-term success.

FUTURE RESEARCH DIRECTIONS

Students are choosing online education at record rates (Babson, 2015; Parker, Lenhart, Moore, 2011) yet approaches to online course design that are retention focused and methods for designing courses with retention and student success in mind are in still in the beginning stages. Research studies focused on online course design methodology and how various approaches impact retention and student success

Figure 5. The benefits of powerful questions

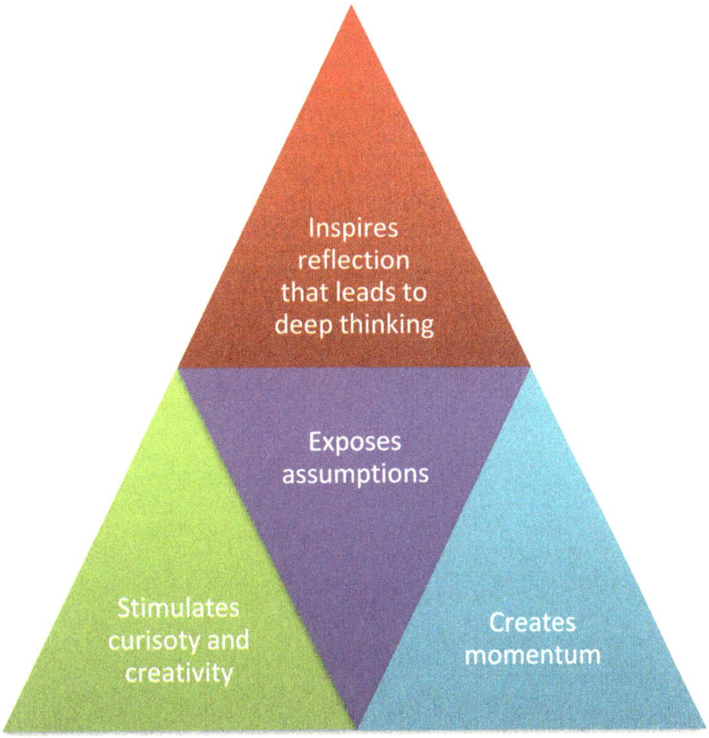

Figure 6. Continuum of powerful questions

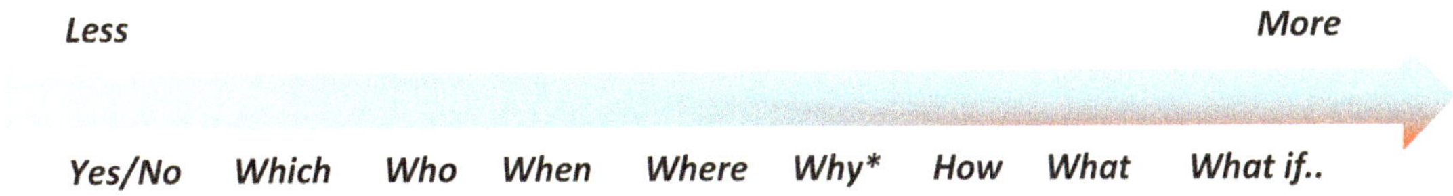

across a range of institution and program type are needed. It will also be important to study and understand how various online course design approaches and methodology impact different student types, different learning preferences and different learning communities. While this author expects that some methods and approaches will be effective with some types of learners and some types of content and subject areas, this is an unexplored area and provides scholars with a range of options and study designs as the field begins to develop and take shape.

CONCLUSION

Students are choosing online education at record rates (Babson, 2015; Parker, Lenhart, Moore, 2011). Consequently, institutions of higher education are offering online courses at record rates and plan to continue to do so into the future (Babson, 2015). Employing online learning systems means universities

Table 2. Examples of powerful questions

Goal of Question	Examples of Powerful Questions
Reflection and Deeper Meaning	What is important (to you) about _____ and why do (should) you care? What has been your major learning insight or discovery so far? What assumptions do we/you need to test or challenge about ____? What is grabbing your attention or focus? What is distracting you from progress?
Expands Possibilities	What would someone with very difference belief/experiences than you say about ____? What are the major dilemmas/opportunities here? What do we/you need to know for you to proceed? What is possible?
Creates Momentum	What opportunities do you foresee? What is the next level of thinking needed? What would it take to create change on this issue? How can we support the necessary next steps?
Simulates Curiosity and Creativity	What is missing from the picture so far? What is it we are not seeing? What do we need more clarity about? What is taking shape? What is emerging? What new connections are you making?

Table 3. Best practices for online student retention

	Guidelines for Design	Guidelines for Faculty
PILLAR ONE BUILDING LEARNING -Backward Design - Assignment Progression	Create an arc to the course. Start with the end in mind and build assignments that will ensure learning.	Carefully craft your responses and feedback. Resist the temptation to be a "sage on the stage".
	Create multiple deliverables that evolve as the course progresses. Give students time to reflect on the material and use the new knowledge and apply to a multi-week assignment.	Feedback and interactions should be student-centered, not lecture based.
	Create enough structure so students feel safe because expectations of deliverables, timelines and behavior are clear and predictable.	Allow students to direct and experience their own learning.
PILLAR TWO CREATE COMMUNITY -Engaged learning communities -Provocative material	Design for students to have authentic interactions with the content and each other.	Encourage and foster quality student-to-student interactions founded on trust and respect that create shared meaning.
	Assign a variety of reading materials with both personal stories and theory based – with contrasting opinions in content.	Encourage students to direct and lead conversations both internally and externally.
	Provide diverse, provocative material that encourages students to look at old paradigms through new lenses.	Be patient, learning is hard and unpredictable; it is courageous work (sometimes the smallest event can create a large change).
PILLAR THREE SUPPORT AUTHENTICITY -Authentic conversations - Open-ended options vs. uniformity	Design at least one assignment integrating one or more of the 'precursors' or evidence of deep learning.	Encourage students to refer to and apply their learning to past or current life experiences.
	Design for students to have authentic interactions with the content, instructor and each other.	Challenge and reward students for confronting and evaluating their current and often hidden assumptions.
PILLAR FOUR FACILITATE THINKING -The art of powerful questions	Design assignments that encourage students to focus their attention, gain deeper insight and meaning, and create forward movement.	Teach and guide through powerful questions that encourage students to reflect inward.
	Design course assignments that employ the powerful questions framework and philosophy	Recall your first experiences learning how to teach and navigate the online classroom and remember that is what your students might be experiencing too.

can access students across time zones, counties, borders and continents. However, the use of technology for online learning is infiltrating higher education and in many cases the technology is moving faster than the instructional design and practices. The first models of online course design were not driven by student engagement and retention. Rather, their focus was based upon convenience by attempting to expose students to material by cramming the content from the face-to- face syllabus into an online course shell. Recent research has proven that early attempts at online course design and instruction yielded high levels of student attrition.

Universities need to take measures which will encourage students to stay enrolled and complete their degrees. A logical place to begin is to improve the student learning experience in the online formats. Palloff and Pratt (2000) argue "technology does not teach students; effective teachers do" (pg. 4). They emphasize the point that technology by itself does not create effective learning; rather it is the effective use of the technology in the design and implementation of the technology. As institutions and educators continue to create and offer learning programs online, they will consequently need to modify their design and facilitation practices. The 21st century learning model requires fresh elements of design that incorporate the newest philosophies of adult student learning. Specifically, retention and student success should be at the core of course design philosophy. In this chapter, we discussed the four pillars for designing and facilitating online learning practices for increasing student learning and retention. Below is a graph that highlights some of the main points to consider for online course design and facilitation.

REFERENCES

Allen, I. E., & Seaman, J. (2013). *Changing course: Ten years of tracking online education in the United States*. Babson Park, MA: Babson Survey Research Group.

Angelino, L. M., Williams, F. K., & Natvig, D. (2007). Strategies to engage online students and reduce attrition rates. *The Journal of Educators Online*, *4*(2), 1–14. doi:10.9743/JEO.2007.2.1

Ausburn, L. J. (2004). Gender and learning strategy differences in non-traditional adult students' design preferences in hybrid distance courses. *Journal of Interactive Online Learning*, *3*(2). Retrieved from http://www.ncolr.org/jiol/issues/pdf/3.2.6.pdf

Babson (2015). 2015 Online Report Card- Tracking Online Education in the United States. Online Learning Consortium. Retrieved from http://www.babson.edu/Academics/faculty/provost/Pages/babson-survey-research-group.aspx

Bandura, A. (1977). *Social learning theory*. New York, NY: General Learning Press.

Bollag, B., & Overland, M. A. (2001). Developing countries turn to distance education. *The Chronicle of Higher Education*, *47*(40), 29–31.

Brock, S. E. (2010). Measuring the importance of precursor steps to transformative learning. *Adult Education Quarterly*, *60*(2), 122–142. doi:10.1177/0741713609333084

Caulfield, J. (2011). *How to Design and Teach a Hybrid Course*. Sterling, VA: Stylus Publishing.

Chickering, A. W., & Ehrmann, S. C. (1996). Implementing the seven principles: Technology as a lever. *AAHE Bulletin*, *49*(2), 3–6.

Cranton, P. (2006). *Understanding and promoting transformative learning: A guide for educators of adults* (2nd ed.). San Francisco, CA: Jossey-Bass.

Fisher-Yoshida, B. (2009). Transformative learning in participative processes that reframe self-identity. In B. Fisher-Yoshida, K. D. Geller, & S. Schapiro (Eds.), *Innovations in transformative learning: Space, culture and the arts* (pp. 43–65). New York, NY: Peter Lang.

Freire, P. (1998a). *Pedagogy of the heart*. New York, NY: Continuum.

Galloway, L. M. (2016). Expanding spaces for transformation: Supporting social justice in a distributed learning environment [Dissertation]. Fielding Graduate University.

Garrison, D. R., & Archer, W. (2000). *A transactional perspective on teaching-learning: A framework for adult and higher education*. Oxford, UK: Pergamon.

Garrison, D. R., & Cleveland-Innes, M. (2003, September). Critical factors in student satisfaction and success: Facilitating student's role adjustment in online communities of inquiry (Invited paper). *Presented to the Sloan Consortium Asynchronous Learning Network Invitational Workshop*, Boston, MA.

Garrison, D. R., & Kanuka, H. (2004). Blended learning: Uncovering its transformative potential in higher education. *The Internet and Higher Education*, *7*(2), 95–105. doi:10.1016/j.iheduc.2004.02.001

Hicks, M., Reid, I., & George, R. (2001). Enhancing on-line teaching: Designing responsive learning environments. *The International Journal for Academic Development*, *6*(2), 143–151. doi:10.1080/713769258

Kegan, R. (2000). What "form" transforms? A constructive-developmental approach to transformative learning. In J. Mezirow et al. (Eds.), *Learning as transformation: Critical perspectives on a theory in progress* (pp. 35–70). San Francisco, CA: Jossey-Bass.

Knowles, M. (1984). *The adult learner: A neglected species*. Houston, TX: Gulf.

Knowles, M. S. (1975). *Self-directed learning*. New York, NY: Association Press.

Knowles, M. S. (1980). *The modern practice of adult education (revised and updated)*. Chicago, IL: Association Press. (Original work published 1970)

Lee, J. E., & Nicolaides, A. (2014). Exploring transformative relationships in a virtual learning environment. In *Proceedings of the Adult Education Research Conference* (pp. 273-278). Kansas State University Libraries, New Prairie Press.

Mezirow, J. (1981). A critical theory of adult learning and education. *Adult Education Quarterly*, *32*(1), 3–24. doi:10.1177/074171368103200101

Mezirow, J. (2009). Transformative learning theory. In J. Mezirow & E. W. Taylor et al. (Eds.), *Transformative learning in practice: Insights from community, workplace, and higher education* (pp. 18–32). San Francisco, CA: Jossey-Bass.

Moore, J. L., Dickson-Deane, C., & Galyen, K. (2011). e-Learning, online learning, and distance learning environments: Are they the same? *The Internet and Higher Education*, *14*(2), 129–135. Retrieved from https://scholar.vt.edu/access/content/group/5deb92b5-10f3-49db-adeb-7294847f1ebc/e-Learning%20 Scott%20Midkiff.pdf doi:10.1016/j.iheduc.2010.10.001

Moore, M., & Kearlsey, G. (2012). *Distance education: A systems view of online learning*. Belmont, CA: Cengage Learning.

Palloff, R. M., & Pratt, K. (2013). *Lessons from the Virtual Classroom* (2nd ed.). San Francisco, CA: Jossey-Bass.

Parker, K., Lenhart, A., & Moore, K. (2011). *The digital revolution and higher education: College presidents, public differ on value of online learning*. Retrieved from http://www.pewinternet.org/~/media// Files/Reports/2011/PIP-Online-Learning.pdf

Rudestam, E. K. (2004). Distributed education and the role of online learning in training professional psychologists. *Professional Psychology, Research and Practice*, *23*(4), 427–432. doi:10.1037/0735-7028.35.4.427

Russell, S. S. (2006). An overview of adult learning processes. *Urologic Nursing*, *26*(5), 349–352, 370. Retrieved from http://www.medscape.com/viewarticle/547417_2 PMID:17078322

Snyder, C. (2008). Grabbing hold of a moving target: Identifying and measuring the transformative learning process. *Journal of Transformative Education*, *6*(3), 159–181. doi:10.1177/1541344608327813

Southern, N. L. (2007). Mentoring for transformative learning: The importance of relationship in creating learning communities of care. *Journal of Transformative Education*, *5*(4), 329–338. doi:10.1177/1541344607310576

Stachowiak, B. (2016, June 2). *Critical Instructional Design with Sean Michael Morris. Teaching in Higher Education Podcast*. Retrieved from http://teachinginhighered.com/podcast/critical-instructional-design/

Tedx Talks. (2013, June). Designing a university for the new millennium: David Helfand at TEDx West Vancouver ED (YouTube video). Retrieved from https://www.youtube.com/watch?v=DZQe73IXZtU

U.S. Department of Education. (2010). Evaluation of Evidence Based Practice in Online Learning. A Meta-Analysis Review of Online Learning Studies. Retrieved from https://www2.ed.gov/rschstat/eval/ tech/evidence-based-practices/finalreport.pdf

Vogt, E., Brown, J., & Issacs, D. (2003). *The Art of powerful questions: Catalyzing insight, innovation, and action*. Mill Valley, CA: Whole Systems Associates.

Wickersham, L., & Dooley, K. (2006). A content analysis of critical thinking skills as an indicator of quality online discussion in virtual learning communities. *Quarterly Review of Distance Education*, *7*(2), 185–193.

Wiggins, G., & McTighe, J. (2005). Understanding By Design. The Association for Curriculum and Development. https://www.amazon.com/Understanding-Design-Grant-Wiggins/dp/1416600353/ref=sr_1_1?ie=UTF8&qid=1488596859&sr=8-1&keywords=wiggins+understanding+by+design#reader_1416600353

Wong, M. L., & Faller, P. (2104). Creating time and space for transformative learning in a cohort-based adult education doctoral program. In *Proceedings of the XI International Transformative Learning Conference*. Teachers College, Columbia University, New York, NY.

Yorks, L., & Kasl, E. (2002). Toward a theory and practice for whole-person learning: Reconceptualizing experience and the role of affect. *Adult Education Quarterly*, *52*(3), 176–192. doi:10.1177/07417136020523002

KEY TERMS AND DEFINITIONS

The Art of Powerful Questions: Thoughtful and provocative questions foster inquiry, understanding, and transfer or application of learning. Thoughtful questions push students to think beyond their present knowledge or understanding. The art of crafting powerful questions can create deep learning experiences for students and can make a critical difference in whether students' minds are opened to or narrowed to possible considerations.

Assignment Progression: Assignment progression means a design of assignments which will have multiple deliverables that evolve as the course progresses. Course assignments are designed to be repurposed and improved upon or added to as the students gain understanding and proficiency.

Backward Design: In "Backward Design" defining the ultimate learning goals of the course is the starting point of online course design for every course. It flips the classic approach, and begins with the end in mind. It considers *what* is to be taught first, and *how* it should be delivered second. *How* the content is delivered should be adapted to the overall purpose or intent of the learning outcomes.

EdX: Harvard and MIT launched the first MOOC platform in 2010, called "edX." It was a joint effort of Harvard and MIT to bring elite virtual education to anyone in the world with an Internet connection.

Four Pillars of Online Course Design: The Four Pillars of Online Course Design are 1) Building Learning, 2) Creating Community, 3) Supporting Authenticity, and 4) Facilitating Thinking.

MOOCs: Massive Open Online Courses or MOOCs. The main idea was that a few star professors would record online lectures to thousands, thereby enabling sharing of their knowledge and enthusiasm for the subject matter with a global audience.

Chapter 15
Realizing Their Potential, Shaping Their Greatness:
Using Self-Efficacy in a Graduate Reading Program to Shape K-6 Student Success

Allyson Leggett Watson
Northeastern State University, USA

Meagan Moreland
Northeastern State University, USA

Melinda Smith
Northeastern State University, USA

ABSTRACT

Research literature in public education often categorizes self-efficacy in order for practitioners of K-12 academic settings to understand how to relate to students. They often view self-efficacy in students' perception, intrinsic motivation and behavior in academic settings. This chapter utilizes the research literature from public education as a bridge to discover the importance of self-actualization of efficacy in developing reading specialists. Too often, students in public education who have been referred to a reading specialist due to struggles with reading, low academic performance or a high dropout risk are missing a level of self-efficacy. In that same context, the authors in this study looked to a specific program which prepares teachers of those students and incorporated rigorous self-efficacy scales to inform and shape practice. This chapter asserts that teachers are paramount in the successful development of an at-risk learner. This chapter provides empirical research centered around two core programs, undergraduate and graduate with specialization in urban education and reading.

DOI: 10.4018/978-1-5225-2998-9.ch015

INTRODUCTION

Self-efficacy, effective learning skills and resilience in retention and student success are critical as a focus for higher education institutions and specialized programs. Bandura (1977) informs us that self-efficacy relies heavily on one's own belief about themselves and how to truly walk in your own belief toward personal attainment. This chapter highlights a regional state university's graduate and undergraduate reading program and identifies the use of self-efficacy to develop and inform pedagogical practices. Within this graduate program, the candidates are serving in public schools in an instructional capacity. The graduate coursework and design centers around preparation for reading specialists and enhanced reading pedagogy for high need learners. The demographic region is closely tied to rural, urban and high poverty K-12 student populations. Moreover, the demographic region includes many underfunded schools and many marginalized and overlooked students.

In this chapter, we classify students as those in the K-12 public and private education system. The term pre-service teacher candidate or reading candidate is used when referring to undergraduate and graduate students within our program. While this chapter focuses mainly on graduate students, we feel it is necessary to highlight the researchers bridge the undergraduate teacher preparation in conjunction with the goals and outcomes that are desired for all students in the program. The goal of this chapter is to delineate an approach to using self-efficacy with graduate and undergraduate students who are studying to be reading specialists. The goal is to so help them understand themselves and their own beliefs prior to shaping and cultivating knowledge in public school classrooms.

The coursework and design of the program center on preparing future teachers and master educators in specialized reading areas. The experiences shared come from the collective work to build self-efficacy in future teachers and the efforts to understand the role teachers' play in the outcomes of K-6th grade students. The research emphasis is twofold. Primarily, the research addresses the notion of teacher preparation in reading which is critical to students served in the elementary school setting. Secondly, the research extends the role of teacher preparation to address the needs of the K-12 students they will serve. By teaching self-efficacy through reflection, our undergraduate and graduate reading program candidates will be more likely to have positive and impactful interactions with their students. They will also be less prone to making assumptions and generalizations, having implicit biases, and to making negative judgments about their students.

BACKGROUND: OUR STUDENTS, OUR STORY

The University. The heritage of our institution was founded on the premise of providing equal education for women regardless of their race or class. We began as a female seminary, with roots stemming from a rich and deep Cherokee tribal tradition. Our institution has provided public education in northeastern Oklahoma since 1909. Our historical accounts trace back to the 1800's when women traditionally were not formally educated. The institutional historian, Dr. Brad Agnew, shares this history in his book, Roots of the Cherokee. The institution records the history on the webpage in the following account:

In 1846, Norland State University was founded as the Cherokee National Female Seminary. This historic link to the Cherokee Nation and Indian Territory make Norland the oldest institution of higher learning in the state. The original Seminary was built in Park Hill south of Tahlequah and destroyed by fire on

Easter Sunday 1887. On May 7, 1889, the Cherokee National Female Seminary was dedicated north of Tahlequah, where it would continue to provide learning opportunities for young women for 20 years. Seminary Hall is now the historic centerpiece of Norland State, and each year on May 7, Descendants of Seminarians gather to observe Seminaries Homecoming in honor of our first students.

Our institution prides itself on the students we prepare for the local and regional workforce. There are many unique language and cultural programs and centers on campus to support and increase knowledge of our Cherokee roots. As an institution serving the largest population of tribal students, we work intentionally to increase retention, recruitment and completion efforts. Due in part to our rural location and the demographic population of students, 33% of the students we serve are non-traditional and come to college after spending years in the business, labor or child care industry. For this reason, the College of Education, has created courses and programs in teacher preparation which meet the needs of non-traditional students and working adults without sacrificing academic rigor or program expectations.

The College of Education. As one of the leading teacher preparation colleges in the state, the College of Education has the largest number of teacher educators. The College of Education prides itself in shaping undergraduate pre-service teachers and graduate reading candidates into impactful and professional educators. Archival data from the university and the College of Education's historical records provide the founding service philosophy:

Our history as a state institution began on March 6, 1909, when the state legislature purchased the Female Seminary and created Norland State Normal School, where the first classes were held on September 14, 1909. A decade later, Norland State Normal School transformed into Norland State Teacher's College, offering a four-year curriculum leading to a bachelor's degree.

The college founding began in one building that still serves as a reminder of the institutions cultural heritage. Today the reach of the institution has expanded across the region and the School of Education now serves students in three distinct metropolitan areas Tahlequah, Muskogee and Broken Arrow. The main campus is in Tahlequah and the College of Education is one of the largest colleges within the institution. Additionally, the most urban branch campus, Broken Arrow, shares the College of Education's mission and focuses on providing graduate and evening classes for non-traditional working adult students.

THE READING CLINIC

The reading clinic provides a space for students to tutor in a supervised practicum experience. Moreover, the reading clinic serves children with reading needs from across the local community. Funding for the clinic was provided by a local tax plan meant to help with economic growth. The reading clinic was completed in 2008 primarily as a result of efforts by the faculty and extensive administrative support. The clinic is a vibrant space designed to serve many stakeholders within the university and local community. Currently, due to the high demand for outreach services, there is a waiting list of over 250 children requesting reading support and assistance.

DEVELOPING THE RESEARCH STUDY

This collaborative evaluation and longitudinal study is the result of over a decade of focused work developing courses within the reading programs. The work has focused on establishing intended academic outcomes and then assessing and reevaluating over time to ensure the intended academic outcomes and the intended impact. The program and its success would not have been possible without extensive collegial faculty collaboration. Faculty members developed a sense of pride in working collaboratively to establish program guidelines and curricular changes that provided the most impactful programs for graduate and undergraduate reading program candidates.

RESEARCH NARRATIVES

One element of our college mission is to be a source for developing teaching professionals to work as future leaders, scholars and developers of human potential. To meet this element of the institutional mission, faculty members in the College of Education are intentional about how they incorporate approaches, practices and strategies that lead our reading program graduates to a path of success with their own students and during their initial teaching assignments. In exploration of this element of the mission the study uses a narrative inquiry methodology.

REVIEW OF THE LITERATURE

Through narrative inquiry, the study seeks to detail and examine the descriptive process of teaching and student impact. Clandinin (2006) gives a clear indication of the importance and the distinctness of narrative as a method of inquiry. In the Handbook on Narrative Inquiry, he states, "there is a distinction between storytelling and the inquiry of the living story" (p. xii). The following narratives provide detailed personal accounts regarding how each of the co-authors identify and cultivate a practice of reflection through self-efficacy in undergraduate and graduate reading program students. These narratives speak to the deep level of thought and consideration the program encourages our developing professionals to have throughout the program. The study identifies key areas within various program courses which allow undergraduate and graduate candidates to envision the population and characteristics of the K-6th grade students they will be teaching. Through the lens of self-reflection, pedagogical practice and new and innovative approaches, the study works to increase student retention in the reading programs and to impact student success in K-6 reading proficiencies.

The importance of pre-service and career teachers being intentional and applying unbiased belief systems that all students can learn and achieve in the classroom cannot be understated. Bandura's (1977) work provides strong and explicit findings that values and belief systems of teachers directly impact into outcomes and student performance. Lee and Johnson (2016) note, "applying the concept of self-efficacy to academic settings, academic self-efficacy is formed by actual performance, experiences, persuasion from others, and physiological reaction" (p. 79). The following passages illustrate the study's recommendations regarding fostering self-efficacy in meaningful and impactful ways.

SHIFTING THE PARADIGM USING SELF-EFFICACY TO DISMANTLE MYTHS IN URBAN EDUCATION

Building on an approach to using reflection through pedagogy begins early in the reading programs as it was clear that the undergraduate pre-service teacher candidates needed to better understand reading subject matter content. In addition to being content proficient, pre-service teacher candidates also needed to understand their own level of efficacy. The School of Education responded to these needs by developing a specific program designed to teach strategies and provide tools for successful teaching of reading in urban school settings. The program focuses on sharing theoretical research and practical applications of what teaching methods work and how they work. Additionally, the program intentionally leads the students through a transparent discovery of understanding the community and the trials and triumphs that the students within the community and school are facing. The program is selective and only students who initiate a desire to work in high poverty schools are accepted. The selected students begin their initial internship and take the program core courses at their school district offices or at their elementary school.

Students are initially given the self-efficacy scale developed by Siwatu, Frazier, Osaghae and Starker (2011). Based on the work of Siwatu, et al, students are taught the importance of being steadfast in developing culturally responsive classrooms. Reflecting on the writings of Siwatu et al. the literature regarding high poverty schools and successful practices led the program to establish a core set of critical resources pertinent to urban education. Siwatu et al. established a groundwork for teacher preparation while keeping the K-12 students in mind. In their 2011 research they shared the following: "preparing pre-service teachers to be successful in the most formidable educational climates and creating a shift in foundational coursework to enhance cultural responsiveness, enriched teaching practice and self-efficacy could potentially impact the future of teacher retention in urban schools" (p. 211).

Utilizing research on self-efficacy for pre-service teachers, the program developed specific pre and post self-efficacy screening for use the urban elementary school reading programs. Study data indicated that the pre-service teachers had a passion to teach in high poverty areas, but that they were not aware that their personal biases and values related to poverty or cultural differences were impacting their teaching efficacy. Kelley, Siwatu, Tost, and Martinez, (2015) inform the literature by providing an important finding. "The core of culturally responsive teaching is utilizing a pedagogy that is inclusive of the students' culture in order to demonstrate utility of the content and to improve academic success" (p. 294). The most recent research in this area has developed out of decades of magnified views around self-efficacy and culturally responsive teaching. The work of Gay, (2000, 2002, and 2004), Ladson-Billings, (2006) and Siwatu, (2007, 2009 and 2011) leads researchers and practitioners to the illuminating and importance need for developing culturally responsive and efficacious teachers. The study findings highlight the importance of combining reflection with pre-service teachers in the urban focused program. Previous research by Watson (2017), summarizes the reflection journey as follows:

The integral part of the development and growth process for the preservice teachers is having the conversation with a faculty facilitator and discussing ideas and reflective thought within the internship and course process. The preservice teachers take anecdotal notes and share them in a weekly journal log. The faculty facilitator selects three to four samples and reads them aloud in class. As a class, the pre-

service teachers then share equally and agree, disagree or offer other suggestions for the shared journal entries. This conversational engagement takes place for an hour at the end of the course day. The set aside reflection time helps the preservice teachers gain a sense of closure after each internship day and develop a community within the classroom. The supportive environment makes difficult conversations less overbearing and more positive. As the preservice teachers delve further into conversations about diversity and culture they become more open to learn. (Watson, 2017)

THEORETICAL CONNECTIONS TO REAL LIFE PRACTICE

The graduate reading candidates begin their program in a foundations course that focuses on immersive study of literacy and learning theory. This introductory course offers an overview of the major theories and models regarding the reading component of literacy learning. Through exposure and experiences with these theories and models, the goal is to provide the candidates with the foundation to enhance their instruction through deeper understanding of these theoretical frameworks. By the end of the course, the candidates are able to articulate why they believe the way they do and how those beliefs are reflected in their classroom practice.

Two signature assignments provided in this course are designed to facilitate achievement of this goal. One of the texts used in the course is *Lenses on Reading* (Tracey & Morrow, 2012). This text provides an in-depth look at major literacy theories and models. Using this text, the candidates reflect on how the reading relates to their own teaching practice (text to self), how it relates to other professional reading they have done (text to text), and how the reading relates to the practice of language arts teaching in general (text to world). The other assignment is a culminating end of the semester project. In this assignment, the candidates design a multigenre/multimodal/multimedia presentation. The presentation must highlight the significance of the chosen theory and the theorists' contributions to our current body of knowledge about reading/language acquisition and its relationship to literacy learning. This presentation must describe how the chosen theory relates to the cognitive, linguistic, motivational, and sociocultural foundations of reading and writing development. Additionally, the presentation must address how the chosen theory is manifested into classroom practice, including word recognition, language comprehension, strategic knowledge, and reading-writing connections (Landis, Umolu, & Mancha, 2010). Successful completion of these two assignments provides the candidates with a comprehensive introductory overview of the theoretical foundations and how understanding these foundational theories can impact the candidate's classroom practice.

The next course the candidates take is literacy assessment. This course builds on the previous class and builds towards a deeper understanding of the individual candidate's theoretical orientation and personal views of reading instruction, along with the role of literacy assessment in the classroom. The first step to helping them delve deeper into what they believe is for them to take the *Deford Theoretical Orientation to Reading Profile* (TORP) (Deford, 1985). According to Deford, "the purpose of using this instrument is to identify a teacher's theoretical framework for literacy teaching and learning" (p 358). The results identify the candidates as phonics driven, skill driven or coming from a holistic orientation about the teaching and learning of reading. This assessment and the outcome help the candidate understand the nature of their self-reported beliefs and how these beliefs provide insight into their orientation towards teaching reading in the classroom.

The next assignment's goal is to utilize all the information gained to this point to determine the candidate's personal philosophy of reading instruction and the role of assessment in the literacy classroom. In this assignment, candidate's write a paper discussing their beliefs and demonstrating proficiency of what they understand about the reading process and how they believe reading should be taught while reflecting on how they arrived at this belief. The hierarch of these assignments has been strategically aligned to allow the candidates to use theoretical research, theoretical orientation, and classroom experiences as a student and as a teacher to reflect on what they believe to be the most effective way to teach students to read. Utilizing this structured approach, the candidates are equipped to articulate and, ultimately, execute their firmly established belief system regarding their personal philosophy of reading instruction and literacy assessment in classroom practice. The goal is to provide the candidates with the confidence and empowerment they need to stand firm in their beliefs and understanding. The goal of the program is for them to be intentional, because they know what is best for their students, and not to be swayed or impacted by what the teacher next door is doing.

BRINGING WORDS TO LIFE THROUGH SELF-REFLECTION TO BUILD SELF-EFFICACY

Graduate reading candidates end their journey with the final two semesters consisting of an applied assessment course and a supervised practicum experience coupled with a capstone research course. Both courses contain content designed to strengthen the self-efficacy of the candidates and to help them transfer that positive efficacy to their current and future students, co-educators, and school administrators. Applied assessment offers the candidates an opportunity to reflect on the impact of word choice. The practicum course offers an assignment meant to assist the candidates with self-evaluation of their teaching coupled with a coaching component. The capstone experience includes embedded reflections that focus both on the program and personal impact of teaching with research woven throughout the final paper assignment.

Applied assessment is a course that includes activities designed to promote self-efficacy. First, candidates are introduced to the text, *Choice Words* (Johnston, 2004). Candidates also complete reflections that are aligned with the text. The premise of *Choice Words* is to inform teachers of their word choices and how word choice impacts their students both emotionally and academically. One example, given in the first chapter, explains how the word choice of a teacher attempting to gently redirect a student caught the attention of the learner without using a harsh tone or harmful words. After reading and reflecting, candidates practice using purposeful, yet gentle, words. Practices like these lead to both increased positive self-efficacy for the candidate and for the student as they are joined in a positive learning environment.

The final practicum course includes an instructional talk tool. For this assignment, candidates are asked to record themselves during a one on one tutoring session with their student, transcribe a section of the lesson, then use a reflective tool to analyze their teaching. The talk tool analysis includes a comparison of two talk tool reflections completed over the course of a year in Applied Assessment and the final practicum course. Candidates note comparisons, similarities, and differences, and identify what changed and why. The reflection tool consists of conceptual and sociocultural functions. The sociocultural function is meant to promote the rich discussion occurring during guided contextual reading. Positive reinforcement facilitates positive self-efficacy (Roskos & Boehlen, 2001).

The final practicum course also includes a literacy coaching component. The graduate program is not meant to only impact the interactions reading specialists have with students, but also to impact the candidate's' ability to become a literacy leader in their district by coaching educators regarding literacy needs and engaging in curricular decisions.

CAPSTONE EXPERIENCE

Graduate candidates complete their capstone in their final semester. The capstone is a culmination of every core reading course and is meant to serve as a final reflection of the graduate reading program. As students journey through their reading courses, they are encouraged to write reflections, to make connections regarding their course work and how it applies in their current classrooms and as future reading specialists. Once the candidates reach the capstone semester, they use the reflections, coupled with research, to create a culminating review. As candidates navigate through the course reflections, they often gain an awareness of the strides they have made as educators. The reading coursework is meant to facilitate a climate of shaping future school leaders in a literacy curriculum capacity. As candidates realize the immense knowledge they have gained and applied, their efficacy is shaped and their understanding of implied and unconscious bias is further developed. As they work towards the end of the capstone course they begin to see themselves as leaders with a sense of responsibility to share knowledge both with peers and children.

FINDINGS

The researchers discovered that grounded theory methodology was the most appropriate theory for determining the phenomenon around the self-efficacy of graduate students who work in the K-12 schools. The theoretical frameworks highlighted throughout the chapter are used to gather data and the implications for this research are helpful in framing the curricular process and pedagogical experiences that the School of Education wants to promote in its graduate programs. The research sample is reflective of the overall cohort graduate student population. The program averages 25-30 graduate full time enrollments. For this study, the researchers identified a purposeful sample of five selected graduate students' capstone reflection paper. The capstone is an in depth personal account of how the specialized curricular program has impacted the teacher candidate, and how they, in turn, plan to use the knowledge, skills and dispositions acquired to impact the K-12 learner.

The small sample allowed the researchers to analyze the most organically developed themes from the student profiles. This method of analysis allowed the researchers to categorize the emerging themes that were developed within the reflection papers. Additionally, using a constant comparative method of data analysis, the study was able to proactively compare data and the detailed notes that the teacher candidates provided in their reflective capstone papers. This approach also allowed the study to develop emerging categories from the data set that was utilized (Creswell, 2007). The open coding process pointed the researchers into a core phenomenon around developing self-efficacy to impact K-12 student potential in reading. The following themes were central to the core phenomenon and aided the researchers in identifying the four emerging themes from the research: knowledge, identity, confidence and reflection.

Figure 1. Core theme developing self-efficacy to impact K-12 student potential in reading

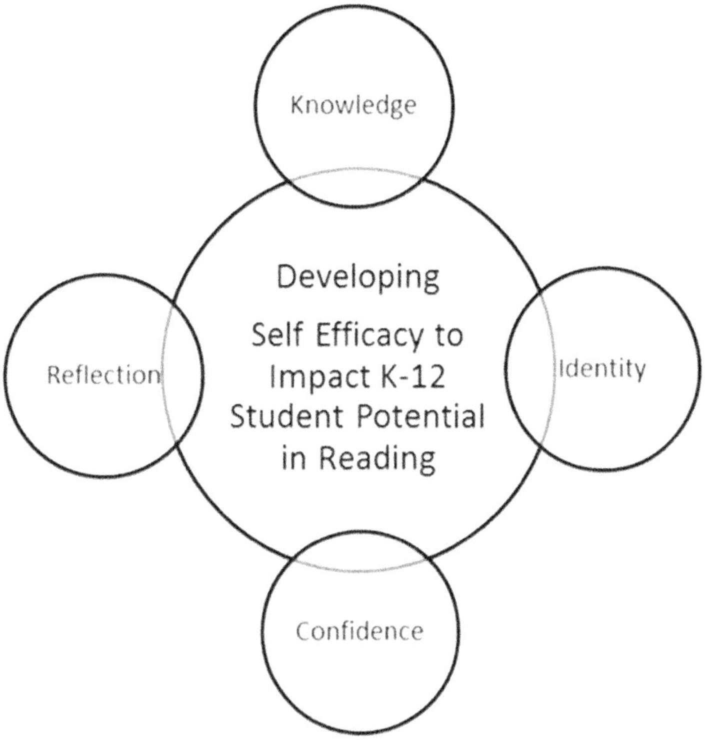

STUDY FINDINGS

One of the research assumptions that was explored through the study was the premise that to build self-efficacy in teacher candidates they must first be knowledge proficient. Through their reflection papers the candidates discussed each core course and the program in detail. Moreover, the teacher candidates provide detailed ideas around how what they were taught directly impacts what they are planning to do in the classroom. The example often used in teaching reading strategies is a question, "How can you teach Johnnie to read?" In the core program courses, our first goal is to ensure that the candidates have the content and pedagogical knowledge to be effective practitioners. The major content reading courses are designed to increase instructional knowledge. Once the teacher candidates develop this critical knowledge, we then take them through strategies of self-actualization, identity, bias, cultural competence, and processing reflection as a developing teacher professional. We have identified that this systematic process allows us to review years of data and cohort research to aid us in assessment and revision of a solid and sought-after teacher education program. Furthermore, the data in this program and the need for impactful reading assistance programs in our demographic area, led to the development of a university-community based clinic to serve K-12 students. The teacher candidates in our program are trained in the knowledge, skills and dispositions throughout the program and then are able to use their course work in the supported and supervised clinic setting that operates as a practicum experience within the program.

The reading clinic on both the Tahlequah and Broken Arrow campuses serves 650 students from the surrounding public-school communities. The students come to campus and our teacher candidates are assigned an individual student or small group to train in reading. The candidates are taught to begin their practicum experiences by identifying the student weaknesses and needs for growth. The reading clinic experiences also assist the teacher candidates in building confidence in their teaching practice. These practicum experiences further provide an opportunity for the students to reflect upon the practices they have learned through the program's coursework.

KNOWLEDGE

Through a systematic analysis of the graduate student reflections, the study uncovered an insight and an unexpected finding. This finding was the identification of the text *Choice Words* (Johnston, 2004) as an embedded class and program resource which positively impacted student learning outcomes and shaped classroom instruction. Resources, such as this text, provide guided instruction for the candidates and help them to consider the non-negotiables of teaching reading and planning for effective instruction. Understanding language positively impacts student learning. It further serves as a pillar of knowledge to highlight the ways that speaking to children impacts their self-efficacy. Words contain the power to motivate, encourage, or discourage students. This highlights the importance of word choice knowledge and demonstrates how careful and thoughtful work choices encourage students to become critical thinkers through instructional language.

The researchers observed graduate candidates discussing how *Choice Words* helps them develop a cognizant awareness of the impact of utilizing targeted questioning and statements (Johnston, 2004). The candidates discussed how targeted questioning allows students to become aware and involved in their own learning. Self-efficacy is positively impacted when children become strategic thinkers. This positively impacts students as they grow and begin to experience the world.

IDENTITY

Through language, teachers assist students as they build an identity and then transfer strategic problem-solving skills from theory into practice. *Choice Words* discusses the influence of the language utilized by teachers as it became a crucial element of the programs ability to deliver effective instruction regarding teaching reading at the K-6 level (Johnston, 2004). During the reflections, some of the candidates discussed the basic concept that knowing their students' names is an important base element in building a secure learning community, promoting individual identity and the importance of the individual. From defining a student's identity to facilitating deeper cognitive thinking, a teacher's language with students impacts more than just the delivery of instruction. Identity can be impacted when teachers facilitate dialogue to increase the students' thought processes. These processes not only impact their learning but also their interactions with their peers and beliefs about themselves.

CONFIDENCE

Building confidence was a recurring theme throughout the graduate students' course and program experiences. Graduate students discussed instructional language, building confidence among their students in the use of language and many ideas for enhancing language choice as it pertains to *Choice Words* (Johnston, 2004). *Choice Words* focuses on and encourages critical thinking through word choice. The graduate students gained hands on experience and developed confidence by providing opportunities to apply the skills and strategies outlined in the *Choice Words* text.

The graduate students also discussed and reflected on the confidence that students gained and the knowledge that they transferred into their classroom experiences. The graduate students utilized reading strategies, new technology in the area of literacy assessments, literacy work stations and a myriad of other resources confidently in their classrooms. Many of our graduate students bloomed as literacy leaders in their school districts and were then able to transfer this knowledge through peer coaching to other teachers and administrators at their schools. Through the reflection study it was evident that the reading program gave the graduate candidates the confidence and positive self-efficacy that they needed to emerge as literacy leaders.

REFLECTION

The instructional talk tool provided ample opportunities for academic reflection. As students arduously transcribed their guided contextual reading work during literacy tutoring, they were able to reflect on the quality and impact of their instruction. This reflection enabled the graduate students to analyze the questions that were posed during literacy tutoring. Reflection of personal teaching provides a valuable tool to aid in improving instruction. Using self-perception inventories encourages personal reflection, as the students can understand how they view themselves as readers, through the lens of how others perceive them. This includes a comparison of themselves to their peers and teacher perception. The case study, at the end of tutoring, provides a comprehensive report of a student (tutee) from the beginning to the end of the semester long practicum. The case study includes reflections on assessment results, teaching strategy effectiveness and future recommendations intended specifically for the individual student. Graduate candidates must reflect as they compile this information to ensure that their case study information is both accurate and helpful as the student transitions out of tutoring.

RECOMMENDATIONS FOR IMPREMENTATION OF SIMILAR PROGRAMS AT OTHER INSTITUTIONS

Overall the study confirmed our belief that the best way to shape greatness in students within school systems is to reach and impact those who will teach them first. Our fundamental goal was to establish and then meaningfully assess how the reading program highlights and teaches best practices, promotes self-efficacy and leads the candidates to content proficiency. The study confirmed that the reading program develops and then shapes curricular expectations in line with the needs of the public and private school students that our graduates serve as reading teachers and reading specialists. Based on the reflections and exit interviews with the candidates, the feedback indicates that the program is effectively shaping the teacher preparation process and leading them to effective classroom instruction and a pathway to success as reading teachers in the community.

In order to implement the findings within other universities, the following guidelines aid in replicating the reading clinic featured in this study:

1. Choose textbooks which promote self-efficacy such as *Choice Words* (Johnston, 2004) and articles such as "Reflection of personal teaching provides a valuable tool to aid in improving instruction" (Roskos & Boehlen, 2001).
2. Create a welcoming reading clinic space. This includes a parent friendly environment, complimentary easily accessible parking, clinical spaces which promote both individualized tutoring and collaboration with peers, an onsite library with ample resources such as leveled readers, classroom supplies, etc.
3. Partner with the community and local school districts so they are aware of clinical services offered.
4. Multidisciplinary partnerships help the reading clinic to stay vibrant. These services may include a partnership with math, counseling services and any other university departments which may help children.
5. Collaborate with other reading clinic faculty across the country in avenues such as the Association of Literacy Educators and Researchers and the Literacy Research Association.
6. Promote an inclusive and culturally diverse curriculum which welcomes students and families from all backgrounds.
7. Practice the art of reflection in every course. At the end of each course, our students reflect over clinical experiences, classroom application, and theory. We use these reflections to modify and change our teaching so the clinic stays dynamic rather than static.

FUTURE RESEARCH DIRECTIONS

The authors reviewed the current literature focusing on new trends in digital and online learning. Based on the research, we concluded that now more than ever, the need for online classes has become evident over the last several years. To address the need for course format options, a future study using the constant comparative method to ascertain an online class format versus the face to face format may be useful in determining self-efficacy. This could include perceptions from graduate students, parents, and children who are tutored in the reading clinic versus being tutored digitally or meeting at another location such as a local library. We feel these implications could further promote the notion of immediate reflection and connection to the materials and information we teach.

Additionally, the suggestions gathered from this study force the authors to consider how desegregated data by campus could inform our practice. We realize this would provide another area where we could benefit and further investigate the overall program impact. The campus in a rural area serves children in the reading clinic who do not have as much access to tutoring services. The urban university setting does still serve children but these children may have access or other options. Both of these future research ideas could help to further determine self-efficacy and how it is impacted when looking at location and course format.

CONCLUSION

The study confirmed the working theory that the reflective process is imperative and vital to developing the most effective teacher practitioners. In order to build self-efficacy through knowledge, identity, confidence and reflection, the practice of reflection should be embedded through both coursework and practicum clinical experiences. The future growth of the reading clinic at Northeastern State University and other institutions wishing to develop effective and impactful reading clinics hinges on deep and longstanding collaboration between faculty members, the institution and the community, a deep commitment to developing reading teachers and a recognition in the importance of investing in the success the program.

REFERENCES

Author. A. A (2017). Developing a practice of cultural awareness in pre-service teachers by promoting positive dialogue around diversity. In Leavitt, L., Wisdom, S. and Leavitt, K. (Eds.), Cultural awareness and competency development in higher education. Hershey, PA: IGI Global Publishing.

Bandura, A. (1977). Self-efficacy: Toward a unifying theory of behavioral change. *Psychological Review*, *84*(2), 191–215. doi:10.1037/0033-295X.84.2.191 PMID:847061

Clandinin, D. J. (Ed.). (2006). *Handbook of narrative inquiry: Mapping a methodology*. Sage Publications.

Creswell, J. (2007). *Qualitative inquiry and research design: Choosing among five approaches*. Sage Publications.

Deford, D. (1985). Validating the construct of theoretical orientation in reading instruction. *Reading Research Quarterly*, *20*(3), 351–367. doi:10.2307/748023

Gay, G. (2000). *Culturally responsive teaching: Theory, research, & practice*. New York, NY: Teachers College Press.

Gay, G. (2002). Preparing for culturally responsive teaching. *Journal of Teacher Education*, *53*(2), 106–116. doi:10.1177/0022487102053002003

Gay, G. (2004). The importance of multicultural education. *Educational Leadership*, *61*, 30–35.

Johnson, K. E., & Golombek, P. R. (2002). *Teachers' narrative inquiry as professional development*. Cambridge University Press.

Johnston, P. H. (2004). *Choice words: How our language affects children's learning*. Portland, Me.: Stenhouse Publishers.

Kelley, H. M., Oginga Siwatu, K., Tost, J. R., & Martinez, J. (2015). Culturally familiar tasks on reading performance and self-efficacy of culturally and linguistically diverse students. *Educational Psychology in Practice*, *31*(3), 293–313. doi:10.1080/02667363.2015.1033616

Ladson-Billings, G. (2006). From the achievement gap to the education debt: Understanding achievement in U.S. schools. *Educational Researcher*, *35*(7), 3–12. doi:10.3102/0013189X035007003

Landis, D., Umolu, J., & Mancha, S. (2010). The Power of Language Experience for Cross-Cultural Reading and Writing. *The Reading Teacher*, *63*(7), 580–589. doi:10.1598/RT.63.7.5

Lee, Y., & Jonson-Reid, M. (2016). The Role of Self-Efficacy in Reading Achievement of Young Children in Urban Schools. *Child & Adolescent Social Work Journal*, *33*(1), 79–89. doi:10.1007/s10560-015-0404-6

Roskos, K., & Boehlen, S. (2001). Enhancing teachers' awareness of their instructional talk in the teaching of reading: Differential features of self-assessment. *Action in Teacher Education*, *22*(4), 59–74. doi:10.1080/01626620.2001.10463030

Siwatu, K. O. (2007). Preservice teachers culturally responsive teaching self-efficacy and outcome expectancy beliefs. *Teaching and Teacher Education*, *23*(7), 1086–1101. doi:10.1016/j.tate.2006.07.011

Siwatu, K. O. (2009). Student teachers' self-efficacy beliefs regarding culturally responsive teaching and their professed classroom practices. *Teacher Education and Practice*, *22*, 323–333.

Siwatu, K. O. (2011). Preservice teachers culturally responsive teaching self efficacy forming experiences: A mixed methods study. *The Journal of Educational Research*, *104*(5), 360–369. doi:10.1080/00220671.2010.487081

Tracey, D. H. (2012). *Lenses on reading: An introduction to theories and models* (2nd ed.). New York, NY: Guilford Press.

KEY TERMS AND DEFINITIONS

Applied Assessment: Applied assessment offers teacher candidates the opportunity to reflect on the impact of word choice. Applied assessment helps teacher candidates with self-evaluation of their teaching coupled with a coaching component. Applied assessment includes embedded reflections that focus program and personal impact of teaching and is interwoven with research.

Culturally Responsive Teaching: The chapter utilizes the definition of culturally responsive teaching developed by Kelley, Siwatu, Tost, and Martinez, (2015). "The core of culturally responsive teaching is utilizing a pedagogy that is inclusive of the students' culture in order to demonstrate utility of the content and to improve academic success" (p. 294). Culturally responsive teaching focuses on the importance of combining reflection with pre-service teachers in the urban focused program.

Pre-Service Teacher Candidate: The term pre-service teacher candidate or reading candidate is used when referring to undergraduate and graduate students within our program.

Self-Efficacy: The chapter utilizes Albert Bandura's definition of self-efficacy from his seminal 1977 work, Self-Efficacy: Toward a Unifying Theory of Behavioral Change. Bandura's definition can be summarized as an individual's confidence in their ability to achieve desired goals and intended outcomes.

Self-Efficacy Scale: The Chapter utilizes the self-efficacy scale developed by Siwatu, Frazier, Osaghae and Starker (2011). Through the use of the self-efficacy scale students are taught the importance of being steadfast in developing culturally responsive classrooms.

Chapter 16
Cognition and the First Year Experience:
How Studying the Mind Equips Students for Learning

Ryan Korstange
Middle Tennessee State University, USA

Kevin Krahenbuhl
Middle Tennessee State University, USA

ABSTRACT

This chapter extrapolates information processing theory and advances in research in memory formation and effective learning practices on to the first-year student. The authors will lay a context for what FYE is, and why it matters, survey advances in memory research, and explore the study practices that students actually use. Finally, the authors posit a structure for organizing FYE courses around the best practices in cognitive and educational psychology so as to change student study behavior and help them increase their academic performance.

INTRODUCTION

Access to higher education has increased dramatically over the last several decades; but, during the same period, completion rates have remained relatively fixed, hovering right around 50% (Eaker & Sells, 2015). Many initiatives have been undertaken to increase student success (Demetriou & Schmitz-Sciborski, 2011). Some of these efforts are co-curricular and are aimed to help students navigate and assimilate into campus culture. Other student success efforts attempt to improve classroom instruction. According to George Kuh in "High Impact Practices," the discussion of retention and student success should not only include co-curricular programming but examine the impact of classroom organization

DOI: 10.4018/978-1-5225-2998-9.ch016

and structure (Kuh, 2008). Furthermore, Kuh identifies First Year Experience (FYE) courses as a "High Impact Practice." FYE courses are not beneficial merely because they are offered; but, they also need to have a demonstrable benefit to learners. This chapter will propose a theoretical foundation for effective FYE courses based on cognitive and educational psychology.

In this chapter, the authors contend that FYE courses become truly effective for students and universities when they provide accurate information about cognitive function and neurobiology, as well as, opportunities to apply this knowledge to develop efficient and effective study strategies that they will use. FYE courses can become beneficial to the university when they increase student retention. However, the class becomes beneficial to students when, and only when, it provides them with essential information which enables them to more effectively and efficiently progress towards the accomplishment of their goals. It is not enough for the FYE course to merely disseminate information or to provide a path for students to develop a connection to the university, rather an effective FYE course must provide opportunities for students to understand and improve their approach to education. Since the retention of students is dependent upon their satisfaction with, and assimilation into the educational institution, the benefit or value of an FYE course is measured by retention and graduation, although satisfied students are certainly retained better (Schreiner & Nelson, 2013).

The need for new retention initiatives is especially evident in light of the academic under-preparedness that typifies incoming first-year students. Moreover, this need is exacerbated by the widening gap between the type of academic activities required in K-12 and Higher Education. A FYE course fits in a unique and influential place in the curriculum to address this gap and help students become successful (Atherton, 2014). Students enroll in FYE courses during their first semester, and the expressed aim of the course is to increase student success. An essential component of an effective FYE course then should include an explicit focus on developing student understanding of how learning occurs, and on the application of effective learning strategies in all of their courses.

This chapter will propose a theoretical foundation for FYE courses grounded in current educational and cognitive psychology and will be organized around several key questions: what is FYE; why does it matter; and, how does learning work? Finally, these questions form the outlines of a FYE curriculum organized around current understandings of cognitive function.

WHAT IS FIRST YEAR EXPERIENCE (FYE) AND WHY DOES IT MATTER?

The transition to college has consistently been acknowledged to be challenging for students (Astin, 1984; Terenzini et al, 1996; Goodman & Pascarella, 2006). Students often experience the challenge of increased autonomy and the unfamiliarity with collegiate instructional style and norms, and in many cases are not prepared for the academic challenge of higher education (Collier & Morgan, 2008). When faced with these challenges, many students perform poorly in their classes and are more likely to drop out of the university (Pascarella & Terenzini, 2005). This challenging transition resulting in low student retention is likely to be exacerbated because of increasing access to higher education for students who learned what success in college requires from their family or previous teachers.

Several influential theories identify reasons as to why students leave college (see Tinto's "Student Integration Model" and Astin's "Theory of Student Involvement"). Both theories effectively assert that students deal with their adjustment as it relates to their academic, social, and personal-emotional integration into the university (Pascarella & Terenzini, 2005). Those students who can adjust to challenges in each of these areas more easily acclimate to the new environment. The varied elements of integration and involvement have been increasingly understood as the degree of student engagement (Kuh, 2003). Student engagement then refers to the amount of time and effort devoted to academic/professional activities by students and the degree to which institutions promote participation in such activities (Kuh, 2003, 2009). Both institutional policies and student expectations and abilities must be understood to improve student transition into any educational institution.

The typical student population has significant differences in demographic makeup, depth of knowledge, and ability to succeed in college. First-generation students, low-income students, and others from underrepresented backgrounds are more likely to come to college academically underprepared (Terenzini et al., 1996). While many students typically come to college with specific goals for an improved opportunity and earning power, they often attempt to find personal relevance in each topic they encounter too quickly, and when they first encounter difficulty in their classes, they often lose motivation. Unfortunately, as noted by Schunk (2008), motivation is an important factor in student learning; and, when students lose motivation, they face an increased challenge to persist in college. Compounding this problem is the fact that many first-year students fail to recognize that their high school coursework did not prepare them for the rigor of college-level coursework until they receive their first mid-term examinations (Tinto, 2000).

Higher retention rates benefit students, who often see a vast increase in their lifelong earnings on average. High retention rates also benefit institutions and society as a whole (Tinto, 1993), and, for these reasons, retention should be improved in all demographic areas. One practice that has actualized potential to help students in this challenging transition are FYE courses. Such courses are intended to support students in the college transition by introducing them to campus resources that can help them achieve their educational and career goals (Jamelske, 2006; Porter & Swing, 2006). They have a wide variation across campuses and depend on institutional goals and student needs (Barefoot, 2005; Pascarella & Terenzini, 2005). In general, these varied FYE courses seek to improve students' academic engagement with the university in two types of integration: structural, which is meeting the institution's academic standards, and normative, which is identifying with the academic system (Braxton & Hirschy, 2005). Another element, which has received wide attention is the deliberate efforts within a FYE course to promote social engagement or the degree of symmetry between an individual student's social expectations and the social milieu of the campus (Strayhorn, 2009).

Although the design and implementation of such FYE seminar courses are diverse, they are increasingly required for students. More than half of all institutions have mandated that every incoming first-year students enroll in some form of FYE course for over a decade now (Tobolowksy & Associates, 2008). This increasing mandate is not surprising, given that grades earned during the first year of college are the single best predictor of student persistence, even after accounting for the input characteristics of students (Pascarella & Terenzini, 2005). Recognizing this impact, George Kuh (2008) described FYE courses as a "High Impact Practice" (HIP).

In this chapter, the authors present a foundation from which FYE courses can intentionally integrate some basic principles of learning with direct implications towards students' self-regulation of learning, and, ultimately, towards student success. These factors make an FYE course truly "high impact." Many FYE courses fall short when attempting to equip students with the tools of learning and knowledge of how it is that learning works. When it comes to preparing students for success in the college environment, helping them monitor their learning is an essential component.

REVIEW OF THE LITERATURE

Before describing how learning happens, the term "learning" must be defined. According to Illeris (2007), learning is "any process that in living organisms leads to permanent capacity change and which is not solely due to biological maturation or aging" (p. 3). Susan Ambrose (2010) and her colleagues define learning in much the same way, emphasizing that learning is a "process, not a product," that it "involves change in knowledge, belief, behaviors or attitudes," and that learning is not "done to students, but rather something students themselves do" (p. 3).

Learning happens in many situations, not just in the classroom, although not all learning is of the same type. Tricot and Sweller (2014) distinguish between biologically primary and secondary forms of learning in which some things seem to be learned naturally and without any teaching (i.e., learning to walk, to talk, etc.,), whereas biologically secondary aspects (i.e., reading) are not learned in the same manner. In this chapter, the researchers are concerned with "biologically secondary learning" as this is where the vast majority of academic learning resides.

Learning is also represented by many different models, in this chapter, the authors will rely on the learning model suggested in information processing theory. Information processing makes sense as a theoretical background for a high impact FYE course for two reasons. First, since the primary goal is to help students envision how information from the classroom environment, passes into their accessible memory, viewing memory creation as a linear process from external and unknown information to internal and known information gives students the foundation for understanding effective study practices and systems. Second, students can more effectively internalize the concepts due to their familiarity with the essential terms and structures from their experience in K-12 education.

Information processing theory models the understanding of human memory on the analogy of the flow of information within a computer. The model starts with the observation that "memory is not a monolithic, unitary entity and that what we label memory, in fact, represents a number of separate but interacting systems" (Schacter & Tulving, 1994, p. 1). The model is imperfect. Notably, it fails to take parallel processing or the simultaneous encoding of traumatic memories into account.

The information flow, as it were, is as follows: a) external stimuli which receive attention from the individual enter the sensory memory briefly and pass quickly into the working memory (Baddeley, 1992); b) from there, information is recalled from the long-term memory to make sense of and to process the new information (Baddeley, 1986; Schacter & Tolving, 1994); and then, c) information that is rehearsed is encoded, stored in the long-term memory. The following chart illustrates the progression of information from outside of the individual to the long-term memory.

An individual encounters information in their "environment." It is important to recognize that any environment is filled with many stimuli, and therefore many streams of information. It might appear that the classroom environment is one in which the content of the class presentation is only one stream of information requiring the students' attention. However, any student could choose to pay attention to the clothing choice of the instructor, the lighting or temperature of the room, a text message from a friend, the cloud formations outside the classroom window. The list could go on, but any learning environment is similarly filled with multiple streams of information.

While in any environment, an individual chooses to direct their attention towards some, but not all, of the available pieces of information, and to disregard others. Illeris (2016) points out the complexity of this choice, "learning implies the integration of two very different processes, namely an external interaction process between the learner and his or her social, cultural or material environment, and an

Figure 1. Information processing theory: A model of the mind

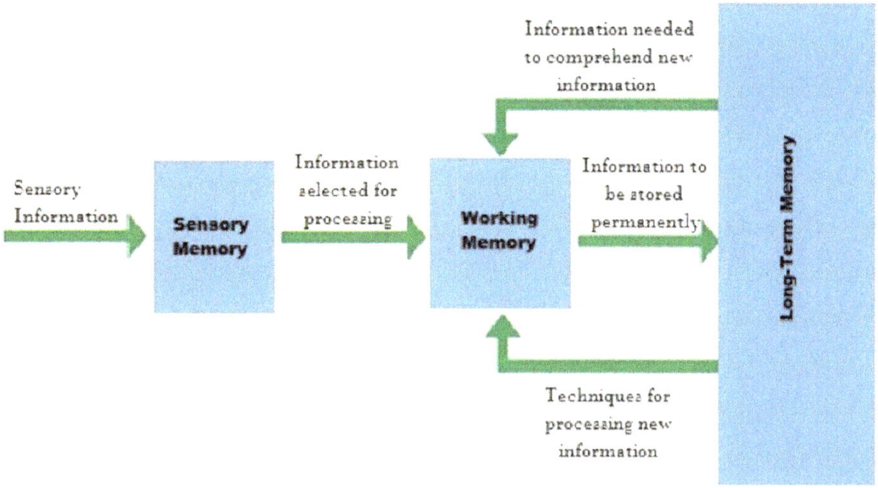

internal psychological process of elaboration and acquisition" (p. 10). The important task in memory is to construct meaning with the new information; and, this depends both upon what the learner already knows (generation, retrieval practice) and their motivation or desire to learn and apply that material (reflection, elaboration).

The pieces of information to which individuals devote their attention pass quickly through the sensory memory and into the working memory (terminology which has now replaced the former designator of "short-term memory"). According to Baddeley (1992), working memory is "...the brain system that provides temporary storage and manipulation of the information necessary for such complex cognitive tasks as language, comprehension, learning, and reasoning" (p. 556). Material held in the working memory can either be learned, i.e., passed into the long-term memory, or forgotten (passed out of the memory completely).

Long-term memory is made up of several components (see the overview of research in Schacter & Tolving 1994, pp. 8-11). Typically, this memory system can be divided into declarative or explicit memory and non-declarative or implicit memory, as illustrated in the following figure:

Non-declarative memory is often referred to as procedural memory, which is considered a "performance-line" system (Hirsh, 1974). Procedural memory exists outside from what Shachter and Tolving (1994) call "cognitive memory" (i.e., declarative memory). This categorization emerges from research on the effects of amnesia (e.g., Knowlton, Ramus, & Squire, 1992; Nissen, Knopman, & Schacter, 1987; Squire, 1994).

Declarative memory is broken down to include the memory of facts (semantic memory) and events (episodic memory) (Tulving, 1993). Although note that some scholars doubt the distinction of these systems (Squire, 1994). Several scholars assert the independent existence of the perceptual representation system (PRS), which is pre-semantic and is used in the identification of words and objects. It mediates "improvement in identifying or processing a stimulus as the result of its having been observed previously" (Gazzaniga, Ivry, & Mangun, 2008), and as such is connected to priming.

Learning, as a process, makes use of these memory systems, and so should be understood in relation to them. The most significant observation is that learning requires cognitive effort (Brown, Roediger

Figure 2. Classification of memory

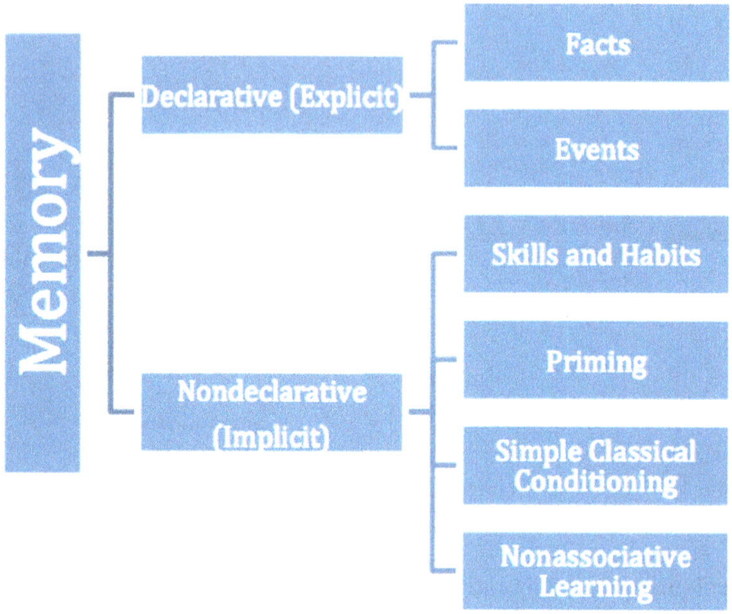

& McDaniel, 2014). Further, the memory system is not a one-way system of moving material from the external into the long-term memory, rather, it is multi-directional. Individual memory systems inform other memory systems. Memories are recalled from the long-term memory into the working memory to help understand or make sense of information as it is being processed.

When one interacts with information from their environment, the information rises into working or short-term memory. This information can be said to be learned when it is moved into one's long-term memory for the purpose of later recall. Willingham (2009) defines memory as the "residue of thought" (p. 54), and in this regard, learning happens when that residue is solidified in the long-term memory.

Learning strategies help at two critical stages in the learning process. First, they help students to direct significant attention to important information from their environment and to connect new information to material that they already know because learning rests on a foundation of prior knowledge (Brown, Roediger, & McDaniel 2014). Second, these strategies help with encoding or storage of information in the long-term memory for later recall. Memory is created through two processes, first "rehearsal," which has two components, maintenance rehearsal and elaborative rehearsal (Reisberg 2010), and encoding. Encoding happens in various ways; but, it is best considered the process of imposing structure on new information and adding to the complex series of connections that creates long-term memories.

HOW DO STUDENTS TRY TO LEARN?

There are clear indications that effective study skills correspond to academic success (Crede & Kuncel, 2008; Hattie, 2009; Prevatt, et. al, 2006), and also clear indications that students do not often use the most effective study techniques (Gurung, 2005). Dunlosky, et. al (2013) evaluated common study strategies with a view towards their effectiveness as demonstrated by cognitive psychology. They found that the study strategies students used most commonly, such as re-reading notes, highlighting notes, etc., were the least effective from the perspective of memory and learning. They further suggest that small changes produce significantly greater results.

Gurung (2005) polled students in introductory history courses, attempting to see which study skills they used in preparing for an exam. He found that the most frequently used techniques were reading notes and the textbook, using mnemonics, rewriting notes, and reviewing highlighted material in the textbook. Of these, only reading notes and the textbook and mnemonics correlated with higher performance on the exam. His conclusion is that "not all techniques were effective. The most effective techniques were often not the ones used most" (p. 240). Compiling general statistics about student study strategies is complicated by two factors. First, students feel compelled to overestimate their academic strategies and to misrepresent the amount of time that they spend on academic tasks (Ehrlinger & Dunning, 2003). Second, studying is an individual process, and effective study methods differ between tasks. Nevertheless, some tendencies amongst student practice can be observed.

Gurung, Weidert, and Jeske, (2010), have put together an assessment tool called the Study Behavior Checklist (SBC), based on Gurung (2005) and Weinstein and Palmer (2002). This assessment attempts to correlate study behaviors with exam performance. The result was that metacognitive strategies were positively correlated with exam scores; and unfortunately, many of the study strategies advocated in many FYE courses are negatively correlated with exam scores (e.g., explaining concepts to a friend, asking for additional study materials, reviewing the chapter after the lecture, highlighting important information in the chapter).

Research also indicates that even effective practices can be utilized incorrectly and that incorrect use mitigates their benefit. Gurung and McCann (2011) have identified several ways that students can modify effective practices, rendering them ineffective (following Gurung 2004). Their description centers around the incorrect use of pedagogical aids, which they describe as the tools provided to the learner by either the instructor or the instructional materials. Students can spend too much time using lists of key terms, chapter summaries, or they can highlight too much text in the notes and reading. In addition, they often spend too much time studying chapter review questions rather than testing their understanding of the material. Study groups can be used ineffectively. Gurung and McCann (2011) point out that student distraction is a huge detriment to effective studying (Gurung 2005; Calderwood, Ackerman, & Conklin, 2014; and Kirschner & Merrienboer, 2013).

When students step foot on their college campus for the first time, they already have an extensive history with "effective" study skills, or so they think. Traditional-aged students will have just graduated from high school months before; and for them, their graduation functions as a validation of the effectiveness of their study practices. However, as Kirschner & Merrienboer (2013) have pointed out, students' preferred methods for learning are often not effective. This observation certainly relates to the so-called "learning styles" students may think that they have, but also to the learning strategies they employ (Ehrlinger & Shain, 2014). Research has clearly demonstrated that students overestimate their knowledge (Ehrlinger & Dunning, 2003), the quality of their performance (Ehrlinger, Johnson, Banner, Dunning, & Kruger, 2008), and the time required for them to master and complete academic work (Buehler, Griffin, & Ross, 1994).

A PROPOSED COGNITIVE FOCUSED FYE FOR STUDENT SUCCESS

Cognitive and educational psychology provide some extremely important theoretical underpinnings for increasing student academic achievement; but, translating these concepts from the sterile environment of the memory lab into the often chaotic reality of the first year student's life takes a bit of effort and does not always work as expected (Daniel & Poole, 2009; Gurung, Weidert, & Jeske, 2010). Cox and Lemon (2016) make it clear that the curriculum, teacher pedagogy, and assessment methods work together to contribute to the effectiveness of FYE courses. Three fundamental cognitive concepts of distributed practice, effortful retrieval, and chunking benefit students in key areas of their learning – at the point of attention, the point of encoding, and the point of recall. These concepts should form the curricular backbone of the FYE course.

First, cognitive science informs a student's learning strategy at the point of attention. The information processing learning model discussed above starts with the assumption that a student only has the potential to learn the things to which they devote their attention. Practically, this means that numerous environmental stimuli escape our notice at any given minute and that those stimuli that are missed will not become learned. On this point, Willingham's (2009) description of memory as a "residue of thought" is incredibly helpful (p. 54). To remember any concept, it first must be noticed, then thought about, and then, thought about some more.

The implications of this observation are that increased attention to significant information will produce increased learning. One method students can use to increase their attention to significant information is to take better notes in class and on the reading material outside of class (Wei, Wang, & Fass, 2014). The goal of note taking is not to record information, but to think about the information so as to start the learning process. Without intentional attention to important external stimuli, learning is not possible (Hembrooke & Gay, 2003; Junco & Cotton, 2012). Effective notes facilitate attention to important information, and so the strategies of active listening, along with meaningful, reflective recording, are essential tools. The Cornell method of note taking or any of its variations is worth special consideration because it adds the benefits of both distributed practice and effortful retrieval.

Second, cognitive science informs learning at the point of encoding, which is the process of storing information in the long-term memory. Human memory is fundamentally associative, that is to say, memory is built by connecting new information to old information. So, memory is more effectively encoded when it is understood to be personally meaningful. Connections can be built between new and old information in a process of reflection. Although it should also be mentioned that a student's prior knowledge may hinder learning (Ambrose, et. al, 2010; Illeris, 2016). Several strategies can be employed to develop better encoding, such as spaced practice and deliberate practice.

Spaced practice is important because it has strong empirical support and yet is usually not the typical way in which students study (Dunlosky, et al., 2013). Carey (2014) asserts that forcing yourself to retrieve information multiple times is one of the most powerful ways of building memory. Correspondingly, he argues that by spacing practice, the quick learning nature of the working memory is mitigated, which builds recall. Typically, students engage in massed practice, commonly referred to as "cramming" for a test; however, research about how we learn as underscored by Carey (2014) has established empirical support for spacing practice out in intervals relative to the length of time away when completing a summative assessment.

A sample illustration of what this might look like is included in Table 1. This method intentionally spaces out study sessions so that students engage in an effortful retrieval of information. One might compare this approach to the commonly utilized approach by most students in which virtually all of their practice occurs the day before or even the day of the test in a "cramming" session. Given the strong evidence by researchers in favor of spaced practice over utilizing the cramming method, it is certainly a strategy that should be encouraged among college students (Brown, Roediger, & McDaniel, 2014; Carey, 2014).

Deliberate practice is a highly structured form of practice in a specific domain that includes the specific goal of improving performance. According to Ericsson, Krampe, and Tesch-Romer (1993), deliberate practice consists of four essential components: a) the learner must be motivated to exert effort to improve his or her performance; b) the practice tasks should take into consideration the students' prior knowledge-base and skill competence so that the study task is specific and understandable; c) students receive immediate informative feedback regarding their performance, such as completing a practice test (Dunlosky et al., 2013); d) students should repeatedly perform the same or similar tasks. Deliberate practice provides strong empirical evidence against the notion that expertise is something inborn and has important implications for student achievement (Dweck, 2006). As Ericsson and his team (1993) note, "expert performance is qualitatively different from normal performance… however, we deny that these differences are immutable, that is, due to innate talent… the differences between expert performers and normal adults reflect a life-long period of deliberate effort to improve in a specific domain" (p. 400).

In many ways, recall matters the most to students, for it is at this point that they either remember the information properly, and pass a test or quiz, or do not, and fail. Recall cannot be simply understood to be the production of information on a test; but rather, recall, particularly when it is effortful, is a key component of any encoding and study strategy (Brown, Roediger, & McDaniel, 2014; Dunlosky et al., 2013). Recall is not a punctiliar activity, that is, one in which students engage at one time, only during the test. Simple strategies can encourage regular effortful recall. Elaborative review is when a student writes a short summary of the class contents from memory after the session has concluded. In addition, students can use practice tests or quizzes (or flash cards) as the main method for information encoding. The key to both of these strategies is that information is recalled, and as it is recalled, it is more thoroughly learned. Effortful retrieval has another benefit for students as it gives them immediate feedback on their progress (Brown, Roediger, & McDaniel, 2014). When students engage in effortful retrieval in any form, they give themselves a convenient data point from which they can more objectively gauge their understanding and acquisition of new information. This data can be used to "shorten the feedback loop," and allows students to understand the effectiveness of their study strategies before they arrive at the exam.

Table 1. Spaced practice to learn for a test

Study Session 1	**Study Session 2**	**Study Session 3**	**Study Session 4**	**TEST**
Study & Learn 50 minutes	Restudy 30 minutes	Restudy 20 minutes	Restudy 20 minutes	
Monday	Wednesday	Friday	Sunday	

As Brown, Roediger and McDaniel (2014) suggest, using effective strategies create "retrieval [that] is harder and feels less productive, but the effort produces longer lasting learning and enables more versatile application of it in later settings" (p. 4). For this reason, students have a hard time seeing the practicality of applying these strategies. For this reason, FYE courses need to be intentionally designed. Effective course design takes students from where they are – in this case, reliance on ineffective study practices – to where the instructor wants them to be – effective and efficient studying (Jones, Noyd, & Sagendorf, 2015). Therefore, the study skills discussed in FYE courses should be structured around the cognitive model developed and demonstrated in Table 2.

As has been demonstrated in the table, providing students with effective and efficient systems for processing new information is beneficial to their becoming successful graduates. A method of organizing a student's approach to studying under the best practices that have emerged from educational and cognitive psychology is proposed by the authors. This model consists of two inter-related systems: a regular system of day-to-day learning accomplished through structured attention and intentional review of class information, and a more focused system to be utilized in preparation for exams.

Table 2. Cognitive focused FYE seminar outline

Unit	Content and Application
Unit I: How Does Learning Work	Content: • Overview of memory systems • High school vs. college learning Application: • Reflection: Articulation of a 'Philosophy of College learning'
Unit II: Learning at the Point of Attention	Content: • Increasing attention by reading assignments before class • Previewing the course syllabus to have context for the session • Effective note taking regulates attention in class • Review notes after class through identifying key questions or elaborative review • Asking good questions in, or after, class Application: • Lab: Review and troubleshoot note taking in other classes • Lab: Review and troubleshoot reading college level assignments • Reflection: Strategy for effective class participation
Unit III: Learning at the Point of Encoding	Content: • Identify the specific tasks that effective learning requires • Determine the appropriate ways of spending time in order to make sure to pass classes • Prioritizing the variety of tasks (academic and otherwise) • Overview of various tools essential for task and time management Application: • Lab: Review tools students are using for managing their time and tasks • Reflection: Identify the regular process of studying that you can use
Unit IV: Learning at the Point of Retrieval	Content: • Spaced practice • Effortful retrieval • Creating and using practice tests • Test-taking strategies Application: • Create a study schedule for mid-term or final exams utilizing spaced practice, deliberate practice, and effortful retrieval • Make one practice exam for one upcoming exam

The main topics covered in a study skills section of many FYE courses thus is marshaled into a complete system of learning, organized around information processing theory, and helps students to learn efficiently.

All of the main content units in this curriculum (Units II, III, and IV) connect essential academic skills and strategies to this cognitive model. Unit II focuses on improving a student's skills at the point of attention through a discussion of strategies that can be employed before and during class. The key here is to provide students with skills in determining which pieces of information to pay attention to, and also to help them find more meaning in the significant information. To this end, this section of the curriculum focuses on developing effective practices of structuring the workflow, such as reading assignments before class, paying attention in class, and doing homework regularly. It also advocates the creation of deliberate and effortful processes of paying attention to information presented to the students. Students are encouraged not to read words, or transcribe class lectures mindlessly, but to think about the information by encouraging them to anticipate exam questions that might be asked about the material they are reading or hearing and noting the questions they anticipate. These questions will be useful for the creation of practice tests later. The method can also move beyond the anticipation of exam questions into evaluative, reflective, and integrative thinking as students advance. Under these systems, the question is no longer simply, "on what might my professor test me?" but becomes, "what is the limit of this information?" or "is this right?" or "how would this change if we included X or Y?" As has been demonstrated above, memory is the residue of thought. Since thinking is crucial to memory creation, then, the stronger the memory base in any domain, the greater is the likelihood of moving towards deeper learning.

Another key to FYE curriculum is its focus on the application of information. Students will not see the benefit of this change in study strategies unless they are encouraged to put the strategy into practice; therefore, assessment focuses on the student application of the principles being, and on student recall of the principles. Therefore, applied projects, not exams, are preferable measures for summative assessment. In addition, students benefit from some formative assessment as they work to acquire these processes, which is accomplished through "lab" sessions. The lab assignments can be extremely simple, such as having students demonstrate and discuss the way they are integrating these skills into their classes. In so doing, the students are forced to think about the material they are learning in their other courses while deliberately practicing these research-based learning strategies. Through a consistent application of these procedures, students may develop routines that support continued use of these strategies beyond the semester in which they participate in the FYE course.

Unit III covers learning at the point of encoding, and this, along with Unit IV, focuses more on what the students typically think of as "studying." This unit incorporates more of the workflow and task processing pieces but subsumes them under a cognitive model. Students need to be reminded that time management is important because learning takes time. The strategies of spaced and deliberate practice do not take more time than other study methods, but they do require that students budget their study time differently. The goal of this unit is for students to think in a "big picture" way about how they spend time and the impact those choices have on their learning. Assessment is conducted by asking students to manage their time, and by identifying a process of studying that they believe will work for them. Feedback on this process is also essential.

The final unit, Unit IV, discusses retrieval practices. The key point is that retrieval must be a regular part of one's study practice and not just something that happens on the exam. Students must be encouraged to build their memory throughout the semester and must understand that recall produces memory. Assessment measures focus on allowing students to utilize the strategies of spaced practice, deliberate practice, and effortful retrieval into practice in their other classes. Students are assigned creating a plan for studying for exams that accommodates these three cognitive concepts.

CONCLUSION

Organizing an FYE course around the development of cognitive tools is important for many reasons. First, it is designed specifically to help students to become self-directed learners. In adopting such a design, the model intentionally shifts the burden of doing the work to students, which is argued to be important for learning (Schunk, 2008) and is furthermore supported in large-scale meta-analyses of such approaches (Hattie, 2009). Second, through the explicit attention paid to these tools of learning, students will have a platform for analyzing and changing their academic progress. This is especially important because researchers have found that learners are typically very poor at assessing their learning (Brown, Roediger, & McDaniel, 2014). In effect, this platform offers learners a means through which they can cultivate a growth mindset (Dweck, 2006) and improve their capacity to self-assess learning. Finally, such an approach pays serious attention to the importance of emphasizing the in-classroom experiences of students as paramount for student retention and success. As Kuh (2008) and Eaker and Sells (2015) have argued, without explicit attention to what students encounter in the classrooms of higher education, any other retention efforts will construct an artificial ceiling for positive impact on student success. These implications of the FYE curriculum are shown in Table 3, along with its corresponding justification that is grounded in empirical research.

While no one would doubt that college is first and foremost a learning experience, the efforts to improve student success have often focused on the peripheral areas of student learning while in college. Until reforms and practices are aimed directly at the locus of learning itself, the classroom and the academic learning requirements of students, the impact will remain marginal. The students who participate in FYE courses organized around developing cognitive tools will be equipped and trained in best practices

Table 3. Positive implications for student success

Implication	Justification for Its Positive Impact on Student Success
Help learners become self-directed	Schunck (2008) and Hattie (2009) advocate for the importance of shifting the burden to students becoming more intrinsically driven when it comes to positively impacting learning
Develop learners' cognitive tools and awareness	Dweck (2006) emphasizes the importance of nurturing a sense of a growth mindset in learners; and Brown, Roediger, & McDaniel (2014) point out that, although learners are typically poor at assessing their learning, an increased awareness of cognitive tools can influence their ability to improve their learning
Emphasis on the in-class challenges faced by students	Eaker & Sells (2015) and Kuh (2008) both emphasize that without explicit attention to what happens in the formal classroom experience of students, any efforts to improve learning are at best, peripheral

for maximizing their learning. They will have an increased capacity to assess their progress, and will furthermore be applying these best practices to courses in their first semester. Given the importance of students' first semester and first-year GPA, these tools and skills will provide immediate impact while offering the potential for extended impact beyond enrollment in the course. Our proposition is that this cognitive focused FYE course is a logical adaptation that has real potential for significantly improving student learning and success in the university.

REFERENCES

Ambrose, S. A., Bridges, M., Lovett, M., DiPietro, M., & Norman, M. (2010). *How learning works: seven research-based principles for smart teaching*. San Francisco, CA: Jossey-Bass.

Astin, A. W. (1984). Student involvement: A developmental theory for higher education. *Journal of College Student Development*, *25*(4), 297–308.

Atherton, M. (2014). Academic Preparedness of First-Generation College Students: Different Perspectives. *Journal of College Student Development*, *55*(8), 824–829. doi:10.1353/csd.2014.0081

Baddeley, A. (1986). *Working memory*. Oxford: Oxford University Press.

Baddeley, A. (1992). Working Memory. *Science*, *255*(5044), 556–559. doi:10.1126/science.1736359 PMID:1736359

Barefoot, B. O. (2005). Current institutional practices in the first college year. In M. Upcraft, J. Gardner, & B. Barefoot (Eds.), *Challenging and supporting the first year student: A handbook for improving the first year of college* (pp. 47–63). San Francisco, CA: Jossey-Bass.

Braxton, J. M., & Hirschy, A. S. (2005). Theoretical developments in the study of college student departure. In A. Seldman (Ed.), *College student retention: Formula for student success* (pp. 61–87). Westport, CT: ACE Praeger.

Brown, P. C., Roediger, H. L., & McDaniel, M. A. (2014). *Make it stick: the science of successful learning*. Cambridge, MA: The Belknap Press of Harvard University Press. doi:10.4159/9780674419377

Buehler, R., Griffin, D., & Ross, M. (1994). Exploring the planning fallacy: Why people underestimate their task completion times. *Journal of Personality and Social Psychology*, *67*(3), 366–381. doi:10.1037/0022-3514.67.3.366

Calderwood, C., Ackerman, P. L., & Conklin, E. M. (2014). What else do college students do while studying? An investigation of multitasking. *Computers & Education*, *75*, 19–29. doi:10.1016/j.compedu.2014.02.004

Carey, B. (2014). *How we learn: the surprising truth about when, where, and why it happens*. New York, NY: Random House.

Collier, P. J., & Morgan, D. L. (2008). Is that paper really due today? Differences in first-generation and traditional college students understandings of faculty expectations. *Higher Education*, *55*(4), 425–446. doi:10.1007/s10734-007-9065-5

Cox, T., & Lemon, M. (2016). A Curricular Intervention for Teaching and Learning: Measurement of Gains of First-Year College Student Learning. *Journal of the Scholarship of Teaching and Learning*, *16*(3), 1–10. doi:10.14434/josotl.v16i3.19268

Crede, M., & Kuncel, N. R. (2008). Study habits, skills, and attitudes: The third pillar supporting collegiate academic performance. *Perspectives on Psychological Science*, *3*(6), 425–454. doi:10.1111/j.1745-6924.2008.00089.x PMID:26158971

Daniel, D. B., & Poole, D. A. (2009). Learning for life: An ecological approach to pedagogical research. *Perspectives on Psychological Science*, *4*(1), 91–96. doi:10.1111/j.1745-6924.2009.01095.x PMID:26158839

Demetriou, C., & Schmitz-Sciborski, A. (2011). Integration, motivation, strengths and optimism: Retention theories past, present and future. In R. Hayes (Ed.), *Proceedings of the 7th National Symposium on Student Retention*, Charleston. (pp. 300-312). Norman, OK: The University of Oklahoma

Dunlosky, J., Rawson, K. A., Marsh, E. J., Nathan, M. J., & Willingham, D. T. (2013). Improving Students Learning with Effective Learning Techniques: Promising Directions from Cognitive and Educational Psychology. *Psychological Science in the Public Interest*, *14*(1), 4–58. doi:10.1177/1529100612453266 PMID:26173288

Dweck, C. (2006). *Mindset: the new psychology of success*. New York, NY: Random House.

Eaker, R., & Sells, D. (2016). *A New Way: Introducing Higher Education to Professional Learning Communities at Work*. Bloomington, IN: Solution Tree.

Ehrlinger, J., & Dunning, D. (2003). How chronic self-views influence (and mislead) estimates of performance. *Journal of Personality and Social Psychology*, *84*(1), 5–17. doi:10.1037/0022-3514.84.1.5 PMID:12518967

Ehrlinger, J., Johnson, K., Banner, M., Dunning, D., & Kruger, J. (2008). Why the unskilled are unaware: Further explorations of (absent) self-insight among the incompetent. *Organizational Behavior and Human Decision Processes*, *105*(1), 98–121. doi:10.1016/j.obhdp.2007.05.002 PMID:19568317

Ehrlinger, J., & Shain, E. A. (2014). How accuracy in students' self-perceptions relates to success in learning. In V.A. Benassi, C.E. Overson, & C.M. Hakala (Eds.), *Applying science of learning in education: Infusing psychological science into the curriculum*. Retrieved from http://teachpsych.org/ebooks/asle2014/index.php

Ericsson, K. A., Krampe, R. T., & Tesch-Romer, C. (1993). The role of deliberate practice in the acquisition of expert performance. *Psychological Review*, *100*(3), 363–406. doi:10.1037/0033-295X.100.3.363

Gazzaniga, M. S., Ivry, R., & Mangun, G. R. (2008). *Cognitive neuroscience: The biology of the mind* (3rd ed.). New York, NY: W.W. Norton.

Goodman, K., & Pascarella, E. T. (2006). First-year seminars increase persistence and retention: A summary of the evidence from how college affects students. *Peer Review, 8*(3), 26-28.

Gurung, R. A. R. (2005). How do students really study (and does it matter)? *Teaching of Psychology*, *32*, 239–241.

Gurung, R. A. R., & McCann, L. (2011) How Should Students Study? Tips, Advice, and Pitfalls. Observer, 24(4). Retrieved March 22, 2017 from http://www.psychologicalscience.org/index.php/publications/observer/ 2011/april-11

Gurung, R. A. R., Weidert, J., & Jeske, A. (2010). Focusing on how students study. *Journal of the Scholarship of Teaching and Learning*, *10*, 28–35.

Gurung, R. R. (2004). Pedagogical Aids: Learning Enhancers or Dangerous Detours? *Teaching of Psychology*, *31*(3), 164–166. doi:10.1207/s15328023top3103_1

Hattie, J. (2009). *Visible learning: A synthesis of over 800 meta-analyses relating to achievement*. London, England: Routledge.

Hembrooke, H., & Gay, G. (2003). The Laptop and the Lecture: The Effects of Multitasking in Learning Environments. *Journal of Computing in Higher Education*, *15*(1), 46–64. doi:10.1007/BF02940852

Hirsh, R. (1974). The hippocampus and contextual retrieval of information from memory: A theory. *Behavioral Biology*, *12*(4), 421–444. doi:10.1016/S0091-6773(74)92231-7 PMID:4217626

Illeris, K. (2007). *How We Learn: Learning and Non-learning in School and Beyond*. London, England: Routledge.

Illeris, K. (2016). *Learning, Development, and Education: from learning theory to education and practice, the selected works of Knud Illeris*. London, England: Routledge.

Jamelske, E. (2006). Measuring the impact of a university first-year experience program on student GPA and retention. *Higher Education*, *57*(3), 373–391. doi:10.1007/s10734-008-9161-1

Junco, R., & Cotton, S. (2012). The relationship between multitasking and academic performance. *Computers & Education*, *59*(2), 505–514. doi:10.1016/j.compedu.2011.12.023

Kirschner, P. A., & van Merrienboer, J. G. (2013). Do learners know best? *Educational Psychologist*, *48*(3), 169–183. doi:10.1080/00461520.2013.804395

Knowlton, B. J., Ramus, S. J., & Squire, L. R. (1992). Intact artificial grammar learning in amnesia: Dissociation of classification learning and explicit memory for specific instances. *Psychological Science*, *3*(3), 172–179. doi:10.1111/j.1467-9280.1992.tb00021.x

Kuh, G. D. (2003). What were learning about student engagement from NSSE. *Change*, *35*(2), 24–32. doi:10.1080/00091380309604090

Kuh, G. D. (2008). *High-impact educational practices: What they are, who has access to them, and why they matter*. Washington, DC: Association of American Colleges and Universities.

Kuh, G. D. (2009). What student affairs professionals need to know about student engagement. *Journal of College Student Development*, *50*(6), 683–706. doi:10.1353/csd.0.0099

Nissen, M. J., Knopman, D. S., & Schacter, D. L. (1987). Neurochemical dissociation of memory systems. *Neurology, 37*(5), 789–794. doi:10.1212/WNL.37.5.789 PMID:3574678

Pascarella, E. T., & Terenzini, P. T. (2005). How College Affects Students: Vol. 2. *A Third. Decade of Research*. San Francisco, CA: Jossey-Bass.

Porter, S. R., & Swing, R. (2006). Understanding how first-year seminars affect persistence. *Research in Higher Education, 47*(1), 89–109. doi:10.1007/s11162-005-8153-6

Prevatt, F., Petscher, Y., Proctor, B. E., Hurst, A., & Adams, K. (2006). The revised Learning and Study Strategies Inventory: An evaluation of competing models. *Educational and Psychological Measurement, 66*(3), 448–458. doi:10.1177/0013164405282454 PMID:26347055

Reisberg, D. (2010). *Cognition: Exploring the Science of the Mind* (4th ed.). New York, NY: Norton.

Schacter, D. L., & Tulving, E. (1994). What are the memory systems of 1994? In D. L. Schacter & E. Tulving (Eds.), *Memory Systems 1994*. Cambridge, MA: MIT Press.

Schreiner, L., & Nelson, D. (2013). The contribution of Student Satisfaction to Persistence. *Journal of College Student Retention, 15*(1), 73–111. doi:10.2190/CS.15.1.f

Schunk, D. H. (2008). Metacognition, self-regulation, and self-regulated learning: Research recommendations. *Educational Psychology Review, 20*(4), 463–467. doi:10.1007/s10648-008-9086-3

Squire, L. R. (1994). Declarative and nondeclarative memory: Mulitiple brian systems support learning and memory. In D. Schacter & E. Tulving (Eds.), *Memory systems 1994*. Cambridge, MA: MIT Press.

Strayhorn, T. L. (2009). An examination of the impact of first-year seminars on correlates of college student retention. *Journal of the First-Year Experience & Students in Transition, 21*(1), 9–27.

Terenzini, P. T., Springer, L., Yaeger, P. M., Pascarella, E. T., & Nora, A. (1996). First-generation college students: Characteristics, experiences, and cognitive development. *Research in Higher Education, 37*(1), 1–22. doi:10.1007/BF01680039

Tinto, V. (1993). *Leaving college: Rethinking the causes and cures of student attrition* (2nd ed.). Chicago, IL: University of Chicago Press.

Tinto, V. (2000). Linking learning and leaving: Exploring the role of the college classroom in student departure. In J. M. Braxton (Ed.), *Reworking the student departure puzzle* (pp. 81–94). Nashville, TN: Vanderbilt University Press.

Tobolowsky, B. F., & Associates. (2008). *2006 National Survey of First-Year seminars: Continuing innovations in the collegiate curriculum* (Monograph No, 51). Columbia, SC: University of South Carolina, National Resource Center for the First-Year Experience and Students in Transition.

Tricot, A., & Sweller, J. (2014). Domain-specific knowledge and why teaching generic skills does not work. *Educational Psychology Review, 26*(2), 265–283. doi:10.1007/s10648-013-9243-1

Tulving, E. (1993). What is episodic memory? *Current Directions in Psychological Science, 2*(3), 67–70. doi:10.1111/1467-8721.ep10770899

Wei, F. F., Wang, Y. K., & Fass, W. (2014). Research Report: An experimental study of online chatting and notetaking techniques on college students cognitive learning from a lecture. *Computers in Human Behavior*, *34*, 148–156. doi:10.1016/j.chb.2014.01.019

Weinstein, C. E., & Palmer, D. R. (2002). *Learning and Study Strategies Inventory (LASSI):User's manual*. Clearwater, FL: H & H Publishing.

Willingham, D. T. (2009). *Why don't students like school?: A cognitive scientist answers questions about how the mind works and what it means for the classroom*. San Francisco, CA: Jossey-Bass.

KEY TERMS AND DEFINITIONS

Deliberate Practice: A highly structured form of practice including increased motivation, incorporation of prior knowledge, receipt of immediate feedback, and repetition.

Encoding: Storage of information in the long-term memory for later recall.

High Impact Practices: Pedagogical strategies proven effective for effective instruction and increased engagement of students from many different backgrounds.

Information Processing Theory: A learning theory that operates on the assumption that the human mind works like a computer, gathering information through the senses, processing it in the working memory, and storing it in the long-term memory.

Long-Term Memory: Large capacity memory store which holds episodic, semantic, and procedural memory for a long period of time.

Spaced Practice: Breaking studying up into several shorter sessions, and spreading those sessions over several days (for example, studying for 30 minutes a day for a week, rather than 4 hours in one day).

Student Integration Model: Vincent Tinto (1975) suggests that whether a student persists or drops out of college is very strongly connected to their level of social integration (friends, personal contact with academics, enjoying being here, etc.) and academic integration (grades, enjoying subjects, identifying with academic norms, etc.).

Theory of Student Involvement: Alexander Astin (1985) theory of student attrition which places emphasis on student "inputs" (previous experience, demographics, etc.), the student's "environment" (all experiences during college), and also a student's "outcomes" (knowledge, attitudes, beliefs of students). This is developed as commonly referred to as the I-E-O Model.

Working Memory: The area of memory where information is temporarily stored and processed.

Chapter 17
Communication as a Retention Factor in Online Course Delivery

Judi Simmons Estes
Park University, USA

ABSTRACT

Online courses have become an integrated aspect of course delivery for institutions of higher education. However, the attrition that occurs in online courses is substantially higher than in face-to-face courses. This chapter considers the influence of student communication factors in both course and degree program retention. An argument is presented related to the role of being intentional in planning and implementing communication strategies within courses and degree programs as a conduit for positively influencing completion. A second argument is that while all college/university personnel have a role in student attention, the role of faculty is significant and requires on-going professional development.

INTRODUCTION

Delivery of courses on-line is becoming commonplace for colleges and universities, partially driven by meeting the needs of many of today's college students who represent what was formally considered a non-traditional population. Rather than attending college right after high school, these students are older, working full-time, have family responsibilities, and need courses that fit into their lifestyle and time availability. These students tend to have a preference for the asynchronous world of distance education where course content is available at their convenience.

In a 2012 survey of chief academic leaders in higher education, 69.1 percent reported that offering courses online is critical to the long-term strategy of the institution (Allen & Seaman, 2013). A complicating factor is that some faculty and students are skeptical that courses offered online are high-quality, engaging, and effective instructionally. However, there is evidence that taking a course online is as effective and sometimes more effective, in meeting student learning outcomes, than face-to-face learning (U.S. Department of Education, 2009). A second complicating factor is that retention is one of the greatest challenges facing online educators and administrators (Lee & Choi, 2011). Retention rates for online courses are cited as being much lower than their face-to-face counterparts (Taylor, Dunn,

DOI: 10.4018/978-1-5225-2998-9.ch017

& Winn, 2015). There has been considerable discussion in the literature about what students want in online programs and factors that contribute to and take-away from student retention; however, one area for which there is a paucity of research and discussion is the role communication plays within course infrastructure and delivery and related impact on student retention.

With attrition rates significantly higher in online courses versus face-to-face courses, the development of models to explain online retention is considered imperative (Boston, Diaz, Gibson, Ice, Richardson & Swan, 2009). The purpose of this chapter is to make an argument that on-going communication touchpoints with students can positively influence retention and that these touchpoints can be embedded in the infrastructure of not only individual courses but also in the infrastructure of online degree delivery.

BACKGROUND

"Lack of persistence in online education and its' consequence of attrition, is an identified problem within the United States and internationally" (Hart, 2012, p. 19). Online learning, e-learning and distance education are each terms used for the process of delivering instruction in a venue other than a traditional brick and mortar setting, most often through a web-based learning management system. Simonson, Smaldino, Albright and Zvacek (2012) defined distance education as "institution based formal education where the learning group is separated, and where interactive telecommunications systems are used to connect learners, resources, and instructors" (p. 32). According to Allen and Seaman (2014), 35.5% of all students have taken at least one online course at an institution of higher education.

"For some institutions, web-based courses have been viewed as a way to attract new students, as well as to provide more convenient education options for students currently enrolled" (Lion, 2011, p. 49). Indeed, there has been an increasing demand (17%) for online courses over face-to-face courses; online program growth is considered a priority at over 80% of major institutions of higher education (Allen & Seaman, 2010). Who are the online learners of today and what factors have been found to influence retention?

Online Learners

In the past "traditional" students were defined as those who attend college full-time immediately after high school and live on campus; however, this group of students represents only 15% of those now attending U.S. colleges; on the other hand, those students who used to be considered "non-traditional" (working parents, veterans and military personnel, and caregivers) represent 85% of U.S. college students (UPCEA, 2014). According to the National Center for Education Statistics (NCES, 2015), there is a "new normal" for students in higher education; 74% of all undergraduate students have at least one nontraditional characteristic. For example, 66% of college students transfer between institutions prior to degree completion; 62% work either full or part time; 43% attend college part-time; 28% have at least one dependent, 35% are enrolled in two-year colleges, and 63% are first generation students (NCES).

Although online courses can be seen as a delivery method offering greater accessibility to education, there is a price associated with this increased access. Students taking courses online may require additional support that may not be required of students in face-to-face classrooms. Student failure can lead to less chance of enrolling in future courses, whereas student success increases these chances (Simonson et al., 2012). Patterson and McFadden (2009) found dropout rates to be six to seven times higher in online programs than in face-to-face course delivery.

Factors Influencing Attrition

Research studies on student retention, in face-to-face classes, began as early as 1926 (Braxton, 2000) and gained prominence in the 1970s primarily through the work of Tinto (1975). Tinto posited that efforts toward student retention need to consider both the academic and social environments of higher education with a focus on engagement. More recently, Rovai (2003) recognized that there is no simple formula for student success in college coursework and that many factors are involved in a student's desire to persist. For example, Rovai created two categories: factors prior to admissions and factors after admission; each influence on-going student decisions relative to whether to stay in or leave a course. In a review of 20 studies, Hart (2012) found the following factors influenced a student's persistence in an online degree program: proximity to graduation, flexibility inherent in asynchronous delivery, desire to complete a degree, GPA, satisfaction and relevance of content, self-efficacy, and self-motivation.

Several researchers identified factors influencing a student's persistence in online courses and/or degree program that relate to communication with the institution and/or faculty: a) quality of interactions and feedback, social connectedness and related presence in a course, and support received (Hart, 2012); b) lack of institutional support, lack of connection between the student and the institution, quality of interaction between the student and faculty, sense of isolation, disconnection, technology, and self-discipline (Heyman, 2010); and c) a lack of personal feedback from the instructor, lack of contact outside of the classroom with the instructor, inaccessibility of the instructor, and unfriendliness of faculty and staff at the institution (O'Brien, 2002). Given that communication plays a role in a student's decision regarding staying or leaving a course or a degree program, briefly reviewing elements of effective communication may provide some insights into developing a course and degree infrastructure that incorporates effective communication.

ELEMENTS OF EFFECTIVE COMMUNICATION

In this section, two forms of communication will be discussed: non-verbal and verbal. Nonverbal communication is communication without words such as facial expressions, touching, body movements, dress, posture, and even spatial distance. Communication with words is verbal and can be spoken or written. Verbal and nonverbal communications are often interactive and interdependent.

Verbal Communication

In a discussion of effective public speaking, Cosmin (2013) identified 7 C's of effective communication applicable in both oral and written communication. Examining these elements of communication can aid in demonstrating the complexity of communication in an online teaching-learning process where the majority of communication is in a written format.

1. **Completeness:** Communication should convey all facts needed to persuade the audience/students of the message content for understanding and discussion.
2. **Conciseness:** Communication needs to be clear and succinct.
3. **Consideration:** Communication must reflect knowledge of the audience and a desire to connect with the audience.
4. **Clarity:** Communication needs to be an organized presentation of the topic, details, and purpose of the message.
5. **Concreteness**: Communication includes supporting facts and examples that strengthen the audience/students understanding of the message.
6. **Courtesy:** Communication is conveyed in a manner that contributes to a reciprocal relationship and sharing of information with the audience/students and faculty.
7. **Correctness:** Communication is evidence-based.

Not only did Cosmin identify elements of effective communication but he also identified rules governing an effective communication process (2013, p. 40). Faculty in an online course could model and invite students to do the following:

- **Demonstrate Tolerance Towards Separate Opinions:** Encourage students to share their viewpoints on topics of discussion, emphasizing that a richness in discussion occurs when a variety of viewpoints are welcomed.
- **Promote Critical Thinking:** Encourage students to examine and share their own thought process with logical arguments, as well as a goal of gaining understanding of others' way of thinking in order to clarify knowledge and improve understanding.
- **Support Respectful Arguments:** Guide students to present clear arguments supporting a certain point of view, through research involving analysis, evaluation, interpretation, and synthesis of facts.
- **Write to Persuade:** Positions taken must convince others that a student's allegations are just, through providing clarity, structure, and valid logical reasoning.
- **Model Concision:** Concision speaks to efficient communication; the faculty member must clearly display every argument, show the impact of the arguments in the discussion and sum up the entire argument, while encouraging students to do the same.
- **Practice Active Listening:** Active listening involves not only listening to and analyze the arguments presented by a student to gain a deeper understanding of the student's position.

The challenge for an institution is how to embed elements of verbal (spoken and written) communication in not only online course delivery, but also in communication with students regarding the processes of completing a degree online. This level of communication is a complex process and requires intentionality in planning and delivery. For example, the seven elements of effective instruction and the six rules of communication provide a thorough checklist for both the faculty providing content and the instructional designer who provides the structure for an online course. Elements of effective communication can guide the instructional design of the course and presentation of content. The rules for the process can guide the process of course delivery and the participation of the faculty member.

Non-Verbal Communication

Common non-verbal communication that can be observed in the face-to-face classroom includes: "reading facial expressions," experience of body space, touching such as hand-shakes, how one sits, stands, walks, gestures, or holds their body. Guerrero and Miller (1998) conducted a literature search to identify nonverbal elements of effective classroom teaching, which led to examining 10-minute sections of video-taped lessons. They selected segments where the instructor was providing course content and demonstrating a variety of behaviors. These segments were shown to undergraduate sophomores and juniors who were asked to rate instructor behaviors and communication competence. Guerrero and Miller found that judgments of the instructor's expressiveness and warmth and general involvement accounted for a large amount of the judgement of the instructor's communicative competence; above all, students favored the instructors that displayed humanness. In a literature review by Lincoln (2008), for the purpose of identifying desirable characteristics of marketing educators, the conclusion was that "many, if not all, of the characteristics students desire in their instructors are either directly or indirectly related to nonverbal communication" (p. 63).

Just how much do non-verbal factors influence communication? There are a range of estimates. For example, Mehrabian (1981) estimates that approximately one-half of all communication is nonverbal communication, while Fromkin and Rodman (1983) suggest that 90 percent of communication is non-verbal. Tubbs and Moss (2006) stated that a nonverbal message functions in one of three ways, either replacing, reinforcing, or contradicting the verbal message of the communicator; "...if a verbal message conflicts with a nonverbal message, the receiver is most likely to use the nonverbal message and discard or downplay the verbal message because nonverbal cues reveal our true intentions and emotions..." (p. 112). Clearly, verbal and non-verbal communication interact with one another in online course delivery, just as it does in a face-to-face course.

Interactions Between Verbal and Non-Verbal Communication

Knapp (1972) identified six ways in which verbal and nonverbal behaviors interact. The examples provided are those that might occur in a face-to-face and an online setting.

1. **Repeating:** Nonverbal behavior sometimes repeats what was said verbally. For example, in a face-to-face classroom, a faculty member might say, "Get into groups of three" and then walk around the room to facilitate the groupings. In an asynchronous online class, the instructor could assign groups of three students or ask student to self-select groups of three.
2. **Contradicting:** Nonverbal behavior sometimes contradict what was said. For example, a student says, "I really want to get an A in this class" but then does not turn in several assignments. This can happen in a face-to-face or online course.
3. **Substituting:** Nonverbal can be used in the place of verbal. For example, if a faculty member has a box inside a door and points to it as a student hands him/her an assignment, then the student may automatically place the assignment in the box with no words exchanged between them. In an online course, non-verbal communication can occur through the use of graphics. For example, an arrow can be used to point to a tab to click to move to the next section.

4. **Complementing:** Nonverbal can elaborate or modify verbal. For example, a student may share with a faculty member that he/she is nervous about presenting in front of the whole class, yet as the presentation unfolds the student visibly becomes more and more relaxed. Similarly, a student in an online course may e-mail the instructor and share that they are having trouble figuring out how to do an assignment and ask for a review prior to submission to support an end goal of turning in an assignment that is high quality.
5. **Accenting:** Nonverbal may accent verbal. For example, in a face-to-face course, a student verbally says "yes" or "no" while shaking their head correspondingly (either side to side or up and down). In an online class, a student may state in a discussion thread that he/she does not know how to do an assignment; then the student fails to hand in the assignment.
6. **Regulating:** Nonverbal behaviors regulate the flow of verbal conversation. For example, a student may ask a faculty member a question during class and then after getting an answer, may turn his/her head to a laptop and begin typing notes. The faculty understands that the student is making notes and goes on with his/her instruction. In an online classroom, a student and faculty member may agree that the student will have contact with the faculty member each Monday to get clarification on assignments, as needed. On Mondays when the student fails to contact the faculty, there is an understanding between them that the student does not want help that week.

The intent of this discussion has been to establish that both verbal and non-verbal communications have multiple influencing factorss that impact the effectiveness of communication in both face-to-face and online courses. The online environment adds some complexity because non-verbal communication is not seen in body language but rather in the inherent structure of the course and actions of both faculty and students as well as perceptions each have of one another.

COMMUNICATION AS A RETENTION FACTOR

It is the premise of this chapter that verbal and non-verbal communications are critical elements in student persistence, retention, and completion of online courses and degree programs. Incomplete or ineffective communication can be a barrier to student persistence in course and degree completion (Aragon & Johnson, 2008). For example, slow feedback, no feedback, late feedback, difficulty contacting faculty or staff, and limited communication with faculty are each experiences that can contribute to a student's sense of frustration which can create a barrier to retention (Bunn, 2004). On the other hand, clear communication with the instructor is a factor used to overcome barriers to persistence and lead the student to success in an online course (Müller, 2008; Park & Choi, 2009). It is noteworthy that technology communication tools for collaboration, social networking, and user-generated content have transformed mainstream culture and how we deliver teacher education and professional development (Foundation for Excellence in Education, 2010).

Research on the effectiveness of online learning is primarily focused on the instructional process used to address course objectives. However, social and relational elements of the online learning environment are equally relevant, including the sense of community that is created, social presence of the students and the instructor, and a sense of connectivity and engagement (Schroeder, Baker, Terras, Mahar, & Chiasson, 2016). It is people who make the online learning environment productive, thus it is important to focus on the human activities more than the capabilities or limitations of technology (Kehrwald, 2008). An examination of communication strategies that are in place for the purpose of guiding and engaging students can be done on two levels: a) communication throughout a course and b) communication throughout a degree program and enrollment to graduation.

Communication Within Course Infrastructure

Intentionally planned strategies for communication encompass the entire student experience. In a course, the student experience involves all contacts from course enrollment to completion. For example, students may need to be shown how to navigate an online classroom; students may need to be guided in order to know what is expected and those expectations need to be clearly articulated in more than one location (Simonson et al., 2012). Without forewarning, students can underestimate the time and effort involved in completing an online course, as well as the skills needed to deal with the technical aspects of online learning (Choi, 2012). Student support services and administrators can address the distance student's specific needs by orienting students to the distance education classroom and format of learning. An orientation can address everything from technical issues and the technical requirements of online learning, to course expectations, etiquette, and course procedures (Ko & Rossen, 2010). Orientations have been demonstrated to be effective. For example, a Nursing program offered a face-to-face orientation session prior to the start of an online class. In another impactful example, a classroom community was created that incorporate the use of follow-up technology that collective reduced attrition from 20% to 1% for the first six weeks of class (Gilbert & Lyons, 2012).

In addition to strategies that provide on-going communication throughout an online course, an additional factor in student course retention is an awareness of the role of social presence. Myoka and Duncan (2007) suggested that a primary objective for online learning has become the creation of social presence, a term that encompasses an environment where learners are at ease and experience comfort in their communication with others. Social presence theory relates to interpersonal communication and looks at the quality of communication within a sequence of interactions (Schroeder, et al., 2016). One study by Richardson and Swan (2003) examined the relationship among students' perceived social presence, their perceived learning, and their satisfaction with their instructor. The study found that the higher the students' perceived social presence in their online courses, the higher they perceived their learning to be and the greater satisfaction they reported with their teacher. These results suggest that faculty must increase the "presence" or level of connectivity students feel in their courses to increase student's perceived learning and overall satisfaction.

Theory to Practice

This author's interest in the role of communication in course retention arose from experiences teaching online and strategies to invite students to engage in a level of critical thinking and discourse that can happen in a face-to-face class as content is discussed, synthesized, and applied to the profession of teaching. From several semesters of developing and delivering course content online, it became clear that online students want a transparent path to engaging with content, peers, and the teacher. Recently, Estes (2015) shared communication strategies that collectively are intended to support steps toward a high retention rate of students in online courses; these strategies include:

1. **Initial Contact:** An instructor reach out to students through e-mail, as soon as a class roster is available, can provide the instructor with the opportunity to provide a self-introduction, set the expectation for active student participation, answer initial questions, and model that the teacher for the course will also be an active participant.
2. **Introductions:** The course instructor can provide a self-introduction that models criteria for students, can respond to every student's introduction, and can begin making connections with and between students.
3. **Discussion Thread:** When students engage in discussions with each other, moderated by the course instructor, the possibilities for collaboration increase. The instructor's role is to demonstrate connections between students' sharing and provide prompts to extend the discussion.
4. **Prompt Feedback:** Instructor feedback allows students to assess existing knowledge, reflect on what students have learned and what they still need to learn; there is a positive correlation between teacher immediacy and student learning and motivation.
5. **Clear and Accessible Guidelines:** Providing a course infrastructure that is consistently used each week provides clarity and accessibility, thus supporting students in focusing on content knowledge acquisition and course participation rather than spending time trying to figure out how to complete assignments and performance expectations.
6. **Open Communication:** It is important for an instructor to create an environment of open communication so that students feel comfortable asking questions and sharing what is and is not working for them. Clearly displaying how students can get in touch with the course teacher by e-mail and phone is critical. Also, having an area that is a "Virtual Café" where students can pose questions of one another can be an effective way for students to answer one another's questions without relying solely on the teacher.
7. **Recognition of Course Completion:** While a grade recognizes course completion, completion becomes more personal when the instructor e-mails students individually to let them know how much they enjoyed the class. This outreach solidifies a connection with a teacher, the content, and contributes to the student's overall positive experience and likelihood to continue enrollment at the institution.
8. **Links to College/University Resources:** Students often need support from resources at the college/university other than the course teacher. Providing links to Student Services, the library, a writing lab, etc. are often contained with the learning management system for the course. However, students may not know about these resources or how to access a link; having the teacher of the course direct the student can be a benefit to some students.

The main idea in supporting student course retention is to provide clear communication throughout the infrastructure of the course. There is nothing more frustrating for a student than getting online in the late evening (after finishing all family obligations for the day) and facing confusion about what to do, how to do it, and where to go to get questions answered. While asynchronous course delivery meets the flexibility needs of students, it can also be a lonely experience when opportunities requiring engagement are not properly provided within the online course. While the instructor for a course has a key role in student course completion, the institution also has a responsibility for creating a system that offers an infrastructure that communicates a clear path from enrollment to graduation for online students.

Communication Within Degree Program Infrastructure

In an online degree program, the student experience spans the time from the initial class enrollment to graduation. There are many factors that impact whether or not the student experience is positive, a) response to initial inquiries into the degree program, b) ease of enrollment in coursework, c) quality of orientation to online learning, d) user-friendly nature of the learning management systems; e) quality of the course instructional design and content; f) opportunities for student engagement with content, peers, and instructors; g) student perception of instructor support; and h) visibility of institutional resources. However, the most important factor may be customer service that demonstrates all personnel on campus are working collaboratively to provide student support. As early as 1985, Noel, Levitz, and Saluri concluded that a caring attitude of college personnel is viewed as the most potent retention force on a campus. Therefore, it comes as no surprise that how institutions work with students may determine student outcomes (Braxton, 2000). "Interaction with faculty, administrative personnel and offices, and other students may also factor into engagement levels. Those levels of engagement are important in face-to-face environments but may be more important in online environments" (Boston, Ice, & Gibson, 2011, p. 8).

In creating a clear path to graduation for online students, touchpoints need to be identified where intentional communication occurs, beginning with recruitment and ending with encouragement to alumni to remain connected to the institution. Transparency is critically important. All employees need to know who to direct students to for information that is unknown to that particular employee. Ease of accessibility is important to students and frustration can arise when a student is told "I don't know who can answer that question." What happens when students don't know next steps and/or do not have easy access to someone who can answer questions is that they lodge complaints or they leave the institution. Either result is problematic. When there is an area where constant student complaints are being received, that is an area that needs immediate attention and an adjustment in the system.

Theory to Practice

The author's exploration of the role of communication in delivery of a fully online course arose after moving into a position that included responding to student inquiries and complaints. Many inquiries and complaints came from students enrolled in a fully online degree program. The inquiries were generally about processes involved in degree completion requirements, particularly: a) applying to the School of Education, b) participating in practicums and internship, and c) completing a portfolio. Other issues that arose included: a) students did not understand that the degree program did not lead to teacher certification and b) frustration from not receiving immediate responses to inquiries. While addressing inquiries and complaints was handled individually, it became clear that a new system needed to be developed with an intentional focus on communication. For example, information relevant to the kinds of inquiries that students were regularly making was not readily available to them online; the inquiry was such that the information needed to be obtained from a specific person on campus; this also resulted in a lack of transparency. In addition, a seamless system between staff Student Success Advisors and a faculty advisor was not in place. During the same two semesters that these issues were tracked and addressed, student enrollment in degree courses was also tracked. Data showed attrition as a problem; approximately 35% of students did not enroll from one semester to another.

Through gathering information from all stakeholders during an academic year, a conclusion that was drawn was that a system needed to be developed where the degree completion process was transparent to students and Student Success Advisors who do initial advising and guidance through the first 60 credit hours of a student's degree program. Development of that systemic plan is in process at this time and involves using the University's learning management system (LMS) to aid in communicating information to students about identified touchpoints. In addition, many steps have been taken that are supportive to the new system. For example, Student Success advisors had reported that students would call and ask about the degree program and what jobs the degree would prepare them for; Student Success Advisors have now been trained to answer these questions and also provided with resources for themselves and potential student inquiries that address these topics.

A plan that is in the process of being implemented begins upon enrollment. Students are given access to a course within the LMS that houses modules containing information and guidelines for each touchpoint of the program. The modules include; a) an orientation to the degree program and the jobs that the programs prepares students for, b) advising and review of the degree audit process, c) using the LMS system for an online class, d) applying to the School of Education, e) practicum placements and guidelines, f) internship and guidelines, g) completing a portfolio, and h) applying for graduation. The student, the adjunct and full-time faculty, and Student Success Advisors (housed within Enrollment services) each have access to the modules. To devise this system required working with the manager of the LMS and an instructional designer, and a faculty person who provided content. Additionally, one Student Success Advisor has been identified to work with all students who inquire about and/or enroll in the online degree program.

Every major area of study will have its own touchpoints. When a student is told what these touchpoints are when being recruited, they will have a better sense of what to expect when enrolling in the program. Informing the student during the recruiting process and again upon enrollment are first steps. There must be continued reminders and guidelines presented throughout the program. Within each module are explicit details about that specific process. For example, the "applying to the School of Education" might address the following content:

- The student may need to apply to the education department at the end of their sophomore year.
- The student fills out an online application process; a link is provided.
- The student needs to pass a state general education exam; a link is provided to exam locations and practice exams.
- Student will request letters of recommendation from two faculty inside the School of Education.
- Student will request one letter from outside of the school of education.
- Student will ask two School of Education to complete a dispositions form; the student will also need to self-complete a dispositions form; with a link provided.

The purpose of sharing this case study is to demonstrate that in one degree program communication was problematic and attrition was greater than in the corresponding face-to-face program. While a cause and effect relationship between communication and attrition was not established, anecdotal evidence suggests that the number of inquiries and complaints have decreased since steps have been put in place to communicate more clearly with students. These steps alone will not be all that is needed to address retention; educating faculty about their role in retention is equally important.

FACULTY ROLE IN RETENTION

The role of a teacher in an online classroom environment is not a new discussion (Keengwe & Kidd, 2010). The National Study on Student Engagement (NCSS, 2015) found that faculty influence students through formal and informal roles, not only as a teacher but also as an advisor and/or mentor. Students tend to decide the merits of an online course according to their perception of the availability and interactivity of the course faculty (Flottemesch, 2000). While the role of faculty in retention was mentioned in an earlier section of this chapter, a more detailed discussion will be presented here, including a discussion of effective faculty attributes, student engagement, and faculty-student relationships.

Effective Faculty Attributes

Tinto (1993) suggested that high rates of retention are marked by relatively high rates of student-faculty interactions. One study investigating the relationship between faculty and students found that students felt energized by their connections and experienced increases in self-esteem after positive interactions with faculty (Schwartz and Holloway, 2012). Reupert, Mayberry, Patrick, and Chittleborough (2009) found that the majority of students need instructors to provide a personal presence by being engaging, approachable, understanding, patient, and passionate about the subject.

Faranda and Clarke (2004) used in-depth student interviews for the purpose of identifying effective and critical faculty attributes; rapport, delivery, fairness, knowledge, credibility, and organization were the top attributes with rapport and delivery as most important. Rapport involved the ability to build relationships with mutual trust and harmony; delivery was described as the way faculty communicate course content, including faculty enthusiasm, positive attitudes, and passion. In a study conducted by Mupinga, Nora, and Yaw (2006), several expectations of online learners were identified, including: regular and prompt communication with professors, prompt feedback on assignments, clear expectations of the professors, and academic rigor. Students' needs that were identified in the Mupinga, et al. (2006) study included: technical help, flexible instructors, course information in advance, sample assignments, additional reference material, and the same course management platform for all online courses.

A perception of availability and interactivity of course faculty has been termed "teacher presence." In a study on the role of teaching presence in developing a learning community, Shea, Li, Swan, and Pickett (2005) found that instructor assistance with student discussions and conversations, as well as the quality of the course design, were important for establishing a clear teaching presence in the online environment and that this presence was positively related to students' perceptions of support and inclusiveness. Shea, Li, and Pickett (2006) reported positive relationships between instructor presence and students' sense of classroom community and overall engagement.

Student Engagement

Student engagement is an important factor in effective instruction whether the course is delivered online or face-to-face. Consider the findings of Arnold and McClure (1996): "Learners retain: 10% of what they read, 20% of what they hear, 30% of what they see, 50% of what they see and hear, 70% of what they say, and 90% of what they say and do." Meaningful academic experiences provide students with relevant accomplishments from which satisfaction is gained and in turn student engagement increases and dropout rates decrease (Park & Choi, 2009). Focusing only on content is not an engaging strategy for students in a face-to-face class or an online class; creating a learning environment that is supportive of student needs is a protective factor for student engagement (Mandernach, 2009). Student engagement equates with a sense of belonging which is a critical component for student success, especially for a first-year student (Pearson, 2012).

As students become engaged in the learning process, attrition appears to decline (Angelino, Williams, & Natvig, 2007). Mandernach (2009) posited that student engagement depends on a number of factors, including faculty personal connections with students and faculty creation of an active online environment. Eengaged students tend to complete a course. It is important that students recognize the importance of understanding what is expected of them in all aspects of a course, including assignment requirements and due dates, and students value feedback on their performance (Brinkerhoff & Koroghlanian, 2007).

Faculty-Student Communication

Catt, Miller, and Schallenkamp (2007) found that while faculty rate "knowledge of the subject" as the most important factor in education, students identified "communication skills" as most relevant. These results were largely congruent with literature pertaining to online learning that indicates a high degree of clarity and communication is essential for student satisfaction (Durrington, Barryhill, & Swafford, 2006). According to Umbach and Wawrzynski (2005), "faculty behaviors and attitudes affect students profoundly, which suggests that faculty members may play the single-most important role in student learning" (p. 176).

Communication can be a barrier in online course delivery, but it can be overcome by clearly setting expectations and explaining unspoken rules and roles to students. Karp and Bork (2012) suggest that high quality interactions and relationships with college personnel, faculty and peers improve student success. A sense of connection can emerge when a student has a relationship with just one key person within the institution and this relationship can positively impact a student's decision to remain in college; "the single most important factor in advising students who are at-risk is helping them to feel that they are cared for by the institution" (Heisserer & Parette, 2002, p. 1).

The results of a literature survey conducted by Terblanché (2015) suggest that the roles of the educator changes in a digital environment; intentionally planned communication enables a faculty member to maintain a degree of control over computer-mediated learning. In short, effective interaction with students requires effective communication. The results of effective communication include building rapport with students, keeping students engaged through a variety of activities, and pursuing strategies that enhance the probability that desired learning outcomes are achieved. It is much easier for students to appreciate the value and importance of learning if they are interacting with faculty who consistently provide interesting and challenging content while building relationships between and with students. While faculty are pivotal to the process of developing and delivering online courses, when Seaman (2011) surveyed faculty it was found that institutions were reported as below average in providing support services for teaching online.

THE INSTITUTIONAL ROLE IN RETENTION

As was discussed earlier, there is a new normal in the demographics of students attending institutions of higher education. Non-traditional students are now the majority of students entering institutions of higher education. To meet the needs of these students, institutions are beginning to examine ways to change and shift practices. For example, today's college students may be working and/or have families which results in a need for non-academic support such as a flexible schedule provided by online courses and other delivery models that allow students to move at their own pace, earn credit for past experience and demonstrated competencies, as well as providing guidance in locating childcare and managing finances (U.S. Department of Education, 2017). An additional consideration is that institutions with the most rigorous admission selection processes have exhibited the highest persistence and retention rates (Tinto, 1993), while institutions with open enrollment generally have the lowest retention rates (Boston, Ice, & Gibson, 2011). Students may need support in navigating the systems of online course delivery and institutional processes. Students representing underserved populations such as students from low-income families, first generation students and English language learners, as well as, students with disabilities often have many support needs (U.S. Department of Education).

The increase in students taking online courses, combined with the research indicating that attrition rates are higher for online courses, is a cause for concern (Hachey, Conway, Wladis, 2013). "As institutions grapple with how to effectively build high-quality online educational programs, they must also reimagine the structure of their organizations" (DelCasino, 2015). Institutions of higher education are finding themselves in a situation where it is fiscally necessary to respond to consumer demand for convenience and accessibility in course delivery; in many colleges and universities, online learners represent increased numbers within the total student count. High attrition rates of online students can result in a noticeable loss of income for institutions of higher education; even a small increase in student retention can result in a significant increase in institutional income. As a result, institutions have developed and refined comprehensive support programs aimed at encouraging and supporting academic and social excellence to assist students in this navigating factors influencing course completion (Nava, 2010).

The Need for a Systemic Approach to Retention

To be most effective, the training available to faculty who teach online must be a part of a systematic institutional retention effort that is sustainable. A well thought-out multifaceted plan must assure that all departments that have contact with or serve students work together and have established methods of communicating with one another. Examining effective methods of communication is not just for students and teachers. Representatives from various departments need to communicate with one another to best meet the needs of the student. There is an internal need for departments to share information and work collaboratively to meet student needs. Technology is continually advancing; tools available to engage students are changing daily including tools available in the adopted learning management system of the institution.

Miller (2008) identified several retention protective factors for online courses including: having students practice using technologies before critical assignments are due; allowing time for social communication within the technological environment; and offering a required activity, with a low point value, early in the course to ensure participation. Another idea by Terrion and Leonard (2010) suggested that a peer mentor "can serve as one source of support to reduce the stress experienced by a younger and less experienced student" (p. 156) which, in turn, has the potential to reduce attrition of the student. While faculty can develop these ideas to meet student needs on their own, the likelihood of ideas being generated will increase through professional development and a supported focus on generating new ideas to meet student needs.

Among faculty there are varied needs for professional development and training and in some cases faculty will need substantial support. For this reason, it is necessary to take instructors' needs into consideration during planning and implementation of training programs. In order to sustain successful online programs, institutions should address the needs of online instructors in a systematic and comprehensive manner and employ different mechanisms to support those who teach online.

Supporting Faculty to Effectively Teach Online

There is a need for preparing faculty to teach online (Almala, 2006). Well-planned and well-organized instructor training programs have been identified as a prerequisite for success in an online setting. Wolf (2006) identified five features for successful training of online faculty: 1) the course was led by an instructor trained to teach online, 2) participants must have computing skills, 3) participants use the course delivery system they will be teaching with, 4) participants have ongoing institutional support, and 5) participants are motivated to teach online. According to Wolf, "effective distance education programs provide ongoing faculty support in the form of mentoring, shadowing, continuing education workshops, or some combination of all these" (p. 58).

There will always be faculty that are new to online teaching that need to be trained, but just as important are the faculty who teach online on an on-going basis who not only need to be challenged to continue developing their skill base, but also supported in having avenues for contributing what they have learned about best practice. The more time faculty spend trying to improve their teaching: a) the more they engage students in discussion, small-group activities, student presentations or performances, and experiential activities; b) interact with students, c) use effective teaching practices, d) have significantly higher learning expectations for their students and e) the less the faculty lecture (NSSE, 2014).

While institutions of higher education typically offer a class for faculty to learn to teach online, this course tends to be focused on learning to use the particular learning management system adopted by the institution than on instructional strategies to engage students. Some institutions may offer a mentor during the first term that the faculty member actually teaches the developed course; however, the focus of the mentor tends to be on following established guidelines rather than instructional effectiveness. Beyond these initial offerings, there may or may not be professional development opportunities provided for faculty to continue to grow and develop both in pedagogy and in practice.

Some methods and techniques used in face-to-face delivery can be transferred to an online setting. However, online instruction does differ from face-to-face instruction; faculty transitioning to an online environment can encounter challenges (Paloff & Pratt, 2005). To address the needs of online students, faculty often need to acquire some new skills, roles, strategies, and techniques not needed in face-to-face course delivery (Varvel, 2007). For example, online teaching requires the instructor to shift "from being content providers to facilitators of student learning…[they] must gain comfort and proficiency in using the Web as the primary teacher-student link, and learn to teach effectively without the visual control provided by direct eye contact" (Smith, Ferguson, & Caris, 2002, p. 62).

Faculty Resistance to Teaching Online

Even though there is still resistance from faculty (who question the validity of online learning), more than 70.8% of chief academic leaders say that online learning must be part of their strategic planning (Allen & Seaman, 2015). As more faculty are needed to teach online, those who are reluctant and those who have few computer skills, may be among those who are invited to teach an online course. Faculty, however, must do more than develop technological skills. In a research study, Roman, Kelsey, and Lin (2010) found that online training programs should emphasize both technological and pedagogical skill development, evaluate participants' training needs prior to the training, and provide ongoing resources and support mechanisms after the training. As was discussed earlier, teaching online requires consideration of pedagogy as a foundation for effective online delivery of instruction.

How do faculty behaviors and attitudes impact students when faculty do not want to teach an online course, but participation is required because of student demand? Can faculty resistance to teaching online be mediated? In a national survey of faculty working within public institutions of higher education, a large percentage (64%), of 10,720 faculty members who have taught online report that it takes more time to teach an online course than a face-to-face course; faculty rate the extra effort needed as a barrier for teaching online (Seaman, 2009). To mediate this concern, institutions can make allowances for first-time online course delivery; for example, there can be a stipend to support the additional time faculty need to spend the first time they develop a course for online delivery.

Resistance of faculty to teaching online can result from real or perceived lack of institutional support (Lion, 2011). "Many faculty, staff and students view online education as devaluing the institutional mission" (DelCasino, 2015.) In order to sustain successful online programs, institutions should address the needs of online instructors in a systematic and comprehensive manner and employ different mechanisms to support instructors who teach online. An instructor's understanding of structuring and facilitating interaction in an online environment requires effective professional development as a foundation for confident and competent online teaching (Cueller, 2002).

Provide On-Going Professional Development Opportunities

To transition successfully from a traditional classroom to an online classroom environment, faculty need not only training, but also institutional support (Taylor & McQuiggan, 2008). "The starting point for any faculty staff development strategy is support from senior leadership..." (Robinson, 1998, p. 35). In a study of 364 institutions of higher education that provide online course offerings, Lion (2011) concluded that institutions that are successful at influencing student learning outcomes and driving faculty performance provide focused supports to help attain those outcomes.

Collom, Dallas, Jong, and Obexer (2001) discuss the process their university went through to develop a framework for professional development. Developing the framework was a process that involved recognizing that faculty would have varying levels of knowledge, skills, and motivation for teaching online. The resulting framework recognizes the developmental process for faculty as they grow in their knowledge, skills, and commitment to the process of teaching online.

- Help faculty to identify their strengths and gaps in knowledge in individual development for online teaching.
- Support faculty in choosing and planning a learning path to further their knowledge and skills.
- Aid administrators to identify broader needs in professional development.
- Assist in teaming developing integrated professional development vs. An approach that is "piecemeal".

Professional development can increase faculty self-efficacy in the process of teaching online, resisting resistance to this delivery model for instruction.

FUTURE RESEARCH DIRECTIONS

Online courses have become an integrated aspect of course delivery for institutions of higher education, yet attrition that occurs in online courses is substantially higher than in face-to-face courses. This chapter suggests that through a system-wide process of intentionally planning transparent and embedded strategies for communicating with students, institutions of higher education can create a conduit for lessening attrition and positively influence course and degree completion. More research, qualitative and quantitative, is needed to examine the role of communication as a retention factor in online course delivery. Following are suggestions for future research directions:

1. Case studies that describe, in detail, experiences with developing system-wide retention initiatives authored by faculty working within institutions of higher education.
2. Investigations identifying and codifying effective communication within the infrastructure of online course design and delivery.
3. Investigations of the relationship between faculty professional development (full-time and adjunct) in online course development and delivery and student retention in online courses.

CONCLUSION

Given that online courses have become an integrated aspect of course delivery for institutions of higher education and that the attrition that occurs in online courses is substantially higher than in face-to-face courses, there is institutional interest in identifying ways to retain online students in courses and degree programs. The purpose of this chapter is to make an argument that on-going communication touchpoints with students can positively influence retention and that these touchpoints can be embedded in the infrastructure of not only individual courses but also in the infrastructure of online degree delivery. Both verbal and non-verbal communication was considered.

An argument is made for institutions to intentionally plan and implement a system-wide focus on communication strategies embedded in the infrastructure of online courses and degree programs that are offered fully online. Inherent in a system-wide communication focus is recognition that each employee at an institution contributes to a student's decision to stay or leave an institution. Faculty play a particularly key role in student retention; faculty need institutional support, including professional development, to effectively teach online in a manner that is a protective factor for student retention. The offering of online courses will continue to increase; examining how institutional communication relates to student retention is a topic worthy of further investigation.

REFERENCES

Allen, I. E., & Seaman, J. (2015). *Grade Level: Tracking Online Education in the United States.* Babson Survey Research Group. Retrieved from http://www.onlinelearningsurvey.com/reports/gradelevel.pdf

Allen, I. E., & Seaman, J. (2013). *Changing course: Ten years of tracking online education in the United States.* Retrieved from http://files.eric.ed.gov/fulltext/ED541571.pdf

Allen, I. E., & Seaman, J. (2010). Learning on demand: Online education in the United States, 2009. Retrieved from http://sloanconsortium.org/publications/survey/pdf/learningondemand.pdf

Almala, A. (2006). The community college leadership perspectives of quality e-learning. *Distance Learning*, *3*, 9–14.

Angelino, L. M., Williams, F. K., & Natvig, D. (2007). Strategies to engage online students and reduce attrition rates. *The Journal of Educators Online*, *4*(2). Retrieved from http://www.thejo.com/

Aragon, S. R., & Johnson, E. S. (2008). Factors influencing completion and noncompletion of community college online courses. *American Journal of Distance Education*, *22*(3), 146–158. doi:10.1080/08923640802239962

Boston, W., Diaz, S., Gibson, A., Ice, D., Richardson, J., & Swan, K. (2009). An exploration of the relationship between indicators of the community inquiry framework and retention in online programs. *Journal of Asynchronous Learning Networks*, *13*(3), 67–83.

Boston, E., Ice, P., & Gibson, A. M. (2011). *Comprehensive assessment of student retention in online learning environments*. Charlestown, West Virginia: School of Arts and Humanities. Retrieved from http://digitalcommons.apus.edu/facultySAH/1

Chickering, A. W., & Gamson, Z. (1999). Development and adaptations of the seven principles for good practice in undergraduate education. *New Directions for Teaching and Learning*, *80*(80), 75–81. doi:10.1002/tl.8006

Braxton, J. M. (Ed.). (2000). *Reworking the student departure puzzle*. Nashville: Vanderbilt University Press.

Brinkerhoff, J., & Koroghlanian, C. M. (2007). Online students' expectations: Enhancing the fit between online students and course design. *Journal of Educational Computing Research*, *36*(4), 383–393. doi:10.2190/R728-28W1-332K-U115

Collom, G., Dallas, A., Jong, R., & Obexer, R. (2002). Six months in a leaky boat: Framing the knowledge and skills needed to teach well online. *Proceedings of ASCILITE '02*, Auckland, New Zealand, December 8-11. Retrieved from http://www.ascilite.org.au/conferences/auckland02/proceedings/papers/181.pdf

Cosmin, I. (2013). Effective communication-an essential step toward public success. *Journal of Media Research.*, *1*(15), 37–42.

DelCasino, V. (2015). Online Program Development in an Era of Rapid Institutional Change (Part 1). Distance and Online Learning. Retrieved from http://evolllution.com/revenue-streams/distance_online_learning/online-program-development-in-an-era-of-rapid-institutional-change-part-1/

Durrington, V. A., Berryhill, A., & Swafford, J. (2006). Strategies for enhancing student interactivity in an online environment. *College Teaching*, *54*(1), 190–193. doi:10.3200/CTCH.54.1.190-193

Estes, J. S. (2015). The pivotal role of online faculty in student engagement and retention. In L. Kyei-Blankson, J. Blankson, E. Ntuli, & C. Agyeman (Eds.), *Handbook of Research on Strategic Management of Interaction, Presence, and Participation in Online Courses*. Hershey, PA: IGI Global.

Faranda, W. T., & Clark, I. (2004). Student observations of outstanding teaching: Implications for marketing educators. *Journal of Marketing Education*, *26*(3), 271–282. doi:10.1177/0273475304268782

Flottemesch, K. (2000). Building effective interaction in distance education. *Educational Technology*, *4*(3), 46–51.

Foundation for Excellence in Education. (2010). *Digital Learning Now*! Tallahassee, FL: Author. Retrieved from www.excelined.org/Docs/Digital%20Learning%20Now%20Report%20FINAL.pdf

Fromkin, V., & Rodman, R. (1983). *An Introduction to Language*. New York: CBS College Publishing.

Gilmore, M., & Lyons, E. (2012). NURSING 911: An orientation program to improve retention of online students. *Nursing Education Perspectives*, *33*(1), 45–47. doi:10.5480/1536-5026-33.1.45 PMID:22416541

Guerrero, L., & Miller, T. A. (1998). Associations between nonverbal behaviors and initial impressions of instructor competence and course content in videotaped distance education courses. *Communication Education*, *47*(1), 30–42. doi:10.1080/03634529809379108

Hachey, A. C., Conway, K. M., & Wladis, C. W. (2013). Community colleges and underappreciated assets: Using institutional data to promote success in online learning. *Online Journal of Distance Learning Administration*, *16*(1), 1–18.

Hart, C. (2012). Factors associated with student persistence in an online program of study: A review of the literature. *Journal of Interactive Online Learning*, *11*(1), 19–42.

Heisserer, D. L., & Parette, P. (2002). Advising at-risk students in college and university settings. *College Student Journal*, *36*, 69–83.

Heyman, E. (2010). Overcoming student retention issues in higher education online programs. *Online Journal of Distance Learning Administration*, *8*(4).

Karp, M. M., & Bork, R. H. (2012). *"They never told me what to expect, so I didn't know what to do": Defining and clarifying the role of a community college student* (Vol. CCRC Working Paper No. 47). New York, NY: Columbia University, Teachers College, Community College Research Center.

Keengwe, J., & Kidd, T. T. (2010). Towards best practices in online learning and teaching in higher education. *MERLOT Journal of Online Learning and Teaching*, *6*(2). Retrieved from http://jolt.merlot.org/vol6no2/keengwe_0610.htm

Kehrwald, B. (2008). Understanding social presence in text-based online learning environments. *Distance Education*, *29*(1), 89–106. doi:10.1080/01587910802004860

Knapp, M. L. (1972). *Nonverbal communication in human interaction*. New York: Holt, Rinehart and Winston, Inc.

Ko, S., & Rossen, S. (2010). *Teaching online: A practical guide* (3rd ed.). New York, NY: Routledge.

Lee, Y., & Choi, J. (2011). A review of online course dropout research: Implications for practice and future research. *Educational Technology Research and Development*, *59*(5), 593–618. doi:10.1007/s11423-010-9177-y

Lincoln, D. J. (2008). Drama in the classroom: How and why marketing educators can use nonverbal communication and enthusiasm to build student rapport. *Marketing Education Review*, *8*(3), 53–65. do i:10.1080/10528008.2008.11489048

Lion, R. W. (2011). A study of performance support in higher education. *Performance Improvement Quarterly*, *24*(1), 49–67. doi:10.1002/piq.20101

Mandernach, B. J. (2009). Three ways to improve student engagement in the online classroom. *Online Classroom*.

Mehrabian, A. (1981). *Silent Messages: Implicit Communication of Emotions and Attitudes*. Belmont, CA: Wadsworth.

Miller, S. K. (2008) *Online, distance learning instructor guidelines to improve student retention*. Maricopa Learning Exchange, Phoenix, AZ. Retrieved from www.mcli.dist.maricopa.edu/mix/slip.php?item-1372

Müller, T. (2008). Persistence of women in online degree-completion programs. *International Review of Research in Open and Distance Learning*, *9*(2), 1–18. doi:10.19173/irrodl.v9i2.455

Mupinga, D. M., Nora, R. T., & Yaw, C. (2006). The learning styles, expectations, and needs of online students. *College Teaching*, *54*(1), 185–189. doi:10.3200/CTCH.54.1.185-189

Mykota, D., & Duncan, R. (2007). Learner characteristics as predictors of online social presence. *Canadian Journal of Education*, *30*(1), 157–170. doi:10.2307/20466630

National Center for Education Statistics. (2015). Demographics and enrollment characteristics of non-traditional undergraduates: 2011-2012. Retrieved from http://nces.ed.gov.pubs2015/2015025.pdf

National Survey of Student Engagement (NCSS). (2014). Center for Postsecondary Research, Indiana University School of Education, Bloomington, IND. Retrieved from http://nsse.iub.edu/NSSE_2014Results/pdfNSSE_2014_Annual_Results.pdf#page=8

Nava, M. E. (2010). *Exploring new paths: The first-year experiences for first-generation college students and the impact of participating in comprehensive programs*. Unpublished doctoral dissertation, University of Texas, Austin.

Noel, L., Levitz, R., & Saluri, D. (1985). *Increasing student retention*. San Francisco: Jossey-Bass.

O'Brien, G. (2002). *Issues Paper 3: A sense of belonging*. Retrieved from http://www.fye.qut.edu.au/documents/FYIEissues&5Fpaper%5F3.pdf

Palloff, R. M., & Pratt, K. (2005). *Collaborating online: Learning together in community*. San Francisco, CA: Jossey-Bass.

Park, J. H., & Choi, H. J. (2009). Factors influencing adult learners' decision to drop out or persist in online learning. *Journal of Educational Technology & Society*, *12*(4), 207–217.

Patterson, B., & McFadden, C. (2009). Attrition in online and campus degree programs. *Online Journal of Distance Learning Administration*, *12*(2). Retrieved from http://www.westga.edu/~distance/ojdla/summer122/patterson112.html

Reupert, A., Maybery, D., Patrick, K., & Chittleborough, P. (2009). The importance of being human: Instructors' personal presence in distance programs. *International Journal of Teaching and Learning in Higher Education*, *21*(1), 47–56.

Richardson, J. C., & Swan, K. (2003). Examining social presence in online courses in relate to students perceived learning and satisfaction. *Journal of Asynchronous Learning Networks*, *1*, 68–98.

Roman, T., Kelsey, K., & Lin, H. (2010). Enhancing online education through instructor skill development in higher education. *Online Journal of Distance Learning Administration*, *8*(1). Retrieved from http://www.westga.edu/~distance/ojdla/winter134/roman_kelsey134.html

Rovai, A. P., & Jordan, H. (2004). Blended learning and sense of community: A comparative analysis with traditional and fully online graduate courses. *International Review of Research in Open and Distance Learning*, *2*(5), http://www.irrodl.org/index.php/irrodl/article/viewArticle/192/274 Retrieved December 21, 2014

Schwartz, H. L., & Holloway, E. L. (2012). Partners in learning: A grounded theory study of relational practice between masters students and professors. *Mentoring & Tutoring: Partnership in Learning*, *20*(1), 115–135. doi:10.1080/13611267.2012.655454

Seaman, J. S. (2009). Online learning as a strategic asset. Volume II: The paradox of faculty voices -Views and experiences with online learning. Results of a National Faculty Survey, Part of the Online Education Benchmarking Study Conducted by the APLU-Sloan National Commission on Online Learning. Association of Public and Land Grant Universities Schroeder.

Schroeder, S., Baker, M., Terras, K., Mahar, P., & Chiasson, K. (2016). Desired and experienced levels of connectivity to an asynchronous, online, distance degree program. *Online Learning*, *20*(3), 244–263.

Shea, P., Li, C. S., Swan, K., & Pickett, A. M. (2005). Developing learning community in online asynchronous college courses: the role of teaching presence. *Journal of Asynchronous Learning Networks*, *9*. Retrieved from http://www.sloan.org/publications/jaln/v9n4/pdf/v9n4_shea.pdf

Simonson, M., Smaldino, S., Albright, M., & Zvacek, S. (2012). *Teaching and learning at a distance: Foundations of distance education* (5th ed.). Boston, MA: Pearson.

Smith, G., Ferguson, D., & Caris, M. (2002). Teaching over the web versus in the classroom: Differences in the instructor experience. *International Journal of Instructional Media*, *29*(1), 61–67.

Taylor, A., & McQuiggan, C. M. (2008). Faculty development programming: If we build it, will they come? *Educause*, *3*, 28–37.

Taylor, J. M., Dunn, M., & Winn, S. K. (2015). Innovative orientation leads to improved success in online courses. *Online Learning*, *19*(4), 112–120. doi:10.24059/olj.v19i4.570

Terblanché, E. (2015). Deciding to teach online: Communication, opportunities and challenges for educators in distance education. *Communicatio*, *41*(4).

Terrion, J., & Leonard, D. (2010). A taxonomy of the characteristics of student peer mentors in higher education: Findings from a literature review. *Mentoring & Tutoring: Partnership in Learning*, *15*(2), 149–164. doi:10.1080/13611260601086311

Tinto, V. (1975). Dropout from higher education: A theoretical synthesis of recent research. *Review of Educational Research*, *45*(1), 89–125. doi:10.3102/00346543045001089

Tinto, V. (1993). *Leaving college: Rethinking the causes and cures of student attrition* (2nd ed.). Chicago: University of Chicago Press.

Tubbs, S., & Moss, S. (2006). *Human communication: Principles and contexts*. New York, NY: McGraw Hill.

Umbach, P. D., & Wawrzynsli, M. R. (2005). Faculty do matter: The role of college faculty in student learning and engagement. *Research in Higher Education*, *46*(2), 153–185. doi:10.1007/s11162-004-1598-1

UPCEA (University Professional and Continuing Education Association). (2014, October). Federal Policy Brief. Retrieved from http://www.upcea.edu/Files/UPCEAPolicyBrief.pdf

U.S. Department of Education. (2017). *Reimagining the role of technology in higher education. A supplement to the National Education Technology Plan*. Washington, DC: Office of Educational Technology.

U.S. Department of Education. (2009). Evaluation of evidence-based practices in online learning: A meta-analysis and review of online learning studies. Washington, D.C. Retrieved from http://www.gibill.va.gov/gi_bill_info/ch33/yellow_ribbon.htm

Wolf, P. (2006). Best practices in the training of faculty to teach online. *Journal of Computing in Higher Education*, *17*(2), 47–78. doi:10.1007/BF03032698

ADDITIONAL READING

Arnold, W. E., & McClure, L. (1996). *Communication Training and Development* (2nd ed.). Prospect Heights, IL: Waveland Press.

Baker, C. (2010). The impact of instructor immediacy and presence for online student affective learning, cognition, and motivation. *Journal of Educators Online*, *7*(1), 1–30. Retrieved from http://www.thejeo.com/Archives/Volume7Number1/BakerPaper.pdf doi:10.9743/JEO.2010.1.2

Bean, J. P., & Metzner, B. S. (1985). A conceptual model of nontraditional undergraduate student attrition. *Review of Educational Research*, *55*(4), 485–540. doi:10.3102/00346543055004485

Bunn, J. (2004). Student persistence in a LIS distance education program. *Australian Academic Research Libraries*, *35*(3), 253–270. doi:10.1080/00048623.2004.10755275

Catt, S., Miller, D., & Schallenkamp, K. (2007). You are the key: Communicate for Learning Effectiveness. *Education*, *127*(3), 369–377.

Cho, M. H. (2012). Online student orientation in higher education: A developmental study. *Educational Technology Research and Development*, *60*(6), 1051–1069. doi:10.1007/s11423-012-9271-4

Collison, G., Elbaum, B., Haavind, S., & Tinker, R. (2000). *Facilitating online learning: Effective strategies for moderators*. Madison, WI: Atwood Publishing.

Dennen, V. P. (2005). From message posting to learning dialogs: Factors affecting learner participation in asynchronous discussion. *Distance Education*, *26*(1), 127–148. doi:10.1080/01587910500081376

Ekstrand, B. (2013). Prerequisites for persistence in distance education. *Online Journal of Distance Learning Administration*, *16*(2). Retrieved from http://www.westga.edu/~distance/ojdla/fall163/ekstrand164.html

Gascoigine, C., & Parnell, J. (2014). Distance education readiness assessments: An overview and application. *Online Journal of Distance Learning Administration*, *17*(4). Retrieved from http://www.westga.edu/~distance/ojdla/winter174/gascoigne_parnell174.html

Jairam, D., & Kiewra, K. A. (2010). Helping students soar to success on computers: An investigation of the SOAR study method for computer-based learning. *Journal of Educational Psychology*, *102*(3), 601–614. doi:10.1037/a0019137

Kitsantas, A. (2013). Fostering college students' self-regulated learning with learning technologies. *Hellenic Journal of Psychology*, *10*, 235–252.

Levy, Y. (2007). Comparing dropouts and persistence in e-learning courses. *Computers & Education*, *48*(2), 185–204. doi:10.1016/j.compedu.2004.12.004

Online Learning Consortium. (2015). *Five pillars of quality online education*. Retrieved from http://onlinelearningconsortium.org/about/quality-framework-five-pillars/

Pearson, M. (2012). Building bridges: Higher degree student retention and counseling support. *Journal of Higher Education Policy and Management*, *34*(2), 187–199. doi:10.1080/1360080X.2012.662743

Rovai, A. P. (2003). In search of higher persistence rates in distance education online programs. *The Internet and Higher Education*, *6*(1), 1–16. doi:10.1016/S1096-7516(02)00158-6

Rovai, A. P., & Downey, J. R. (2010). Why some distance education programs fail while others succeed in a global environment. *The Internet and Higher Education*, *13*(3), 141–147. doi:10.1016/j.iheduc.2009.07.001

Shea, P., Li, C. S., & Pickett, A. (2006). A study of teaching presence and student sense of learning community in fully online and web-enhanced college courses. *The Internet and Higher Education*, *9*(3), 175–190. doi:10.1016/j.iheduc.2006.06.005

Stanford-Bowers, D. (2008). Persistence in online classes: A study of perceptions among stakeholders. *Journal of Online Learning and Teaching*, *4*(1), 37–50.

KEY WORDS AND DEFINITIONS

Asynchronous: In an online environment, communication is primarily in a written format occurring at a time determined by each respondent, rather than according to a synchronized time on a clock.

Attrition: In academia, attrition refers to the loss of student enrollment; a reduction in the number of enrolled students which results in a loss of revenue.

Distance Education: One of the terms used for the delivery model of providing education in a setting other than a brick and mortar building, typically through a web-based learning management system; other terms include online learning and e-learning.

Learning Management System: A learning management system (LMS) used in higher education is a web-based technology used to deliver content; generally providing a way for an instructor to plan, implement, and assess an instructional- learning process that occurs between instructor and student.

National Study on Student Engagement (NSSE): Since 2000, the NSSE has annually collected College Survey Report results from colleges and universities; results pertain to how undergraduate students spend their time and what they gain from college attendance.

Online Learning: Online learning is a term that refers to the process of gaining informing through web-based, internet based sources. The term is also referred to as e-learning, virtual learning, net-based learning; a wide variety of technology is encompassed by the term online learning; in this chapter the term is used for whole course internet delivery.

Retention: Within a college setting, student retention refers to continued participation in a course or program until completion.

Student Engagement: Student engagement involves, not only active participation, but also critical thinking, synthesis, and application of content to real-life experiences. Highly engaged students tend to feel a connection to their institutions of higher education and remain committed to completion of individual courses within their degree plan.

Compilation of References

Abele, C., Penprase, B., & Ternes, R. (2013). A closer look at academic probation and attrition: What courses are predictive of nursing student success? *Nurse Education Today*, *33*(3), 258–261. doi:10.1016/j.nedt.2011.11.017 PMID:22177474

ACT. (2013). *ACT Engage College Domains and Subscales.* Retrieved from http://www.act.org/content/dam/act/unsecured/documents/EngageCollegeHandout.pdf

Adelman, C., Ewell, P., Gaston, P., & Schneider, C. G. (2014). *The Degree Qualifications Profile.* Lumina Foundation. Retrieved from http://www.luminafoundation.org/publications/The_Degree_Qualifications_Profile.pdf

Adelman, C. (1999). *Answers in the toolbox: Academic intensity, attendance patterns, and bachelor's degree attainment (Document No. PLLI 1999-8021).* Washington, DC: U.S. Department of Education, Office of Educational Research and Improvement.

Adelman, C. (2006). *The toolbox revisited: Paths to degree completion from high school through college.* Washington, DC: U.S. Department of Education.

Aguirre, A. Jr. (2000). *Women and Minority Faculty in the Academic Workplace: Recruitment, Retention, and Academic Culture. ASHE-ERIC Higher Education Report, 27(6).*

Ai, A. L., Cascio, T., Santangelo, L. K., & Evans-Campbell, T. (2005). Hope, meaning, and growth following the September 11, 2001, terrorist attacks. *Journal of Interpersonal Violence*, *20*(5), 523–548. doi:10.1177/0886260504272896 PMID:15788553

Aiken-Wisneiwski, S. (Ed.). (2010). *Guide to Assessment in Academic Advising* (2nd ed.). Manhattan, KS: The National Academic Advising Association.

Alalshaikh, S. (2015). Cultural Impacts on distance learning, online learning styles, and design. *Quarterly Review of Distance Education*, *16*(3), 67–75.

Alexander, C., Kohnke, M., & Naginey, A. (2011) *Undergraduate and graduate retention: Two concepts, one outcome.* Retrieved 28 March 2017 from www.earth.callutheran.edu/fact_book/documents

Allen, I. E., & Seaman, J. (2010). Learning on demand: Online education in the United States, 2009. Retrieved from http://sloanconsortium.org/publications/survey/pdf/learningondemand.pdf

Allen, I. E., & Seaman, J. (2013). *Changing course: Ten years of tracking online education in the United States.* Retrieved from http://files.eric.ed.gov/fulltext/ED541571.pdf

Allen, I. E., & Seaman, J. (2015). *Grade Level: Tracking Online Education in the United States.* Babson Survey Research Group. Retrieved from http://www.onlinelearningsurvey.com/reports/gradelevel.pdf

Allen, I. E., & Seaman, J. (2015, February). *Grade level: Tracking online education in the United States 2015*. Retrieved from http://www.onlinelearningsurvey.com/reports/ gradechange.pdf

Allen, I. E., & Seaman, J. (2013). *Changing course: Ten years of tracking online education in the United States*. Babson Park, MA: Babson Survey Research Group.

Allen, I., & Seaman, J. (2011). *Going the Distance: Online Education in the USA 2011*. Wellesley, MA: Babson Survey Research Group.

Allen, S. J., Miguel, R. F., & Martin, B. A. (2014). Know, see, plan, do: A model for curriculum design in leadership development. *SAM Advanced Management Journal*, *79*(2), 26.

Alliger, G. M., & Dwight, S. A. (2000). A meta-analytic investigation of the susceptibility of integrity tests to coaching and faking. *Educational and Psychological Measurement*, *60*(1), 59–72. doi:10.1177/00131640021970367

Allport, G. (1971). *Personality: A psychological interpretation*. London: Constable.

Almala, A. (2006). The community college leadership perspectives of quality e-learning. *Distance Learning*, *3*, 9–14.

Altback, P., & Knight, J. (2007). The internationalization of higher education: Motivations and Realities. *Journal of Studies in International Education*, *11*(3-4), 290–306. doi:10.1177/1028315307303542

Ambrose, S. A., Bridges, M., Lovett, M., DiPietro, M., & Norman, M. (2010). *How learning works: seven research-based principles for smart teaching*. San Francisco, CA: Jossey-Bass.

American Association of Colleges of Nursing (AACN). (2007). Alliance for Nursing Accreditation Statement on Distance Education Policies.

American Association of Colleges of Nursing (AACN). (2015). New AACN Data Confirm Enrollment Surge in Schools of Nursing.

American Council on Education. (2011). Report of the blue-ribbon panel on global engagement strength through global leadership and engagement: U.S. higher education in the 21st Century. Campus internationalization. Retrieved from http://www.acenet.edu/news-room/Documents/2011-CIGE-BRPReport.pdf

American Council on Education. (2014). *Campus internationalization*. Retrieved from http://www.acenet.edu/higher-education/topics/Pages/Campus-Internationalization.aspx

American Institutes for Research. (2010). *College dropouts cost cash-strapped states billions*. Retrieved from http://www.air.org/news/press-release/college-dropouts-cost-cash-strapped-states-billions

American Psychological Association. (2016). *Depression*. Retrieved from http://www.apa.org/topics/depression/

Anderson, E. (1997). *Chip*. Iowa City, IA: Academic Advising for Student Success and Retention. Noel-Levitz.

Angelino, L. M., Williams, F. K., & Natvig, D. (2007). Strategies to engage online students and reduce attrition rates. *The Journal of Educators Online*, *4*(2). Retrieved from http://www.thejo.com/

Angelino, L. M., Williams, F. K., & Natvig, D. (2007). Strategies to engage online students and reduce attrition rates. *The Journal of Educators Online*, *4*(2), 1–14. doi:10.9743/JEO.2007.2.1

Antonio, A. L. (2004). The influence of friendship groups on intellectual self-confidence and educational aspirations in college. *The Journal of Higher Education*, *75*(4), 446–471. doi:10.1353/jhe.2004.0019

Apergis, N., & Pekka-Economou, V. (2010). Incentives and female entrepreneurial activity: Evidence from panel firm level data. *International Advances in Economic Research*, *16*(4), 371–387. doi:10.1007/s11294-010-9277-9

Aragon, S. R., & Johnson, E. S. (2008). Factors influencing completion and noncompletion of community college online courses. *American Journal of Distance Education, 22*(3), 146–158. doi:10.1080/08923640802239962

Aranda, P. (2015, September). In most MBA programs, the diversity trend is down. *The Consortium.* http://cgsm.org/in-most-mba-programs-the-diversity-trend-is-down/

Armstrong, L. (2007). Competing in the global higher education marketplace: Outsourcing, twinning, and franchising. *New Directions for Higher Education, 140*(140), 131–138. doi:10.1002/he.287

Arnau, R. C., Rosen, D. H., Finch, J. F., Rhudy, J. L., & Fortunato, V. J. (2007). Longitudinal Effects of Hope on Depression and Anxiety: A Latent Variable Analysis. *Journal of Personality, 75*(1), 43–64. doi:10.1111/j.1467-6494.2006.00432.x PMID:17214591

Arnett, J. J. (2000). Emerging adulthood: A theory of development from the late teens through the twenties. *The American Psychologist, 55*(5), 469–480.

Asia Society Center for Global Education. (2016). *South Korean education reforms.* Retrieved from http://asiasociety.org/global-cities-education-network/south-korean-education-reforms

Asian Society. (2016). South Korean Education Reforms. Retrieved July 7, 2016 from http://asiasociety.org/global-cities-education-network/south-korean-education-reforms

Association of American Colleges and Universities. (2011). *The LEAP vision for learning: Outcomes, practices, impact, and employers' views.* Washington, DC: Author.

Association of American Colleges and Universities. (2014). Essential learning outcomes. Retrieved from https://www.aacu.org/leap/essential-learning-outcomes

Astin, A. W. (1984). Student involvement: A developmental theory for higher education. *Journal of College Student Development, 25*(4), 297–308.

Astin, A. W. (1984). Student involvement: A developmental theory for higher education. *Journal of College Student Personnel, 25*(2), 297–308.

Astin, A. W. (1984). Student involvement: A developmental theory for higher education. *Journal of College Student Personnel, 25*, 297–308.

Astin, A. W. (1990). *Assessment for excellence: The philosophy and practice of assessment and evaluation in higher education.* New York: Macmillan.

Astin, A. W. (1993). Diversity and multiculturalism on the campus: How are students affected? *Change: The Magazine of Higher Learning, 25*(2), 44–49. doi:10.1080/00091383.1993.9940617

Astin, A. W. (1993). *What matters in college? Four critical years revisited.* San Francisco, CA: Jossey Bass.

Astin, A. W. (1993). *What Matters in College?* San Francisco: Jossey-Bass.

Astin, A. W. (1993). *What matters in college?: Four critical years revisited* (Vol. 1). San Francisco: Jossey-Bass.

Atherton, M. (2014). Academic Preparedness of First-Generation College Students: Different Perspectives. *Journal of College Student Development, 55*(8), 824–829. doi:10.1353/csd.2014.0081

Attract Project. (2012). Retrieved from http://attractproject.org/

Ausburn, L. J. (2004). Gender and learning strategy differences in non-traditional adult students' design preferences in hybrid distance courses. *Journal of Interactive Online Learning*, *3*(2). Retrieved from http://www.ncolr.org/jiol/issues/pdf/3.2.6.pdf

Austin, A. W. (1999). Student involvement: A development theory for higher education. *Journal of College Student Development*, *40*(5), 518–529.

Author. A. A (2017). Developing a practice of cultural awareness in pre-service teachers by promoting positive dialogue around diversity. In Leavitt, L., Wisdom, S. and Leavitt, K. (Eds.), Cultural awareness and competency development in higher education. Hershey, PA: IGI Global Publishing.

Babson (2015). 2015 Online Report Card- Tracking Online Education in the United States. Online Learning Consortium. Retrieved from http://www.babson.edu/Academics/faculty/provost/Pages/babson-survey-research-group.aspx

Baddeley, A. (1986). *Working memory*. Oxford: Oxford University Press.

Baddeley, A. (1992). Working Memory. *Science*, *255*(5044), 556–559. doi:10.1126/science.1736359 PMID:1736359

Baldacchino, D. R., & Galea, P. (2012). Student nurses' personality traits and the nursing profession: Part 2. *British Journal of Nursing (Mark Allen Publishing)*, *21*(9), 530–535. doi:10.12968/bjon.2012.21.9.530 PMID:22585266

Ball, S., Maguire, M., & Macrae, S. (1997). The Post-16 Education Market: Ethics, interests and survival. *Paper presented at BERA Conference*, York, UK. September 11-14.

Ball, S. (1997). Policy sociology and critical social research: A personal review of recent education policy and policy research. *British Educational Research Journal*, *23*(3), 257–274. doi:10.1080/0141192970230302

Bamber, J., & Tett, L. (2001). Ensuring integrative learning experiences for non-traditional students in Higher education. *Journal of Widening Participation and Lifelong Learning*, *3*(1), 8–18.

Bandura, A. (1977). Self-efficacy: Toward a unifying theory of behavioral change. *Psychological Review*, *84*(2), 191–215. doi:10.1037/0033-295X.84.2.191 PMID:847061

Bandura, A. (1977). *Social learning theory*. New York, NY: General Learning Press.

Bandura, A. (1986). *Social foundations of thought and action: A social cognitive theory*. Englewood Cliffs, NJ: Prentice Hall.

Bandura, A. (1997). *Self-efficacy: The exercise of control*. New York: Freeman.

Bandura, A., & Locke, E. A. (2003). Negative self-efficacy and goal effects revisited. *The Journal of Applied Psychology*, *88*(1), 87–99. doi:10.1037/0021-9010.88.1.87 PMID:12675397

Bang, H., & Kim, J. (2016). Korean and American teachers' praising styles and teaching practices. *Mid-Western Educational Researcher*, *28*(1), 28–54.

Barefoot, B. (2000). The first year experience: Are we making it any better? *About Campus: Enriching the Student Learning Experience*, *4*(6), 12–18.

Barefoot, B. O. (2005). Current institutional practices in the first college year. In M. Upcraft, J. Gardner, & B. Barefoot (Eds.), *Challenging and supporting the first year student: A handbook for improving the first year of college* (pp. 47–63). San Francisco, CA: Jossey-Bass.

Baron, N. (2010). The forgotten underrepresented minority. *Huffington Post*. http://www.huffingtonpost.com/noah-baron/the-forgotten-underrepres_b_620995.html

Bart, C. (1998). Mission Matters. *The CPA Journal*, *68*(8), 56–57.

Bart, C. K. (1997). Sex, Lies, and Mission Statements. *Business Horizons*, *40*(6), 9–18. doi:10.1016/S0007-6813(97)90062-8

Bataille, G., & Brown, B. (2006). *Faculty career paths: Multiple routes to academic success and satisfaction*. Wesport, CT: Praeger Publishers.

Bauer, K. (2011). Training women for success: An evaluation of entrepreneurship training programs in Vermont, USA. *Journal of Entrepreneurship Education*, *14*, 1.

Baum, S., Ma, J., & Payea, K. (2010). *Education pays 2010: The benefits of higher education for individuals and society*. Retrieved March 28, 2017 from www.advocacy.collegeboard.org

Baumgartner, L. M. (2001). An update on transformational learning. *New Directions for Adult and Continuing Education*, *2001*(89), 15–24. doi:10.1002/ace.4

Baum, S., Ma, J., & Payea, K. (2010). *Education Pays 2010: The Benefits of Higher Education for Individuals and Society*. College Board Trends in Higher Education Series.

Bean, J. P. (1980). Dropouts and turnover: The synthesis and test of a casual model of student attrition. *Research in Higher Education*, *12*(2), 155–187. doi:10.1007/BF00976194

Bean, J. P. (1983). The application of a model of turnover in work organizations to the student attrition process. *Review of Higher Education*, *6*(2), 129–148. doi:10.1353/rhe.1983.0026

Bean, J. P. (1985). Interaction effects based on class level in an explanatory model of college student dropout syndrome. *American Educational Research Journal*, *22*(1), 35–64. doi:10.3102/00028312022001035

Bean, J. P. (2005). Nine themes of college student retention. In A. Seidman (Ed.), *College student retention* (pp. 215–243). Westport: Praeger Publishers.

Bean, J. P., & Metzner, B. S. (1985). A conceptual model of nontraditional undergraduate student attrition. *Review of Educational Research*, *55*(4), 485–540. doi:10.3102/00346543055004485

Bean, J., & Eaton, S. (2000). A psychological model of college student retention. In J. Braxton (Ed.), *Reworking the departure puzzle: New theory and research on college student retention* (pp. 48–61). Nashville, TN: Vanderbilt Universities Press.

Benjamin, L. (1997). *Black women in academy: Promises and perils*. Gainesville, Fl: University Press of Florida.

Bennet, C., & Okinaka, A. (1990). Factors Related to Persistence Among Asian, Black, Hispanic, and White Undergraduates at a Predominantly White University: Comparison Between First and Fourth Year Cohorts. *The Urban Review*, *22*(1), 33–59. doi:10.1007/BF01110631

Bennett, R. E. (2010). Cognitively based assessment of, for, and as learning: A preliminary theory of action for summative and formative assessment. *Measurement: Interdisciplinary Research and Perspectives*, *8*(2-3), 70–91. doi:10.1080/15366367.2010.508686

Berger, B. J. (2001). Understanding the Organizational Nature of Student Persistence: Empirically-Based Recommendations for Practice. *Journal of College Retention*, *3*(1), 3–21. doi:10.2190/3K6A-2REC-GJU5-8280

Berger, J. B., & Lyon, S. C. (2005). Past to present: A historical look at retention. In A. Seidman (Ed.), *College student retention* (pp. 1–29). Westport, CT: Praeger Publishers.

Berg, G. (2010). *Low-income students and perpetuation of inequality: Higher education in America*. Burlington, VT: Ashgate Pub.

Berlin, L.G., Furstenberg, F.F., & Waters, M.C. (2010). Introducing the Issue. *The Future of Children*, *20*(1), 3–17. doi:10.1353/foc.0.0042 PMID:20064752

Berry, J. W. (2005). Acculturation: Living successfully in two cultures. *International Journal of Intercultural Relations*, *29*(6), 697–712. doi:10.1016/j.ijintrel.2005.07.013

Berry, R. M. (1997). *Direct Selling: From door-to-door to network marketing*. Butterworth Heinemann.

Bettinger, E. P., & Baker, R. B. (2013). The effects of student coaching: An evaluation of a randomized experiment in student advising. Retrieved from http://www.insidetrack.com/wp-content/uploads/2014/09/aera-educational-evaluation-policy-analysis.pdf

Bettinger, E., & Baker, R. (2011). *The effects of student coaching in college: An evaluation of a randomized experiment in student mentoring (No. w16881)*. National Bureau of Economic Research. doi:10.3386/w16881

Bhardwaj, B. (2014). Impact of education and training on performance of women entrepreneurs: A study in emerging market context. *Journal of Entrepreneurship in Emerging Economies*, *6*(1), 38–52. doi:10.1108/JEEE-05-2013-0014

Bikowski, D., & Kessler, G. (2002). Making the most of the discussion boards in the ESL classroom. *TESOL Journal*, *11*, 27–30.

Black, R. (2016). Coaching to Increase Retention in Online Degree Completion Programs, a Promising Practice Study. Retrieved from https://rossier.usc.edu/files/2015/06/Ruth_Claire_Black.pdf

Black, R. (2016). Coaching to Increase Retention in Online Degree Completion Programs. Retrieved from http://digitallibrary.usc.edu/cdm/compoundobject/collection/p15799 coll40/ id/298507/rec/2

Black, R. (2016). Coaching to increase retention in online degree completion programs. *USC*. Retrieved from, http://digitallibrary.usc.edu/cdm/compoundobject/collection/p15799coll40/id/298507/rec/2

Black, R. C. (2016). *Coaching to increase retention in online degree completion programs*. Retrieved from http://digitallibrary.usc.edu/cdm/compoundobject/collection/p15799coll40/id/298507/rec/2

Blair, J. (2015). Predicting success in nursing programs. *Research in Higher Education Journal*, *28*, 1.

Bollag, B., & Overland, M. A. (2001). Developing countries turn to distance education. *The Chronicle of Higher Education*, *47*(40), 29–31.

Borchert, M. A. E. (2005). *Master's education: A guide for faculty and administrators. A policy statement* (2nd ed.). Washington, DC: Council of Graduate Schools.

Borkowski, J. G., Carr, M., Relliger, E., & Pressley, M. (1990). Self-regulated cognition: Interdependence of metacognition, attributions, and self-esteem. In B. Jones & L. Idol (Eds.), *Dimensions of Thinking and Cognitive Instruction* (Vol. 1). Hillsdale, NJ: Erlbaum.

Bosch, W. C., Hester, J. L., MacEntee, V. M., MacKenzie, J. A., Morey, T. M., Nichols, J. T., & Young, R. R. et al. (2008). Beyond lip service: An operational definition of learning-centered college. *Innovative Higher Education*, *33*(2), 83–98. doi:10.1007/s10755-008-9072-1

Boston, E., Ice, P., & Gibson, A. M. (2011). *Comprehensive assessment of student retention in online learning environments*. Charlestown, West Virginia: School of Arts and Humanities. Retrieved from http://digitalcommons.apus.edu/facultySAH/1

Boston, W., Diaz, S., Gibson, A., Ice, D., Richardson, J., & Swan, K. (2009). An exploration of the relationship between indicators of the community inquiry framework and retention in online programs. *Journal of Asynchronous Learning Networks*, *13*(3), 67–83.

Bowles, S., Gintis, H., & Osborne, M. (2001). The determinants of earnings: A behavioral approach. *Journal of Economic Literature*, *39*(4), 1137–1176. doi:10.1257/jel.39.4.1137

Bradley, T., & Lomicka, L. (2000). A case study of learner interaction in technology-enhanced language learning environments. *Journal of Educational Computing Research*, *22*(3), 347–368. doi:10.2190/TCUA-3YUV-B1P5-26P3

Brahm, G. (2006). Students with staying power. *Nacubo*. Retrieved from http://www.nacubo.org/Business_Officer_Magazine/Magazine_Archives/June_2006/Students_With_Staying_Power.html

Brandtstädter, J., & Renner, G. (1990). Tenacious goal pursuit and flexible goal adjustment: Explication and age-related analysis of assimilative and accommodative strategies of coping. *Psychology and Aging*, *5*(1), 58–67. doi:10.1037/0882-7974.5.1.58 PMID:2317302

Brandtstädter, J., & Rothermund, K. (2002). The life-course dynamics of goal pursuit and goal adjustment: A two-process framework. *Developmental Review*, *22*(1), 117–150. doi:10.1006/drev.2001.0539

Braxton, J. M., Hirschy, A. S., & McClendon, S. A. (2004). Toward understanding and reducing college student departure. ASHE-ERIC Higher Education Report, 30(3).

Braxton, J. M. (Ed.). (2000). *Reworking the student departure puzzle*. Nashville: Vanderbilt University Press.

Braxton, J. M., & Hirschy, A. S. (2005). Theoretical developments in the study of college student departure. In A. Seldman (Ed.), *College student retention: Formula for student success* (pp. 61–87). Westport, CT: ACE Praeger.

Braxton, J. M., Hirschy, A. S., & McClendon, S. A. (2004). *Understanding and reducing college student departure* (A. J. Kezar, Ed.). (Vol. 30). San Francisco, CA: Wiley Periodicals, Inc.

Brinkerhoff, J., & Koroghlanian, C. M. (2007). Online students' expectations: Enhancing the fit between online students and course design. *Journal of Educational Computing Research*, *36*(4), 383–393. doi:10.2190/R728-28W1-332K-U115

Brock, S. E. (2010). Measuring the importance of precursor steps to transformative learning. *Adult Education Quarterly*, *60*(2), 122–142. doi:10.1177/0741713609333084

Bronstein, S. (2007). Supplemental instruction: supporting persistence in barrier courses [Thesis Ed.D.]. University of Massachusetts.

Brown, C., & Bartee, R. (2009). *The broken cisterns of African American education: academic performance and achievement in the post-Brown era*. Charlotte, NC: Information Age Pub.

Brownell, J., & Swaner, L. (2012). Five High-Impact Practices: Research on Learning Outcomes, Completion, and Quality. *Peer Review*, *14*(3), 29.

Brown, P. C., Roediger, H. L., & McDaniel, M. A. (2014). *Make it stick: the science of successful learning*. Cambridge, MA: The Belknap Press of Harvard University Press. doi:10.4159/9780674419377

Bryson, C., & Hand, L. (2007). The Role of engagement in inspiring teaching and learning. *Innovations in Education and Teaching International*, *44*(4), 349–363. doi:10.1080/14703290701602748

Buchmann, C., & DiPrete, T. A. (2006). The growing female advantage in college completion: The role of family background and academic achievement. *American Sociological Review*, *71*(4), 515–541. doi:10.1177/000312240607100401

Buehler, R., Griffin, D., & Ross, M. (1994). Exploring the planning fallacy: Why people underestimate their task completion times. *Journal of Personality and Social Psychology*, *67*(3), 366–381. doi:10.1037/0022-3514.67.3.366

Burger, E. B., & Starbird, M. (2012). *The 5 elements of effective thinking*. Princeton, NJ: Princeton University Press.

Burns, P. (2006). *Success in college: From C's in high school to A's in college*. Lanham, MD: Rowman & Littlefield Education.

Burrus, J., Elliott, D., Brenneman, M., Markle, R., Carney, L., Moore, G., . . . Roberts, R. (2013). *Putting and keeping students on track: Toward a comprehensive model of college persistence and goal attainment.* (ETS Research Report No. RR-13-14). Princeton, NJ: Educational Testing Service.

Cabrera, A. F., Castaneda, M. B., Nora, A., & Hengstler, D. (1992). The convergence between two theories of college persistence. *The Journal of Higher Education*, *63*(2), 143–164. doi:10.2307/1982157

Calderwood, C., Ackerman, P. L., & Conklin, E. M. (2014). What else do college students do while studying? An investigation of multitasking. *Computers & Education*, *75*, 19–29. doi:10.1016/j.compedu.2014.02.004

Calhoun, S. L., & Mayes, S. D. (2005). Processing speeds in children with clinical disorders. *Psychology in the Schools*, *42*(4), 333–343. doi:10.1002/pits.20067

Campaign Brief Asia. (2012). The Global Media Intelligence Report. Retrieved from http://www.campaignbriefasia.com/2012/10/10/eMarketer%20Global%20Media%20Intelligence%20Report%20Exec%20Summary.pdf

Campbell, N. (2007). Bringing ESL students out of their shells: Enhancing participation through online discussion. *Business Communication Quarterly*, *70*(1), 37–43. doi:10.1177/108056990707000105

Campbell, S., Nutt, C., Robbins, R., Kirk-Kuwaye, M., & Higa, L. (2005). *NACADA guide to assessment in academic advising*. Manhattan, KS: National Academic Advising Association.

Campbell-Whatley, G. D. (2003). Recruiting and retaining of culturally and linguistically diverse groups in special education: Defining the problem. *Teacher Education and Special Education*, *26*(4), 255–263. doi:10.1177/088840640302600402

Carey, B. (2014). *How we learn: the surprising truth about when, where, and why it happens*. New York, NY: Random House.

Carnevale, A. P. (2012, July 2). *The great sorting*. Retrieved from http://chronicle.com/article/The-Great-Sorting/132635/

Carnevale, A. P., & Rose, S. J. (2015). *The economy goes to college: The hidden promise of higher education in the post-industrial service economy.* Retrieved from https://cew.georgetown.edu/wp-content/uploads/EconomyGoesToCollege.pdf

Carnevale, A. P., Jayasundera, T., & Cheah, B. (2012). *The college advantage: Weathering the economic storm.* Retrieved from https://cew.georgetown.edu/wp-content/uploads/2014/11/CollegeAdvantage.ExecutiveSummary.081412.pdf

Carnevale, A. P., Smith, N., & Strohl, J. (2013). *Recovery: Job growth and education requirements through 2020.* Retrieved from https://cew.georgetown.edu/wp-content/uploads/2014/11/Recovery2020.FR_.Web_.pdf

Carnevale, A.P., Smith, N., & Strohl, J. (2014). Recovery: Job growth and education requirements through 2020 (executive summary). Georgetown University. Retrieved from https://cew.georgetown.edu/wp-content/uploads/2014/11/Recovery 2020.ES_.Web_.pdf

Carr, S. (1999). *Distance-Learning group blends offerings of 2- and 4-Year colleges*. The Chronicle of Higher Education. Retrieved from: www.chonicle.com

Carr, J., Walker, W., Carr, M., & Fulwood, D. (2012, May). Reflect for success: Use of mentor recommendations to help failing students. *British Journal of Community Nursing*, *17*(5), 226–228. doi:10.12968/bjcn.2012.17.5.226 PMID:22584401

Carroll, J. (2015). *Tools for Teaching in an Educationally Mobile World*. London: Routledge.

Carter, M., & Fuller, C. (2015). Symbolic interactionism. *Sociopedia*. doi:10.1177/205684601561

Casner-Lotto, J., & Benner, L. (2006). *Are they really ready to work? Employers' perspectives on the basic knowledge and applied skills of new entrants into the 21st century workforce*. New York, NY: The Conference Board, Inc. Retrieved from http://www.p21.org/storage/documents/FINAL_REPORT_PDF09-29-06.pdf

Cassuto, L. (2015, June 1). The degree for quitters and losers. *The Chronicle of Higher Education*. Retrieved March 16, 2017 from www.chronicle.com/article/The-Degree-for-Quitters

Catalano, R. F., Berglund, M. L., Ryan, J. A., Lonczak, H. S., & Hawkins, J. D. (2002). Positive youth development in the United States: Research findings on evaluations of positive youth development programs. *Prevention & Treatment*, *5*(1), 15a.

Caulfield, J. (2011). *How to Design and Teach a Hybrid Course*. Sterling, VA: Stylus Publishing.

Chandra, S., Shiffman, S., Scharf, D. M., Dang, Q., & Shadel, W. G. (2007). Daily smoking patterns, their determinants, and implications for quitting. *Experimental and Clinical Psychopharmacology*, *15*(1), 67–80. doi:10.1037/1064-1297.15.1.67 PMID:17295586

Chapelle, C. (2001). *Computer applications in second language acquisition*. Cambridge University Press. doi:10.1017/CBO9781139524681

Charles, M. (2007). Language matters in global communication: Article based on ORA lecture, October 2006. *Journal of Business Communication*, *44*(3), 260-282.

Chau, J. (2012). More students are enrolled in college and on financial aid, annual report shows. The Chronicle of Higher Education. Retrieved March 3 2012, from http://chronicle.com/article/More-Students-Are_Enrolled-in/131320/

Chen, H., & Lattica, L. (2008). Conceptualizing engagement: Contributions of faculty to Student Engagement in Engineering. *The Journal of Engineering Education*, *97*(3), 339–352. doi:10.1002/j.2168-9830.2008.tb00983.x

Chester, A., & Gwynne, G. (2006). Online teaching: Encouraging collaboration through anonymity. *Journal of Computer-Mediated Communication*, 4.

Chickering, A. W., & Ehrmann, S. C. (1996). Implementing the seven principles: Technology as a lever. *AAHE Bulletin*, *49*(2), 3–6.

Chickering, A. W., & Gamson, Z. (1999). Development and adaptations of the seven principles for good practice in undergraduate education. *New Directions for Teaching and Learning*, *80*(80), 75–81. doi:10.1002/tl.8006

Chickering, A., & Reisser, L. (1993). *Education and identity. San*. Francisco: Jossey-Bass Publishers.

Child Trends Data Bank. (2015). Parental Expectations for Their Children's Academic Attainment. Retrieved from https://www.childtrends.org/wp-content/ uploads/2015/10/115_Parental_Expectations.pdf

Choi, I. C. (2008). The impact of EFL testing on EFL education in Korea. *Language Testing*, *25*(1), 39–62. doi:10.1177/0265532207083744

Cho, M. J., Nam, J. J., & Suh, G. H. (1998). Prevalence of symptoms of depression in a nationwide sample of Korean adults. *Psychiatry Research*, *81*(3), 341–352. doi:10.1016/S0165-1781(98)00122-X PMID:9925185

Christopherson, K. M. (2007). The positive and negative implications of anonymity in Internet social interactions: On the Internet, nobody knows you're a dog. *Computers in Human Behavior*, *23*(6), 3038–3056. doi:10.1016/j.chb.2006.09.001

Cinoğlu, H., & Arıkan, Y. (2012). Self, identity and identity formation: From the perspectives of three major theories. *Journal of Human Sciences*, *9*(2), 1115–1131.

Clandinin, D. J. (Ed.). (2006). *Handbook of narrative inquiry: Mapping a methodology*. Sage Publications.

Clayton, K., Blumberg, F., & Auld, D. (2010). The relationship between motivation, learning strategies and choice of environment whether traditional or including an online component. *British Journal of Educational Technology*, *41*(3), 349–364. doi:10.1111/j.1467-8535.2009.00993.x

Cohen, K. E. (2012). *Persistence of master's students in the United States: Development and testing of a conceptual model* [Doctoral dissertation].

Collier, P. J., & Morgan, D. L. (2008). Is that paper really due today? Differences in first-generation and traditional college students understandings of faculty expectations. *Higher Education*, *55*(4), 425–446. doi:10.1007/s10734-007-9065-5

Collier-Reed, B., Wolmarans, N., & Smit, R. (2010, October). The impact of NSC mathematics on student performance in mathematics in first-year engineering programmes: Where does the gap lie? *Paper presented at the STEM, Mind the Gap Conference, Academy of Science of South Africa, Cape Town.* Retrieved January 19, 2017 from http://uct.academia.edu/BrandonCollierReed/Papers/362566/

Collins, M., & Berge, Z. (1995). Introduction: Computer-mediated communications and the online classroom in higher education. In Z.L. Berge & M.P. Collins (Eds.), Computer mediated communication and the online classroom (Vol. 2, pp. 1–10). New Jersey: Hampton Press.

Collins, J. (2001). *Good to great: Why some companies make the leap...and others don't.* New York: HarperCollins Publishers Inc.

Collom, G., Dallas, A., Jong, R., & Obexer, R. (2002). Six months in a leaky boat: Framing the knowledge and skills needed to teach well online. *Proceedings of ASCILITE '02,* Auckland, New Zealand, December 8-11. Retrieved from http://www.ascilite.org.au/conferences/auckland02/proceedings/papers/181.pdf

Comacchio, A., & Scapolan, A. (2004). The adoption process of corporate e-learning it Italy. *Education + Training*, *46*(6/7), 315–325. doi:10.1108/00400910410555222

Compton, W. C., & Hoffman, E. (2013). *Positive psychology: The science of happiness and flourishing*. Belmont: Wadsworth Cengage Learning.

Conley, D. (2007). *Towards a more comprehensive conception of College Readiness. A report prepared for the Bill and Melinda Gates Foundation.* Eugene, OR: Educational Policy Improvement Center. Retrieved January 1, 2017 from http://www.smallschoolsproject.org/PDFS

Conley, C. S. (2015). SEL in higher education. In J. A. Durlak, C. E. Domitrovich, R. P. Weissberg, & T. P. Gullotta (Eds.), *Handbook of social and emotional learning: Research and practice* (pp. 197–212). New York: Guilford.

Conley, P. A., & Hamlin, M. L. (2009). Justice-learning: Exploring the efficacy with low-income, first-generation college students. *Michigan Journal of Community Service Learning*, *16*(1), 47–58.

Conrad, C. F., Duren, K. M., & Haworth, J. G. (1998). Students perspectives on their master's degree experiences: Disturbing the conventional wisdom. *New Directions for Higher Education*, *26*(1), 65–76. doi:10.1002/he.10106

Cosmin, I. (2013). Effective communication-an essential step toward public success. *Journal of Media Research.*, *1*(15), 37–42.

Coughlin, E. K. (1994). *The emergence of the 'global city'*. The Chronicle of Higher Education. Retrieved from: www.chonicle.com

Council for the Advancement of Standards in Higher Education. (2011). *The role of counseling services.* Washington, DC: Author. Retrieved from http://collegecounseling.org/docs/CAS.pdf

Council for the Advancement of Standards. (2013). Academic Advising Programs: CAS Standards and Guidelines. Retrieved from http://standards.cas.edu/getpdf.cfm?PDF=E864D2C4-D655-8F74-2E647CDECD29B7D0

Council of Graduate Schools. (2013). *Completion and attrition in STEM master's programs: Pilot study findings.*

Covington, M. V. (2000). Goal theory, motivation, and school achievement: An integrative review. *Annual Review of Psychology*, *51*(1), 171–200. doi:10.1146/annurev.psych.51.1.171 PMID:10751969

Cox, T., & Lemon, M. (2016). A Curricular Intervention for Teaching and Learning: Measurement of Gains of First-Year College Student Learning. *Journal of the Scholarship of Teaching and Learning*, *16*(3), 1–10. doi:10.14434/josotl.v16i3.19268

Cranton, P. (2006). *Understanding and promoting transformative learning: A guide for educators of adults* (2nd ed.). San Francisco, CA: Jossey-Bass.

Crede, M., & Kuncel, N. R. (2008). Study habits, skills, and attitudes: The third pillar supporting collegiate academic performance. *Perspectives on Psychological Science*, *3*(6), 425–454. doi:10.1111/j.1745-6924.2008.00089.x PMID:26158971

Credé, M., Roch, S. G., & Kieszczynka, U. M. (2010). Class attendance in college a meta-analytic review of the relationship of class attendance with grades and student characteristics. *Review of Educational Research*, *80*(2), 272–295. doi:10.3102/0034654310362998

Credé, M., Tynan, M. C., & Harms, P. D. (2016). Much ado about Grit: A meta-analytic synthesis of the grit literature. *Journal of Personality and Social Psychology*. doi:10.1037/pspp0000102 PMID:27845531

Creswell, J. W. (2009). *Research Design: Qualitative, Quantitative, and Mixed Methods Approaches* (3rd ed.). Thousand Oaks, CA: Sage.

Creswell, J. W. (2012). *Educational research: Planning, conducting, and evaluating quantitative and qualitative research* (4th ed.). Boston, MA: Pearson Education, Inc.

Creswell, J. W. (2013). *Qualitative inquiry and research design: Choosing among five approaches* (3rd ed.). Thousand Oaks, California: Sage Publications, Inc.

Crockett, D.S., & American Coll. *Testing Program.* (1988). Academic Advising Audit: An Institutional Evaluation and Analysis of the Organization and Delivery of Advising Services.

Crookston, B. B. (1972). A developmental view of academic advising as teaching. *Journal of College Student Personnel*, *13*, 12–17.

Dacanay, A. P., Vaughn, S., Orr, M., Andre, J., & Mort, K. (2015). Factors contributing to nursing distance education student success. *Journal of Nursing Education and Practice*, *5*(2), 34.

Daehwan, N. (2003). The eclectic development of the Neo-Confucianism and statecraft from the 18th to the 19th century. *Korea Journal*, *43*(4), 87–112.

Daniel, B. (2015). Big Data and analytics in higher education: Opportunities and challenges: The Value of Big Data in Higher Education. *British Journal of Educational Technology*, *46*(5), 904–920. doi:10.1111/bjet.12230

Daniel, D. B., & Poole, D. A. (2009). Learning for life: An ecological approach to pedagogical research. *Perspectives on Psychological Science*, *4*(1), 91–96. doi:10.1111/j.1745-6924.2009.01095.x PMID:26158839

Darryl, T. R. (2010). The influence of student involvement with campus life on the retention of African American students enrolled at Public Historically Black University. Retrieved from http://repository.lib.ncsu.edu/ir/bitstream/1840.16/5067/1/etd.pdf

Davenport, R. (2005). A new shade of BIG BLUE IBM reinvents its workforce. *T + d*, *59*(5), 34.

Davidson, C., & Wilson, K. (2013). Reassessing Tinto's concepts of social and academic integration in student retention. *Journal of College Student Retention*, *15*(3), 329–346. doi:10.2190/CS.15.3.b

Davis, K. (1949). *Human Society*. New York, NY: Macmillan.

Davismoon, S. (2016). Supporting Transition, Transforming Culture, Improving Retention and Student Success Through Personal Tutoring. *Paper presented at the HEA Conference 'Ready for Retention: Effective Systems for Transition and Student Success' at University of South Wales.*

DeCapua, A., & Wintergerst, A. (2004). *Crossing cultures in the language classroom*. Ann Habor, MI: The University of Michigan Press.

Deci, E., & Ryan, R. (1985). *Intrinsic motivation and self-determination in human behavior*. New York: Plenum Publishing. doi:10.1007/978-1-4899-2271-7

Deford, D. (1985). Validating the construct of theoretical orientation in reading instruction. *Reading Research Quarterly*, *20*(3), 351–367. doi:10.2307/748023

Deil-Amen, R., & Rosenbaum, J. E. (2003). The social prerequisites of success: Can college structure reduce the need for social know-how? *The Annals of the American Academy of Political and Social Science*, *586*(1), 120–143. doi:10.1177/0002716202250216

DelCasino, V. (2015). Online Program Development in an Era of Rapid Institutional Change (Part 1). Distance and Online Learning. Retrieved from http://evolllution.com/revenue-streams/distance_online_learning/online-program-development-in-an-era-of-rapid-institutional-change-part-1/

DeMarco, T. (1982). *Controlling software projects: management, measurement & estimation*. New York, NY: Yourdon Press, c.

Dembo, M. H., & Seli, H. (2008). *Motivation and learning strategies for college success: A focus on self-regulated learning*. New York, NY: Routledge Taylor.

Demetriou, C., & Schmitz-Sciborski, A. (2011). Integration, motivation, strengths and optimism: Retention theories past, present and future. In R. Hayes (Ed.), *Proceedings of the 7th National Symposium on Student Retention*, Charleston. (pp. 300-312). Norman, OK: The University of Oklahoma

Dennis, J. M., Phinney, J. S., & Chuateco, L. I. (2005). The role of motivation, parental support, and peer support in the academic success of ethnic minority first-generation college students. *Journal of College Student Development*, *46*(3), 223–236. doi:10.1353/csd.2005.0023

Desruisseaux, P. (1991). *Congress approves program to support overseas study*. The Chronicle of Higher Education. Retrieved from www.chonicle.com

Desruisseaux, P. (1999). *What parents should know about study-abroad programs.* The Chronicle of Higher Education. Retrieved from www.chonicle.com

Desruisseaux, P. (1999a). *15% rise in American students abroad shows popularity of non-European destinations.* The Chronicle of Higher Education. Retrieved from www.chonicle.com

Desruisseaux, P. (1999b, December 10). *Foreign Students Continue to Flock to the U.S.* The Chronicle of Higher Education. Retrieved from www.chonicle.com

Desruisseaux, P. (2000,). *As exchanges lose a political rationale, their role is debated.* The Chronicle of Higher Education. Retrieved from www.chonicle.com

Dewar, R. (2002). Case study: Space station robot embeds Ada. *COTS Journal, 3.* Retrieved from http://www.adacore.com/uploads/customers/CaseStudy_SpaceArm.pdf

DeWitz, S. J., Woolsey, M. L., & Walsh, W. B. (2009). College student retention: An exploration of the relationship between self-efficacy beliefs and purpose in life among college students. *Journal of College Student Development, 50*(1), 19–34. doi:10.1353/csd.0.0049

Diener, E. (2000). Subjective Well Being: The science of happiness and a proposal for a national index. *The American Psychologist, 55*(1), 34–43. doi:10.1037/0003-066X.55.1.34 PMID:11392863

Diener, E., Ng, W., Harter, J., & Arora, R. (2010). Wealth and happiness across the world: material prosperity predicts life evaluation, whereas psychosocial prosperity predicts positive feeling. *Journal of Personality and Social Psychology, 99*(1), 52–61. doi:10.1037/a0018066 PMID:20565185

Dilchert, S., Ones, D. S., Viswesvaran, C., & Deller, J. (2006). Response distortion in personality measurement: Born to deceive, yet capable of providing valid self-assessments? *Psychological Science, 48*(3), 209.

Drake, J. K., Jordan, P., & Miller, M. A. (Eds.). (2013). *Academic advising approaches: Strategies that teach students to make the most of college.* John Wiley & Sons.

Drasgow, F., Chernyshenko, O. S., & Stark, S. (2010). 75 years after Likert: Thurstone was right! *Industrial and Organizational Psychology: Perspectives on Science and Practice, 3*(4), 465–476. doi:10.1111/j.1754-9434.2010.01273.x

Duckworth, A. (2016). *Grit: Passion, Perseverance, and The Science Of Success.* Scribner.

Duckworth, A. L., Gendler, T. S., & Gross, J. J. (2016). Situational strategies for self-control. *Perspectives on Psychological Science, 11*(1), 35–55. doi:10.1177/1745691615623247 PMID:26817725

Duckworth, A. L., Peterson, C., Matthews, M. D., & Kelly, D. R. (2007). Grit: Perseverance and passion for long-term goals. *Journal of Personality and Social Psychology, 92*(6), 1087–1101. doi:10.1037/0022-3514.92.6.1087 PMID:17547490

Duckworth, A. L., White, R. E., Matteucci, A. J., Shearer, A., & Gross, J. J. (2016). A stitch in time: Strategic self-control in high school and college students. *Journal of Educational Psychology, 108*(3), 329–341. doi:10.1037/edu0000062 PMID:27158155

Duckworth, A. L., & Yeager, D. S. (2015). Measurement matters: Assessing personal qualities other than cognitive ability for educational purposes. *Educational Researcher, 44*(4), 237–251. doi:10.3102/0013189X15584327 PMID:27134288

Duff, P. (2002). Pop culture and ESL students: Intertextuality, identity, and participation in classroom discussions. *Journal of Adolescent & Adult Literacy, 45*(6), 482–487.

Dumbrigue, C., Moxley, D., & Najor-Durack, A. (2001). *Keeping students in higher education: Successful practices and strategies for retention.* Routledge.

Dunlosky, J., Rawson, K. A., Marsh, E. J., Nathan, M. J., & Willingham, D. T. (2013). Improving Students Learning with Effective Learning Techniques: Promising Directions from Cognitive and Educational Psychology. *Psychological Science in the Public Interest*, *14*(1), 4–58. doi:10.1177/1529100612453266 PMID:26173288

Durlak, J. A., Weissberg, R. P., Dymnicki, A. B., Taylor, R. D., & Schellinger, K. B. (2011). The impact of enhancing students social and emotional learning: A meta-analysis of school-based universal interventions. *Child Development*, *82*(1), 405–432. doi:10.1111/j.1467-8624.2010.01564.x PMID:21291449

Durrington, V. A., Berryhill, A., & Swafford, J. (2006). Strategies for enhancing student interactivity in an online environment. *College Teaching*, *54*(1), 190–193. doi:10.3200/CTCH.54.1.190-193

Dweck, C. (2006). *Mindset: the new psychology of success*. New York, NY: Random House.

Dweck, C. S. (2008). *Mindset: The new psychology of success*. Random House Digital, Inc.

Dwyer, M. M., & Peters, C. K. (2004). The benefits of study abroad. *Transitions abroad, 37*(5), 56-58.

Dwyer, B. (2006). Framing the effect of multiculturalism on diversity outcomes among students at historically black colleges and universities. *Educational Foundations*, *20*(1-2), 37–59.

Eaker, R., & Sells, D. (2016). *A New Way: Introducing Higher Education to Professional Learning Communities at Work*. Bloomington, IN: Solution Tree.

Eaton, S. B., & Bean, J. P. (1995). An approach/avoidance behavioral model of college student attrition. *Research in Higher Education*, *36*(6), 617–645. doi:10.1007/BF02208248

Eccles, J. S., & Wigfield, A. (2002). Motivational beliefs, values, and goals. *Annual Review of Psychology*, *53*(1), 109–132. doi:10.1146/annurev.psych.53.100901.135153 PMID:11752481

Education First. (2013). South Korea. Retrieved from http://www.ef.edu/epi/spotlight/south-korea/

Edwards, J. (2007). Challenges and opportunities for the internationalization of higher education in the coming decade: Planned and opportunistic initiatives in American Institutions. *Journal of Studies in International Education*, *11*(3/4), 373–381. doi:10.1177/1028315307303920

Egeland, J., Rund, B. R., Sundet, K., Landre, N. I., Asbjornsen, A., Lund, A., & Hugdahl, K. et al. (2003). Attention profile in schizophrenia compared with depression: Differential effects of processing speed, selective attention and vigilance. *Acta Psychiatrica Scandinavica*, *108*(4), 276–284. doi:10.1034/j.1600-0447.2003.00146.x PMID:12956828

Ehrlinger, J., & Shain, E. A. (2014). How accuracy in students' self-perceptions relates to success in learning. In V.A. Benassi, C.E. Overson, & C.M. Hakala (Eds.), *Applying science of learning in education: Infusing psychological science into the curriculum*. Retrieved from http://teachpsych.org/ebooks/asle2014/index.php

Ehrlinger, J., & Dunning, D. (2003). How chronic self-views influence (and mislead) estimates of performance. *Journal of Personality and Social Psychology*, *84*(1), 5–17. doi:10.1037/0022-3514.84.1.5 PMID:12518967

Ehrlinger, J., Johnson, K., Banner, M., Dunning, D., & Kruger, J. (2008). Why the unskilled are unaware: Further explorations of (absent) self-insight among the incompetent. *Organizational Behavior and Human Decision Processes*, *105*(1), 98–121. doi:10.1016/j.obhdp.2007.05.002 PMID:19568317

Eisenberger, R. (1992). Learned industriousness. *Psychological Review*, *99*(2), 248–267. doi:10.1037/0033-295X.99.2.248 PMID:1594725

Elliot, E. S., & Dweck, C. S. (1988). Goals: An approach to motivation and achievement. *Journal of Personality and Social Psychology*, *54*(1), 5–12. doi:10.1037/0022-3514.54.1.5 PMID:3346808

Elliott, K. M., & Healy, M. A. (2001). Key factors influencing student satisfaction related to recruitment and retention. *Journal of Marketing for Higher Education*, *10*(4), 1–11. doi:10.1300/J050v10n04_01

Engle, J., & Tinto, V. (2008). *Moving beyond access: College success for low-income, first-generation college students.* The Pell Institute. Retrieved from http://files.eric.ed.gov/fulltext/ED504448.pdf

Engle, L., & Engle, J. (2003). Study abroad levels: Toward a classification of program types. *Frontiers: The Interdisciplinary Journal of Study Abroad*, *9*, 1–20.

Equal Employment Opportunity Commission. (1963). The Equal Pay Act of 1963. Retrieved from https://www.eeoc.gov/laws/statutes/epa.cfm

Ericsson, K. A., Krampe, R. T., & Tesch-Romer, C. (1993). The role of deliberate practice in the acquisition of expert performance. *Psychological Review*, *100*(3), 363–406. doi:10.1037/0033-295X.100.3.363

Eskreis-Winkler, L., Duckworth, A. L., Shulman, E. P., & Beal, S. (2014). The grit effect: Predicting retention in the military, the workplace, school and marriage. *Frontiers in Psychology*, *5*, 36. doi:10.3389/fpsyg.2014.00036 PMID:24550863

Esmieu, P., & Martinez, A. (2014). *Expanding the HBCU legacy: Enrolling and supporting the rising Latino population.* US: Palgrave Macmillan. doi:10.1057/9781137480415.0012

Estes, J. S. (2015). The pivotal role of online faculty in student engagement and retention. In L. Kyei-Blankson, J. Blankson, E. Ntuli, & C. Agyeman (Eds.), *Handbook of Research on Strategic Management of Interaction, Presence, and Participation in Online Courses.* Hershey, PA: IGI Global.

Evans, N. J., Forney, D. S., & Guido-DiBrito, F. (1998). *Student development in college: Theory, research, and practice.* San Francisco: John Wiley & Sons, Inc.

Evans, N. J., Forney, D. S., Guido, F. M., Patton, L. D., & Renn, K. A. (2010). *Student development in college: theory, research, and practice.* San Francisco: Jossey-Bass.

Facione, P. A., Sánchez, C. A., Facione, N. C., & Gainen, J. (1995). The disposition toward critical thinking. *The Journal of General Education.*

Faranda, W. T., & Clark, I. (2004). Student observations of outstanding teaching: Implications for marketing educators. *Journal of Marketing Education*, *26*(3), 271–282. doi:10.1177/0273475304268782

Farkas, K. (2014, October 8). Adults who re-enroll in college often don't graduate. *Northeast Ohio Media Group.* Retrieved from http://www.cleveland.com/metro/index.ssf/2014/10/ adults_who_re-enroll_in_colleg.html

Farrington, C. A., Roderick, M., Allensworth, E., Nagaoka, J., Keyes, T. S., Johnson, D. W., & Beechum, N. O. (2012). *Teaching adolescents to become learners. The role of noncognitive factors in shaping school performance: A critical literature review.* Chicago: University of Chicago Consortium on Chicago School Research.

Feldman, D. B., & Dreher, D. E. (2012). Can hope be changed in 90 minutes? Testing the efficacy of a single-session goal-pursuit intervention for college students. *Journal of Happiness Studies*, *13*(4), 745–759. doi:10.1007/s10902-011-9292-4

Fickel, L. (2015, May 1). What's in a terrible name? *U.S. News.* Retrieved from https://www.usnews.com/opinion/knowledge-bank/2015/05/01/non-cognitive-skills-are-important-but-have-a-terrible-name

Fidishun, D. (2000, April). Andragogy and technology: Integrating adult learning theory as we teach with technology. In *Proceedings of the 2000 Mid-South Instructional Technology Conference.*

Fisher-Yoshida, B. (2009). Transformative learning in participative processes that reframe self-identity. In B. Fisher-Yoshida, K. D. Geller, & S. Schapiro (Eds.), *Innovations in transformative learning: Space, culture and the arts* (pp. 43–65). New York, NY: Peter Lang.

Fleming, J. (2012). *Enhancing minority student and retention and academic performance: What we can learn from program evaluations*. San Francisco, CA: Jossey-Bass.

Flemming, J. (2004). The significance of historically black colleges for high achievers: correlates of standardized test scores in African American students. In M. Brown & K. Freeman (Eds.), *Black colleges: new perspectives on policy and practice* (pp. 29–53). Westport, CT: Praeger.

Flottemesch, K. (2000). Building effective interaction in distance education. *Educational Technology*, *4*(3), 46–51.

Fong, C. J., Davis, C. W., Kim, Y., Kim, Y. W., Marriott, L., & Kim, S. (2016). Psychosocial factors and community college student success a meta-analytic investigation. *Review of Educational Research*, *87*(2), 388–424.

Foundation for Excellence in Education. (2010). *Digital Learning Now*! Tallahassee, FL: Author. Retrieved from www.excelined.org/Docs/Digital%20Learning%20Now%20Report%20FINAL.pdf

Freire, P. (1998a). *Pedagogy of the heart*. New York, NY: Continuum.

Fromkin, V., & Rodman, R. (1983). *An Introduction to Language*. New York: CBS College Publishing.

Frost, S. H. (1989). Academic responsibility: Can it be taught? *NACADA Journal*, *9*(2), 17–24. doi:10.12930/0271-9517-9.2.17 PMID:2593029

Gabriel, K. F. (2008). *Teaching unprepared students: Strategies for promoting success and retention in higher education*. Sterling, VA: Stylus Publishing, LLC.

Gallien, L., & Peterson, M. (2006). *Instructing and mentoring the African American college student: Strategies for success in higher education*. Boston: Pearson/Allyn and Bacon.

Galloway, L. M. (2016). Expanding spaces for transformation: Supporting social justice in a distributed learning environment [Dissertation]. Fielding Graduate University.

Gallup (2013). State of the global workplace. Retrieved from http://www.gallup.com/services/176735/state-global-workplace.aspx

Gardner, J. (2003). as quoted by Joe Cuseo on the First-Year Assessment List FYA-LIST@listserv.SC.EDU, 02/05/03.

Garing, M. T. (1992). Intrusive academic advising. *New Directions for Community Colleges*, *21*(2), 97–104.

Garrison, D. R., & Cleveland-Innes, M. (2003, September). Critical factors in student satisfaction and success: Facilitating student's role adjustment in online communities of inquiry (Invited paper). *Presented to the Sloan Consortium Asynchronous Learning Network Invitational Workshop*, Boston, MA.

Garrison, D. R., & Archer, W. (2000). *A transactional perspective on teaching-learning: A framework for adult and higher education*. Oxford, UK: Pergamon.

Garrison, D. R., & Cleveland-Innes, M. (2005). Facilitating cognitive presence in online learning: Interaction is not enough. *American Journal of Distance Education*, *19*(3), 133–148. doi:10.1207/s15389286ajde1903_2

Garrison, D. R., & Kanuka, H. (2004). Blended learning: Uncovering its transformative potential in higher education. *The Internet and Higher Education*, *7*(2), 95–105. doi:10.1016/j.iheduc.2004.02.001

Gasman, M. (n. d.). The changing face of Historically Black Colleges and Universities. University of Pennsylvania. Retrieved from http://www.gse.upenn.edu/pdf/cmsi/Changing_Face_HBCUs.pdf

Gasman, M., & Baez, B. (2008). *Understanding minority-serving institutions*. Albany: Sate University of New York Press.

Gaudelli, W. (2006). Convergence of technology and diversity: Experiences of two beginning teachers in web-based distance learning for Global/Multicultural education. *Teacher Education Quarterly*, *33*(1), 97–116.

Gay, G. (2000). *Culturally responsive teaching: Theory, research, & practice*. New York, NY: Teachers College Press.

Gay, G. (2002). Preparing for culturally responsive teaching. *Journal of Teacher Education*, *53*(2), 106–116. doi:10.1177/0022487102053002003

Gay, G. (2004). The importance of multicultural education. *Educational Leadership*, *61*, 30–35.

Gazzaniga, M. S., Ivry, R., & Mangun, G. R. (2008). *Cognitive neuroscience: The biology of the mind* (3rd ed.). New York, NY: W.W. Norton.

Geert Hofstede. (2017). Retrieved March 20, 2017, from https://geert-hofstede.com/south-korea.html

Georgia Learning Outcomes of Students Studying Abroad Research Initiative. (2010.) Retrieved from: http://glossari.uga.edu/datasets/pdfs/FINAL.pdf

Gilbert, S., So, D., Russell, T. M., & Wessel, T. (2006). Racial identity and psychological symptoms among African Americans attending a historically black university. *Journal of College Counseling*, *9*(2), 111–122. doi:10.1002/j.2161-1882.2006.tb00098.x

Gilmore, M., & Lyons, E. (2012). NURSING 911: An orientation program to improve retention of online students. *Nursing Education Perspectives*, *33*(1), 45–47. doi:10.5480/1536-5026-33.1.45 PMID:22416541

Girves, J. E., & Wemmerus, V. (1988). Developing models of graduate student degree progress. *The Journal of Higher Education*, *59*(2), 163–189. doi:10.2307/1981691

Glazer-Raymo, J. S. (2005). Professionalizing graduate education: The master's degree in the marketplace. *ASHE Higher Education Report*, *31*(4).

Gloria, A. M., Kurpius, S. E. R., Hamilton, K. D., & Willson, M. S. (1999). African American students' persistence at a predominantly white university: Influence of social support, university comfort, and self-beliefs. *Journal of College Student Development*, *40*, 257–268.

Goldberg, L. R. (1990). An alternative description of personality: The Big-Five factor structure. *Journal of Personality and Social Psychology*, *59*(6), 1216–1229. doi:10.1037/0022-3514.59.6.1216 PMID:2283588

Goldberg, L. R. (1992). The development of markers for the Big-Five factor structure. *Psychological Assessment*, *4*(1), 26–42. doi:10.1037/1040-3590.4.1.26

Goleman, D. (2004). What makes a leader? *Harvard Business Review*, *82*(1), 1–11. PMID:10187249

Goodman, A. E. (1999). *America is devaluing international exchanges for students and scholars*. The Chronicle of Higher Education. Retrieved from www.chonicle.com

Goodman, K., & Pascarella, E. T. (2006). First-year seminars increase persistence and retention: A summary of the evidence from how college affects students. *Peer Review*, *8*(3), 26-28.

Gosling, S. D., Augustine, A. A., Vazire, S., Holtzman, N., & Gaddis, S. (2011). Manifestations of personality in online social networks: Self-reported Facebook-related behaviors and observable profile information. *Cyberpsychology (Brno), 14*(9), 483–488. PMID:21254929

Grabmeier, J. (2006). Better grades and greater incentives help explain why women outpace men in college degrees. Retrieved from http://researchnews.osu.edu/archive/ womcolge.htm

Green, D. F., Jr. (2016). Expanding the dialogue on writing assessment at HBCUs: Foundational assessment concepts and legacies of Historically Black Colleges and Universities. Retrieved from http://www.ncte.org/library/NCTEFiles/Resources/Journals/CE/0792nov2016/CE0792Expandingpword.pdf

Guardado, M., & Shi, L. (2007). ESL students experience of online peer feedback. *Computers and Composition, 24*(4), 443–461. doi:10.1016/j.compcom.2007.03.002

Guerrero, L., & Miller, T. A. (1998). Associations between nonverbal behaviors and initial impressions of instructor competence and course content in videotaped distance education courses. *Communication Education, 47*(1), 30–42. doi:10.1080/03634529809379108

Gurung, R. A. R., & McCann, L. (2011) How Should Students Study? Tips, Advice, and Pitfalls. Observer, 24(4). Retrieved March 22, 2017 from http://www.psychologicalscience.org/index.php/publications/observer/ 2011/april-11

Gurung, R. A. R. (2005). How do students really study (and does it matter)? *Teaching of Psychology, 32*, 239–241.

Gurung, R. A. R., Weidert, J., & Jeske, A. (2010). Focusing on how students study. *Journal of the Scholarship of Teaching and Learning, 10*, 28–35.

Gurung, R. R. (2004). Pedagogical Aids: Learning Enhancers or Dangerous Detours? *Teaching of Psychology, 31*(3), 164–166. doi:10.1207/s15328023top3103_1

Habley, W. R. (1994). Key Concepts in Academic Advising. In Summer Institute on Academic Advising Session Guide. National Academic Advising Association, Kansas State University, Manhattan, KS.

Habley, W. R. (2008). Using assessment to improve academic advising. *Paper presented at the annual Noel Levitz National Conference on Student Recruitment, Marketing and Retention*, Chicago, IL.

Habley, W. R. (1983). Organizational structures for academic advising: Models and implications. *Journal of College Student Personnel*, 1983, 24535–24540.

Habley, W. R. (1997). Organizational Models and Institutional Advising Practices. *NACADA Journal, 17*(2), 39–44. doi:10.12930/0271-9517-17.2.39

Habley, W. R., Bloom, J. L., & Robbins, S. (2012). *Increasing persistence: Research-based strategies for college student success*. John Wiley & Sons.

Habley, W. R., & McCauley, M. E. (1987). The relationship between institutional characteristics and the organization of advising services. *NACADA Journal, 7*(1), 27–39. doi:10.12930/0271-9517-7.1.27 PMID:2951508

Hachey, A. C., Conway, K. M., & Wladis, C. W. (2013). Community colleges and underappreciated assets: Using institutional data to promote success in online learning. *Online Journal of Distance Learning Administration, 16*(1), 1–18.

Hale, F. (2006). *How black colleges empower black students: Lessons for higher education*. Sterling, VA: Stylus Publishing.

Harper, S. R. (2012). *Black male student success in higher education: A report from the National Black Male College Achievement Study*. Philadelphia: University of Pennsylvania. Retrieved from http://works.bepress.com/cgi/viewcontent.cgi?article=1049&context=sharper

Hart, C. (2012). Factors associated with student persistence in an online program of study: A review of the literature. *Journal of Interactive Online Learning*, *11*(1), 19–42.

Hartley, D. E., & Cone, J. (2000). *On-demand learning: Training in the new millennium*. Amherst, Mass: HRD Press.

Hartman, K., Neuwirth, C. M., Kiesler, S., Cochran, C., Palmquist, M., & Zubrow, D. (1995). Patterns of social interaction and learning to write: Some effects of network technologies. In Z.L. Berge & M.P. Collins, (Eds.), Computer mediated communication and the online classroom (pp. 47–78). New Jersey: Hampton Press.

Hattie, J. (2009). *Visible learning: A synthesis of over 800 meta-analyses relating to achievement*. London, England: Routledge.

Heckman, J. J., & Rubinstein, Y. (2001). The importance of noncognitive skills: Lessons from the GED testing program. *The American Economic Review*, *91*(2), 145–149. doi:10.1257/aer.91.2.145

Hedayati, M., & Khazaei, M. (2014). An Investigation of the relationship between depression, meaning in life and adult hope. *Procedia: Social and Behavioral Sciences*, *114*(21), 598–601. doi:10.1016/j.sbspro.2013.12.753

Heisserer, D. L., & Parette, P. (2002). Advising at-risk students in college and university settings. *College Student Journal*, *36*, 69–83.

Hembrooke, H., & Gay, G. (2003). The Laptop and the Lecture: The Effects of Multitasking in Learning Environments. *Journal of Computing in Higher Education*, *15*(1), 46–64. doi:10.1007/BF02940852

Heo, J., Oh, J., Subramanian, S. V., Kim, Y., & Kawachi, I. (2014). Addictive Internet Use among Korean Adolescents: A National Survey. *PLoS ONE*, *9*(2), e87819. doi:10.1371/journal.pone.0087819 PMID:24505318

Heyman, E. (2010). Overcoming student retention issues in higher education online programs. *Online Journal of Distance Learning Administration*, *8*(4).

Hicks, M., Reid, I., & George, R. (2001). Enhancing on-line teaching: Designing responsive learning environments. *The International Journal for Academic Development*, *6*(2), 143–151. doi:10.1080/713769258

Hidi, S., & Harackiewicz, J. M. (2001). Motivating the academically unmotivated: A critical issue for the 21st century. *Review of Educational Research*, *70*(2), 151–180. doi:10.3102/00346543070002151

Hill, C. (2010). *Faculty focus special report: Strategies for increasing online student retention and satisfaction*. Retrieved from http://www.facultyfocus.com/free-reports/strategies-for-increasing-online-student-retention-and-satisfaction/

Hill, C. (2017). 11 tips for improving retention of distance learning students. Retrieved from http://cdn.facultyfocus.com/wp-content/uploads/2015/02/Strategies-for-Increasing-Online-Student-Retention.pdf

Himelhoch, C. R., Nichols, A., Ball, S. R., & Black, L. C. (1997, November). A comparative study of the factors which predict persistence for African American students at historically Black institutions and predominantly White institutions. *Paper presented at the annual meeting of the Association for the Study of Higher Education*, Albuquerque, NM.

Hinkel, E. (1990). *Culture is second language teaching and learning*. New York, NY: Cambridge University Press.

Hirsh, R. (1974). The hippocampus and contextual retrieval of information from memory: A theory. *Behavioral Biology*, *12*(4), 421–444. doi:10.1016/S0091-6773(74)92231-7 PMID:4217626

Hogan, S. J., & Coote, L. V. (2013). Organizational culture, innovation, and performance: A test of Schein's model. *Journal of Business Research*. doi:10.1016/j.jbusres.2013.09.007

Homburg, C., & Pflesser, C. (2000). A multiple-layer model of market-oriented organizational culture: Measurement issues and performance outcomes. *JMR, Journal of Marketing Research*, *37*(4), 449–462. doi:10.1509/jmkr.37.4.449.18786

Hooper, J., McEwen, M., & Mancini, M. E. (2013). A Regulatory Challenge: Creating a Metric for Quality RN-to-BSN Programs. *Journal of Nursing Regulation*, *4*(2), 34–38. doi:10.1016/S2155-8256(15)30156-3

Horstmanshof, L., & Zimitat, C. (2007). Future time orientation predicts academic engagement among first-year university students. *The British Journal of Educational Psychology*, *77*(3), 703–718. doi:10.1348/000709906X160778 PMID:17908382

Horvath, Z. (2014). History of the South Korean flag. Retrieved April 10, 2014 from http://www.crwflags.com/fotw/flags/kr_hist.html#1856

Hser, M. P. (2005). Campus internationalization: A study of American universities' internationalization efforts. *International Education*, *35*(1), 35–48.

Hu, E. (2015, April 15). The All-Work, No-Play Culture Of South Korean Education. Retrieved March 27, 2017 from http://www.npr.org/sections/parallels/2015/04/15/393939759/the-all-work-no-play-culture-of-south-korean-education

Hudzik, J. (2011). *Comprehensive internationalization: From concept to action.* NAFSA: Association of International Educators. Retrieved from http://www.nafsa.org

Hunter, M. S., & White, E. R. (2004). Could fixing academic advising fix higher education? *About Campus: Enriching the Student Learning Experience*, *9*(1), 20–25.

Hu, S. (2011). Reconsidering the relationship between student engagement and persistence in college. *Innovative Higher Education*, *36*(2), 97–106. doi:10.1007/s10755-010-9158-4

ICEF. (2012, June 288 countries leading the way in online education. Retrieved from http://monitor.icef.com/2012/06/8-countries-leading-the-way-in-online-education/

Illeris, K. (2007). *How We Learn: Learning and Non-learning in School and Beyond.* London, England: Routledge.

Illeris, K. (2016). *Learning, Development, and Education: from learning theory to education and practice, the selected works of Knud Illeris.* London, England: Routledge.

Ingraham, E. C., & Peterson, D. L. (2004). Assessing the impact of study abroad on student learning at Michigan State University. *Frontiers: The Interdisciplinary Journal of Study Abroad*, *10*, 83–100.

Institute of Education Sciences (IES) Website. (2014). *What works clearinghouse.* Retrieved from http://ies.ed.gov/ncee/wwc/

Institute of Medicine. (2011). *Future of Nursing: Leading Change, Advancing Health.* Washington, DC: The National Academies Press.

Irving, L. M., Snyder, C. R., Cheavens, J., Gravel, L., Hanke, J., Hilberg, P., & Nelson, N. (2004). The Relationships Between Hope and Outcomes at the Pretreatment, Beginning, and Later Phases of Psychotherapy. *Journal of Psychotherapy Integration*, *14*(4), 419–443. doi:10.1037/1053-0479.14.4.419

Irving, L. M., Snyder, C. R., & Crowson, J. J. Jr. (1998). Hope and the negotiation of cancer facts by college women. *Journal of Personality*, *66*(2), 195–214. doi:10.1111/1467-6494.00009 PMID:9529663

Jamelske, E. (2006). Measuring the impact of a university first-year experience program on student GPA and retention. *Higher Education*, *57*(3), 373–391. doi:10.1007/s10734-008-9161-1

James, M. (2000). Culture in ESL instruction: An analytic framework. *TESOL Canada Journal, 17*(2), 36–49.

Jeffreys, M. R. (2007). Nontraditional students' perceptions of variables influencing retention: A multi-site study. *Nurse Educator, 32*(4), 161–167. doi:10.1097/01.NNE.0000281086.35464.ed PMID:17627208

Jeon, B., & Kim, J. (2012). Development and validation of a teachers praise disposition scale. *Journal of Fisheries and Marine Sciences Education, 24*(5), 627–642. doi:10.13000/JFMSE.2012.24.5.627

Johnson, K. E., & Golombek, P. R. (2002). *Teachers' narrative inquiry as professional development.* Cambridge University Press.

Johnston, P. H. (2004). *Choice words: How our language affects children's learning.* Portland, Me.: Stenhouse Publishers.

Jones, E. (2017). Problematising and reimagining the notion of international student experience. *Studies in Higher Education, 42*(5), 933–943. doi:10.1080/03075079.2017.1293880

Jordan, Z. (2012). In A. Inoue, M. Poe, & P. Lang (Eds.), *Students' Right, African American English, and Writing Assessment: Considering the HBCU. In Race and Writing Assessment* (pp. 97–110).

Junco, R., & Cotton, S. (2012). The relationship between multitasking and academic performance. *Computers & Education, 59*(2), 505–514. doi:10.1016/j.compedu.2011.12.023

Karim, M. N., & Behrend, T. S. (2013). Controlling engagement: The effects of learner control on engagement and satisfaction. In C. Wankel & P. Blessinger (Eds.), *Increasing Student Engagement and Retention in e-learning Environments: Web 2.0 and Blended Learning Technologies (175 – 197).* Bingley, UK: Emerald Group Publishing Limited. doi:10.1108/S2044-9968(2013)000006G005

Karp, M. M., & Bork, R. H. (2012). *"They never told me what to expect, so I didn't know what to do": Defining and clarifying the role of a community college student* (Vol. CCRC Working Paper No. 47). New York, NY: Columbia University, Teachers College, Community College Research Center.

Kasbeek, D. (2013). *Reframing retention strategy for institutional improvement.* San Francisco, CA: Jossey-Bass Publishing.

Kashima, Y., Yamaguchi, S., Kim, U., Choi, S., Gelfand, M. J., & Yuki, M. (1995). Culture, gender, and self: A perspective from individualism-collectivism research. *Journal of Personality and Social Psychology, 69*(5), 925–937. doi:10.1037/0022-3514.69.5.925 PMID:7473038

Katal, A., Wazid, M., & Goudar, R. H. (2013). Big data: Issues, challenges, tools and good practices. *Presented at the International Conference on Contemporary Computing* (IC3), Noida, India. Retrieved from https://doi.org/http://ieeexplore.ieee.org/document/6612229/

Kautz, T., Heckman, J. J., Diris, R., Ter Weel, B., & Borghans, L. (2014). *Fostering and measuring skills: Improving cognitive and non-cognitive skills to promote lifetime success (No. w20749).* National Bureau of Economic Research.

Keengwe, J., & Kidd, T. T. (2010). Towards best practices in online learning and teaching in higher education. *MERLOT Journal of Online Learning and Teaching, 6*(2). Retrieved from http://jolt.merlot.org/vol6no2/keengwe_0610.htm

Kegan, R. (2000). What "form" transforms? A constructive-developmental approach to transformative learning. In J. Mezirow et al. (Eds.), *Learning as transformation: Critical perspectives on a theory in progress* (pp. 35–70). San Francisco, CA: Jossey-Bass.

Kehrwald, B. (2008). Understanding social presence in text-based online learning environments. *Distance Education, 29*(1), 89–106. doi:10.1080/01587910802004860

Kelley, H. M., Oginga Siwatu, K., Tost, J. R., & Martinez, J. (2015). Culturally familiar tasks on reading performance and self-efficacy of culturally and linguistically diverse students. *Educational Psychology in Practice*, *31*(3), 293–313. doi:10.1080/02667363.2015.1033616

Kelly, A. P., & Schneider, M. (2012). *Getting to graduation: The completion agenda in higher education.* Baltimore, MD: John Hopkins University Press.

Kenny, A., Kidd, T., Nankervis, K., & Connell, S. (2011). Mature age students access, entry, and success in nurse education: An action research study. *Contemporary Nurse*, *38*(1-2), 106–118. doi:10.5172/conu.2011.38.1-2.106 PMID:21854242

Kern-Bowen, J., & Gardner, R. (2010). Creating campus community for graduate students through programs, services, and facilities. *The Bulletin, 78*(2). Retrieved March 28, 2017 from www.acui.org

Kezar, A., & Kinzie, J. (2006). Examining the ways Institutions create student engagement: The role of mission. *Journal of College Student Development*, *47*(2), 149–173. doi:10.1353/csd.2006.0018

Kift, N., Nelson, K., & Clarke, J. (2010). Transition Pedagogy: A third generation of FYE – A case study of policy and practice for the higher education sector. *International Journal of the First Year Experience in Higher Education*, *1*(1), 1–20. doi:10.5204/intjfyhe.v1i1.13

Kim, J. (2013, November 12). How to cure South Korea's English fever? *The Wall Street Journal.* Retrieved from http://blogs.wsj.com/korearealtime/2013/11/12/how-to-cure-south-koreas-english-fever/

Kirk, G., & Nolan, S. (2010). Nonprofit Mission Statement Focus and Financial Performance. *Nonprofit Management & Leadership*, *20*(4), 473–490. doi:10.1002/nml.20006

Kirschner, P. A., & van Merrienboer, J. G. (2013). Do learners know best? *Educational Psychologist*, *48*(3), 169–183. doi:10.1080/00461520.2013.804395

Klingbeil, N., Mercer, R., Rattan, K., Raymer, M., & Reynolds, D. (2004, June). Rethinking engineering mathematics education: A model for increased retention, motivation and success in engineering. *Paper presented at the American Society for Engineering Education Annual Conference & Exposition*, Salt Lake City, Utah.

K-MOOC. (2017). Retrieved March 27, 2017, from http://www.kmooc.kr/

Knapczyk, D. R., Frey, T. J., & Wall-Marencik, W. (2005). An evaluation of web conferencing in online teacher preparation. Teacher Education and Special Education. *The Journal of the Teacher Education Division of the Council for Exceptional Children*, *28*(2), 114–124. doi:10.1177/088840640502800205

Knapp, M. L. (1972). *Nonverbal communication in human interaction.* New York: Holt, Rinehart and Winston, Inc.

Knight, J. (1993). Internationalization management strategies and issues. *International Education Magazine*, *9*(6), 21–22.

Knight, J. (1997). Internationalization of higher education: A conceptual framework.

Knight, J. (2001). Internationalization. *Journal of Studies in International Education*, *5*(3), 228–243. doi:10.1177/102831530153004

Knight, J. (2003). Updating the definition of internationalisation. *Industry and Higher Education*, *33*, 2.

Knight, J. (2004). Internationalization remodeled: Definition, approaches, and rationales. *Journal of Studies in International Education*, *8*(1), 5–31. doi:10.1177/1028315303260832

Knowles, M. (1984). *The adult learner: A neglected species.* Houston, TX: Gulf.

Knowles, M. L., Holton, E. F., & Swanson, R. A. (2011). *The adult learner* (7th ed.). Houston, TX: Gulf.

Knowles, M. S. (1975). *Self-directed learning*. New York, NY: Association Press.

Knowles, M. S. (1980). *The modern practice of adult education (revised and updated)*. Chicago, IL: Association Press. (Original work published 1970)

Knowlton, B. J., Ramus, S. J., & Squire, L. R. (1992). Intact artificial grammar learning in amnesia: Dissociation of classification learning and explicit memory for specific instances. *Psychological Science*, *3*(3), 172–179. doi:10.1111/j.1467-9280.1992.tb00021.x

Koh, Y. J. (2003). Neo-Confucianism as the dominant ideology in Joseon. *Korea Journal*, *43*(4), 59–86.

Ko, S., & Rossen, S. (2010). *Teaching online: A practical guide* (3rd ed.). New York, NY: Routledge.

Kot, F. C. (2014). The impact of centralized advising on first-year academic performance and second-year enrollment behavior. *Research in Higher Education*, *55*(6), 527–563.

Krause, K., & Coates, H. (2008). Students Engagement in first-year university. *Assessment & Evaluation in Higher Education*, *33*(5), 493–505. doi:10.1080/02602930701698892

Kuh, G. (1990, November). Influence of collegiate environments on student learning. *Paper presented at the Association for the Study of Higher Education Annual Meeting*, Portland, Oregon.

Kuh, G. D., Cruce, T. M., Shoup, R., Kinzie, J., & Gonyea, R. M. (2008). Unmasking the effects of student engagement on first-year college grades and persistence.

Kuh, G. D., Kinzie, J., Buckley, J. A., Bridges, B. K., & Hayek, J. C. (2006, July). What matters to student success: A review of the literature. *Commissioned report for the national symposium on postsecondary student success: spearheading a dialog on student success*. Retrieved from http://nces.ed.gov/npec/pdf/kuh_team_report.pdf

Kuh, G. (1997). The student learning agenda: Implications for academic advisors. *NACADA Journal*, *17*(2), 7–12. doi:10.12930/0271-9517-17.2.7 PMID:10538182

Kuh, G. (2003). How are we doing at engaging students? *About Campus: Enriching the Student Learning Experience*, *8*(1), 9–16.

Kuh, G. (2012). High-impact educational practices: What they are, who has access to them, and why they matter. *Peer Review*, *14*(3), 29.

Kuh, G. D. (2003). What were learning about student engagement from NSSE. *Change*, *35*(2), 24–32. doi:10.1080/00091380309604090

Kuh, G. D. (2008). *High-impact educational practices: What they are, who has access to them, and why they matter*. Washington, DC: Association of American Colleges and Universities.

Kuh, G. D. (2009). What student affairs professionals need to know about student engagement. *Journal of College Student Development*, *50*(6), 683–706. doi:10.1353/csd.0.0099

Kuh, G. D., Gonyea, R. M., & Williams, J. M. (2005). What students expect from college and what they get. In T. Miller, B. Bender, & J. Schuh et al. (Eds.), *Promoting reasonable expectations: Aligning student and institutional views of the college experience* (pp. 34–64). San Francisco: Jossey-Bass.

Kuh, G. D., Kinzie, J., Schuh, J. H., & Whitt, E. J. (2010). *Student success in college: Creating conditions that matter*. San Francisco, CA: Jossey-Bass.

Kuh, G. D., & Love, P. G. (2000). A cultural perspective on student departure. In J. M. Braxton (Ed.), *Reworking the student departure puzzle* (pp. 196–212). Nashville: Vanderbilt University Press.

Kuh, G. D., Schuh, J., Whitt, E., Andreas, R., Lyons, J., & Strange, C. et al.. (1991). *Involving colleges: Successful approaches to fostering student learning and personal development outside the classroom*. San Francisco: Jossey-Bass.

Kuh, G. D., & Whitt, E. J. (1988). ASHE-ERIC Higher Education Report: Vol. 1. *The Invisible Tapestry. Culture in American Colleges and Universities*. Washington, D.C.: Association for the Study of Higher Education.

Kuh, G., Kinzie, J., Buckley, J., Bridges, B., & Hayek, J. (2007). Piecing together the student success puzzle: Research, propositions and recommendations. *ASHE Higher Education Report*, *32*(5), 1–182.

Kuh, G., & O'Donnell, K. (2013). Ensuring quality and taking high-impact practices to scale. *Peer Review*, *15*(2), 32.

Kuijpers, M., Meijers, F., & Gundy, C. (2011). The relationship between learning environment and career competencies of students in vocational education. *Journal of Vocational Behavior*, *78*(1), 21–30. doi:10.1016/j.jvb.2010.05.005

Kyllonen, P. C., Lipnevich, A. A., Burrus, J., & Roberts, R. D. (2014). *Personality, motivation, and college readiness: A prospectus for assessment and development. Educational Testing Service Research Report No: RR-14-06*. Princeton, NJ: Educational Testing Service.

Ladson-Billings, G. (2006). From the achievement gap to the education debt: Understanding achievement in U.S. schools. *Educational Researcher*, *35*(7), 3–12. doi:10.3102/0013189X035007003

Landefeld, T. (2009). *Mentoring and diversity: Tips for students and professionals for developing and maintaining a diverse community*. Dordrecht, New York: Springer.

Landis, D., Umolu, J., & Mancha, S. (2010). The Power of Language Experience for Cross-Cultural Reading and Writing. *The Reading Teacher*, *63*(7), 580–589. doi:10.1598/RT.63.7.5

Las Vegas Education Review Journal. (n. d.). 60 percent of job openings will require a college education. Retrieved from http://www.reviewjournal.com/news/education/2018-60-percent-job-openings-will-require-college-education

Lassithiotaki, A. (2011). rural women and entrepreneurship: A case study in Heraklion Crete prefecture, Greece. *Journal of Developmental Entrepreneurship*, *16*(2), 269–284. doi:10.1142/S1084946711001835

Lawrence, J. (2005). Re-conceptualising attrition and retention: Integrating theoretical, research and student perspectives', *Studies in Learning. Evaluation Innovation and Development*, *2*(3), 16–33.

Leary, M., & Hoyle, R. (2009). *Handbook of individual differences in social behavior*. New York: Guilford Press.

Lee, J. E., & Nicolaides, A. (2014). Exploring transformative relationships in a virtual learning environment. In *Proceedings of the Adult Education Research Conference* (pp. 273-278). Kansas State University Libraries, New Prairie Press.

Lee, Y. H. (2016). ICT as a key engine for development: good practices and lessons learned from Korea. Retrieved April 13, 2017 from siteresources.worldbank.org/INTEGOVERNMENT/Resources/NoteKoreaICT.doc

Lee, B., Yoon, J., & Lee, I. (2009). Learners acceptance of e-learning in South Korea: Theories and results. *Computers & Education*, *53*(4), 1320–1329. doi:10.1016/j.compedu.2009.06.014

Leeds, E., Campbell, S., Baker, H., Ali, R., Brawley, D., & Crisp, J. (2013). The impact of student retention strategies: An empirical study. *International Journal of Management in Education*, *7*(1/2), 22–43. doi:10.1504/IJMIE.2013.050812

Lee, J. K., & Yi, C. G. (2002). *Korean higher education: A Confucian perspective. 22*. Jimoondang Publishing Co.

Lee, S. U., Fraser, B. J., & Fisher, D. L. (2003). Teacher-student interactions in Korean high school science classrooms. *International Journal of Science and Mathematics Education*, *1*(1), 67–85. doi:10.1023/A:1026191226676

Lee, Y., & Choi, J. (2011). A review of online course dropout research: Implications for practice and future research. *Educational Technology Research and Development*, *59*(5), 593–618. doi:10.1007/s11423-010-9177-y

Lee, Y., & Jonson-Reid, M. (2016). The Role of Self-Efficacy in Reading Achievement of Young Children in Urban Schools. *Child & Adolescent Social Work Journal*, *33*(1), 79–89. doi:10.1007/s10560-015-0404-6

Levitz, R. (2001). What's Working Right Now in Student Retention! Building Blocks for Retention Success – the Basis for Recruiting Success. *Prepared for the National Conference for Student Retention.*

Li, K., Zelenka, R., Buonaguidi, L., Beckman, R., Casillas, A., Crouse, J., & Robbins, S. et al. (2013). Readiness, behavior, and foundational mathematics course success. *Journal of Developmental Education*, *37*(1), 14.

Lim, D. H. (2004). Cross cultural difference in online learning motivation. *Educational Media International*, *41*(2), 163–175. doi:10.1080/09523980410001685784

Lincoln, D. J. (2008). Drama in the classroom: How and why marketing educators can use nonverbal communication and enthusiasm to build student rapport. *Marketing Education Review*, *8*(3), 53–65. doi:10.1080/10528008.2008.11489048

Lindley, M. K., Ashwill, R., Cipher, D. J., & Mancini, M. E. (2017). Expanding capacity with an accelerated online BSN program. *Journal of Professional Nursing*, *33*(1), 5–10. doi:10.1016/j.profnurs.2016.03.010 PMID:28131147

Lindsay, B., & Justiz, M. (2001). *The quest for equity in higher education: Toward new paradigms in an evolving affirmative action era*. Albany: State University of New York Press.

Lion, R. W. (2011). A study of performance support in higher education. *Performance Improvement Quarterly*, *24*(1), 49–67. doi:10.1002/piq.20101

Lipsman, M. (2000). *Globalization isn't good or inevitable*. The Chronicle of Higher Education. Retrieved from www.chonicle.com

Liu, M. C. (July, 2011). Investing in higher education for Latinos: trends in Latino college access and success. Denver, CO: *National Conference of State Legislatures*.

Lloyd, K. M., Tienda, M., & Zajacova, A. (2001). Trends in educational achievement of minority students since Brown v. Board of Education. In C. Snow (Ed.), *Achieving high educational standards for all: Conference summary* (pp. 147–182). Washington, DC: National Academy Press.

Loftin, C., Newman, S., Dumas, B. P., Gilden, G., & Bond, M. L. (2012). Perceived barriers to success for minority nursing students: an integrative review. International Scholarly Research Network Nursing.

Long, M. C., Conger, D., & Iatarola, P. (2012). Effects of high school course-taking on secondary and postsecondary success. *American Educational Research Journal*, *49*(2), 285–322. doi:10.3102/0002831211431952

Lotkowski, V. A., Robbins, S. B., & Noeth, R. J. (2004). The role of academic and non-academic factors in improving college retention. *ACT.org*. Retrieved from https://www.act.org/research/policymakers/pdf/college_retention.pdf

Lotkowski, V. A., Robbins, S. B., & Noeth, R. J. (2004). *The Role of Academic and Non-Academic Factors in Improving College Retention. ACT Policy Report*. American College Testing ACT Inc. doi:10.1037/e420492008-001

Louhiala-Salminen, L., Charles, M., & Kankaanranta, A. (2005). English as a lingua franca in Nordic corporate mergers: Two case companies. *English for Specific Purposes*, *24*(4), 401–421. doi:10.1016/j.esp.2005.02.003

Louhiala-Salminen, L., & Kankaanranta, A. (2012). *Language as an issue in international internal communication: English or local language? if English, what English?* Public.

Lounsbury, J. W., Fisher, L. A., Levy, J. J., & Welsh, D. P. (2009). Investigation of character strengths in relation to the academic success of college students. *Individual Differences Research*, *7*(1), 52–69.

Lumina Foundation. (2015). A stronger nation through higher education. Retrieved from http://www.luminafoundation.org/publications

Lumina Foundation. (2016). Report. Retrieved from http://strongernation.luminafoundation.org/report/2016/

Mabizela, M. (1994). Voices from first year students at UWC. In B. Leibowitz & M. Walker (Eds.), *AD Dialogues 3* (pp. 23–46). Bellville, South Africa: University of the Western Cape.

MacCann, C., Duckworth, A. L., & Roberts, R. D. (2009). Empirical identification of the major facets of conscientiousness. *Learning and Individual Differences*, *19*(4), 451–458. doi:10.1016/j.lindif.2009.03.007

Maddux, J., & Gosselin, J. (2015). *Self-Efficacy*. Oxford Bibliographies in Psychology; doi:10.1093/obo/9780199828340-0088

Magolda, B. (2004). Self-authorship as the common goal. In *Learning partnerships: Theory and models of practice to educate for self-authorship.*

Malmgren, J., & Galvin, J. (2008). Effects of study abroad participation on student graduation rates: A study of three incoming freshman cohorts at the University of Minnesota, Twin Cities. *NACADA Journal*, *28*(1), 29–42. doi:10.12930/0271-9517-28.1.29

Mancini, M. E., Ashwill, J., & Cipher, D. J. (2015). A comparative analysis of demographic and academic success characteristics of on-line and on-campus RN-to-BSN students. *Journal of Professional Nursing*, *31*(1), 71–76. doi:10.1016/j.profnurs.2014.05.008 PMID:25601248

Mandernach, B. J. (2009). Three ways to improve student engagement in the online classroom. *Online Classroom.*

Mantovani, G. (2000). *Exploring borders: Understanding culture and psychology*. London: Routledge.

Markle, R. E., Olivera-Aguilar, M., Jackson, T., Noeth, R., & Robbins, S. (2013). *Examining evidence of reliability, validity, and fairness for SuccessNavigator.* (ETS Research Report 13-12). Princeton, NJ: Educational Testing Service.

Markle, R. E., Wang, J. Z., Sullivan, T., & Russell, J. A. (2015, April). Myths and realities: Effort and response distortion in low-stakes, self-report assessments of noncognitive skills. *Paper presented to the annual meeting of the American Educational Research Association*, Denver, CO.

Markle, R., Brenneman, M., Jackson, T., Burrus, J., & Robbins, S. (2013). *Synthesizing frameworks of higher education student learning outcomes* (Research Report No. RR-13-22). Princeton, NJ: Educational Testing Service. doi:10.1002/j.2333-8504.2013.tb02329.x

Marton, F., & Saljo, R. (1976). On qualitative differences in learning: Outcome and process. *The British Journal of Educational Psychology*, *46*(1), 4–11. doi:10.1111/j.2044-8279.1976.tb02980.x

Mascaro, N., & Rosen, D. (2005). Existential Meanings Role in the Enhancement of Hope and Prevention of Depressive Symptoms. *Journal of Personality*, *73*(4), 985–1013. doi:10.1111/j.1467-6494.2005.00336.x PMID:15958142

Mazzarol, T., & Soutar, G. N. (2002). Push-pull factors influencing international student destination choice. *International Journal of Educational Management*, *16*(2/3), 82–90. doi:10.1108/09513540210418403

McAfee, A., & Brynjolfsson, E. (2012). Big data: The management revolution. *Harvard Business Review*, *90*(10), 59–68. PMID:23074865

McCabe, L. (2001). Globalization and internationalization: The impact on education abroad programs. *Journal of Studies in International Education*, *5*(2), 138–145. doi:10.1177/102831530152004

McCollough, M. A., & Gremler, D. D. (1999). Guaranteeing Student Satisfaction: An Exercise in Treating Students as Customers. *Journal of Marketing Education*, *21*(2), 118–130. doi:10.1177/0273475399212005

McComb, M. (1994). Benefits of computer-mediated communication in college courses. *Communication Education*, *43*(2), 159–170. doi:10.1080/03634529409378973

McEnroe-Petitte, D. M. (2011, April). Impact of faculty caring on student retention and success. *Teaching and Learning in Nursing*, *6*(2), 80–83. doi:10.1016/j.teln.2010.12.005

McGillin, V. A. (2003). The role of evaluation and reward in faculty advising. In G. L. Kramer & G. L. Kramer (Eds.), *Faculty advising examined: Enhancing the potential of college faculty as advisors* (pp. 88–124). Bolton, MA, US: Anker Publishing Company.

McInnis, C., & James, R. (1995). *First Year on Campus: Diversity in the Initial Experiences of Australian Undergraduates*. Canberra: AGPS.

McLanahan, S., Haskins, R., Paxson, C., & Sawhill, I. (2010). Transition to Adulthood. *The Future of Children*, *20*(1).

Megginson, L. A. (2008). RN-BSN education: 21st century barriers and incentives. *Journal of Nursing Management*, *16*(1), 47–55. PMID:18211335

Mehrabian, A. (1981). *Silent Messages: Implicit Communication of Emotions and Attitudes*. Belmont, CA: Wadsworth.

Merriam, S. B. (2001). Andragogy and self-directed learning: Pillars of adult learning theory. *New Directions for Adult and Continuing Education*, *(* 89), 3-14. doi:10.1002/ace.3

Merriam-Webster.com. (2017). Beliefs. Retrieved April 18, 2017, from https://www.merriam-webster.com/dictionary/beliefs

Merrill, K. C., Braskamp, D. C., & Braskamp, L. A. (2012). Assessing individual's global perspective. *Journal of College Student Development*, *53*(2), 356–360. doi:10.1353/csd.2012.0034

Mezirow, J. (1981). A critical theory of adult learning and education. *Adult Education Quarterly*, *32*(1), 3–24. doi:10.1177/074171368103200101

Mezirow, J. (2009). Transformative learning theory. In J. Mezirow & E. W. Taylor et al. (Eds.), *Transformative learning in practice: Insights from community, workplace, and higher education* (pp. 18–32). San Francisco, CA: Jossey-Bass.

Miller, S. K. (2008) *Online, distance learning instructor guidelines to improve student retention*. Maricopa Learning Exchange, Phoenix, AZ. Retrieved from www.mcli.dist.maricopa.edu/mix/slip.php?item-1372

Miller, T. E. (2005). Student persistence and degree attainment. In T. Miller, B. Bender, & J. Schuh et al. (Eds.), *Promoting reasonable expectations: Aligning student and institutional views of the college experience* (pp. 122–139). San Francisco: Jossey-Bass.

Monitor, I. C. E. F. (2014, January 23). *High performance, high pressure in South Korea's education system*. Retrieved from http://monitor.icef.com/2014/01/high-performance-high-pressure-in-south-koreas-education-system/

Moore, J. L., Dickson-Deane, C., & Galyen, K. (2011). e-Learning, online learning, and distance learning environments: Are they the same? *The Internet and Higher Education*, *14*(2), 129–135. Retrieved from https://scholar.vt.edu/access/content/group/5deb92b5-10f3-49db-adeb-7294847f1ebc/e-Learning%20Scott%20Midkiff.pdf doi:10.1016/j.iheduc.2010.10.001

Moore, M., & Kearlsey, G. (2012). *Distance education: A systems view of online learning*. Belmont, CA: Cengage Learning.

Müller, T. F. (2007). *Factors affecting persistence of women in online degree-completion programs* [Doctoral dissertation].

Müller, T. (2008). Persistence of women in online degree-completion programs. *International Review of Research in Open and Distance Learning*, *9*(2), 1–18. doi:10.19173/irrodl.v9i2.455

Mupinga, D. M., Nora, R. T., & Yaw, C. (2006). The learning styles, expectations, and needs of online students. *College Teaching*, *54*(1), 185–189. doi:10.3200/CTCH.54.1.185-189

Mykota, D., & Duncan, R. (2007). Learner characteristics as predictors of online social presence. *Canadian Journal of Education*, *30*(1), 157–170. doi:10.2307/20466630

NACADA National Survey of Academic Advising. (2011). Advisor Load Statistics. Retrieved from http://www.nacada.ksu.edu/Portals/0/Clearinghouse/documents/Chapter%206%20-%20Professional%20Advisor%20Load%20-%20FINAL.pdf

NAFSA. Association of International Educators. (2010). Members of the Teaching, Learning, and Scholarship Knowledge Community 2009 Task Force on Assessment and Evaluation. Retrieved from http://www.nafsa.org/uploadedFiles/NAFSA_Home/Resource_Library_Assets/Publications_Library/Assess%20and%20Eval%20in%20IE.pdf

NAFSA. Association of International Educators. (2015). International Education Professional Competencies. Retrieved from http://www.nafsa.org/Professional_Resources/Career_Center/IE_Competencies/International_Education_Professional_Competencies/

NASPA Student Affairs Administrators in Higher Education. (2014). *National study of non-first-time students shows disturbing completion rates.* [News Release]. Retrieved from NASPALowestGradRatesNonFirstTimeOct2013.pdf

National Center for Education Statistics. (2011). HBCU enrollment rates component. Retrieved from nces.ed.gov/ipeds/tablefiles/tableDocs/IPEDS201112Tablesdoc.xlsx

National Center for Education Statistics. (2014, May). *The condition of education. Institutional retention and graduation rates for undergraduate students.* [Data file]. Retrieved from http://nces.ed.gov/programs/coe/indicator_cva.asp

National Center for Education Statistics. (2015). Demographics and enrollment characteristics of nontraditional undergraduates: 2011-2012. Retrieved from http://nces.ed.gov.pubs2015/2015025.pdf

National Center for Education Statistics. (2016). Status and trends in the education and ethnic groups 2016. Retrieved from https://nces.ed.gov/pubs2016/2016007.pdf

National Center for Education Statistics. (n. d.). Historically Black Colleges and Universities. Retrieved from https://nces.ed.gov/fastfacts/display.asp?id=667

National Center for Educational Statistics. (2016). *The Condition of Education 2016* (NCES 2016-144).

National Center for Higher Education Management Systems. (2016). *Persistence and completion: Graduation rates.* [Data file]. Retrieved from http://www.higheredinfo.org/dbrowser/index.php?submeasure=27&year=2009&level=nation&mode=data&state=0#/-1/

National Council of State Boards of Nursing. (2009). Innovations in education regulation report: Background and literature review. Retrieved May 15, 2016 from https://www.ncsbn.org/Innovations_Report.pdf

National League of Nursing. (2016). Retention rates in RN programs. Retrieved May 15, 2016 from http://www.nln.org/newsroom/nursing-education-statistics/retention-rates-in-rn-programs

National Research Council. (1999). How people learn: Brain, mind, experience, and school. Washington, D.C: National Academy Press.

National Research Council. (2012). *Education for Life and Work: Developing Transferable Knowledge and Skills in the 21st Century*. Washington, DC: The National Academies Press.

National Survey of Student Engagement (NCSS). (2014). Center for Postsecondary Research, Indiana University School of Education, Bloomington, IND. Retrieved from http://nsse.iub.edu/NSSE_2014Results/pdfNSSE_2014_Annual_Results.pdf#page=8

Nava, M. E. (2010). *Exploring new paths: The first-year experiences for first-generation college students and the impact of participating in comprehensive programs*. Unpublished doctoral dissertation, University of Texas, Austin.

Nemeth, S. (2013). Advising Assessment Implementation: Moving from Fearful to Formative. *Presented at the annual Noel Levitz National Conference on Student Recruitment, Marketing and Retention*, Chicago, IL.

Nguyen, L. V. (2008). Technology-enhanced EFL: Syllabus design and materials development. *English Language Teaching*, *1*(2), 135–142. doi:10.5539/elt.v1n2p135

Nichols, M. (2010). Student perceptions of support services and the influence of targeted interventions on retention in distance education. *Distance Education*, *31*(1), 93–113. doi:10.1080/01587911003725048

Nicpon, M. F., Huser, L., Blanks, E. H., Sollenberger, S., Befort, C., & Robinson-Kurpius, S. E. R. (2007). The relationship of loneliness and social support with college freshmen's academic performance and persistence. *Journal of College Student Retention*, *8*(3), 345–358. doi:10.2190/A465-356M-7652-783R

Nielsen Korea. (n. d.). Retrieved March 20, 2017 from http://www.koreanclick.com/english/top10/mobile_web_ranking.html

Nissen, M. J., Knopman, D. S., & Schacter, D. L. (1987). Neurochemical dissociation of memory systems. *Neurology*, *37*(5), 789–794. doi:10.1212/WNL.37.5.789 PMID:3574678

Nistor, N., & Neubauer, K. (2010). From participation to dropout: Quantitative participation patterns in online University courses. *Computers & Education*, *55*(2), 663–672. doi:10.1016/j.compedu.2010.02.026

Noel, L., Levitz, R., & Saluri, D. (Eds.) (1985). Increasing student retention. San Francisco, CA: Jossey-Bass.

Noel, L., Levitz, R., & Saluri, D. et al. (1985). *Increasing Student Retention*. San Francisco, CA: Jossey-Bass.

Nora, A. (1987). Determinants of Retention Among Chicano College Students: A Structural Model. *Research in Higher Education*, *26*(1), 31–59. doi:10.1007/BF00991932

O' Shea, S., & Stone, C. (2014). The hero's journey: Stories of women returning to education. *The International Journal of the First Year in Higher Education, 5*(1), 79. doi:http://dx.doi.org.libproxy1.usc.edu/10.5204/intjfyhe.v5i1.186

O' Shea, S., & Stone, C. (2014). The hero's journey: Stories of women returning to education. *The International Journal of the First Year in Higher Education, 5*(1), 79. Retrieved from http://dx.doi.org.libproxy1.usc.edu/10.5204/intjfyhe.v5i1.186

O'Banion, T. (1972). An academic advising model. *Junior College Journal*, *42*(6), 62–64.

O'Brien, G. (2002). *Issues Paper 3: A sense of belonging*. Retrieved from http://www.fye.qut.edu.au/documents/FYIEissues&5Fpaper%5F3.pdf

Obama, B. (2010). *Remarks by the President on Higher Education and the Economy at the University of Texas at Austin.* Retrieved from https://obamawhitehouse.archives.gov/the-press-office/2010/08/09/remarks-president-higher-education-and-economy-university-texas-austin

OConnor, M. C., & Paunonen, S. V. (2007). Big Five personality predictors of post-secondary academic performance. *Personality and Individual Differences*, *43*(5), 971–990. doi:10.1016/j.paid.2007.03.017

OECD. (2013). *Suicide rates* [Data file]. Retrieved from https://data.oecd.org/healthstat/suicide-rates.htm

OECD. (2016). *Population with tertiary education* [Data file]. Retrieved from https://data.oecd.org/eduatt/population-with-tertiary-education.htm

OECD. (2016). *Population with tertiary education.* Retrieved July 6, 2016, from https://data.oecd.org/eduatt/population-with-tertiary-education.htm

OECD. (2016). *Suicide rates* [Data file]. Retrieved from https://data.oecd.org/healthstat/suicide-rates.htm

Offir, B., Lev, Y., & Bezalel, R. (2008). Surface and deep learning processes in distance education: Synchronous versus asynchronous systems. *Computers & Education*, *51*(3), 1172–1183. doi:10.1016/j.compedu.2007.10.009

Ohayon, M. M., & Hong, S. C. (2006). Prevalence of major depressive disorder in the general population of South Korea. *Journal of Psychiatric Research*, *40*(1), 30–36. doi:10.1016/j.jpsychires.2005.02.003 PMID:15878179

Ohio State University Diversity Action Plan 2007-2012. (2012). Retrieved from https://www.osu.edu/diversityplan/index.php

Olivera-Aguilar, M., Rikoon, S. H., & Robbins, S. B. (2017). Noncognitive skill profiles and their relationships with early college academic outcomes. *The Journal of Higher Education*, *88*(2), 234–257.

Oliveri, M. E., & Markle, R. (2017). *Continuing a culture of evidence: Expanding skills in higher education* (Research Report No. RR-17-09). Princeton, NJ: Educational Testing Service. doi:10.1002/ets2.12137

Oliver, M. L., Smith, A. W., & Wilson, K. R. (1989). Supporting successful black students: Personal, organizational, and institutional factors. *Nat. J. Soc.*, *3*, 199–221.

Olsen, C. L., Green, M. F., & Hill, B. A. (2005). *Building a strategic framework for comprehensive internationalization.* Washington, DC: American Council on Education.

Olson, C. L., Evans, R., & Shoenberg, R. F. (2007). *At home in the world: Bridging the gap between internationalization and multicultural education.* Washington, DC: American Council on Education.

Olson, C. L., Green, M. F., & Hill, B. A. (2006). *A handbook for advancing comprehensive internationalization: What institutions can do and what students should learn.* Washington, DC: American Council on Education.

Ortiz, A., & Santos, S. (2009). *Ethnicity in college: Advancing theory and improving diversity practices on campus.* Sterling, VA: Stylus Publishing.

Oswald, F. L., Schmitt, N., Kim, B. H., Ramsay, L. J., & Gillespie, M. A. (2004). Developing a biodata measure and situational judgment inventory as predictors of college student performance. *The Journal of Applied Psychology*, *89*(2), 187–207. doi:10.1037/0021-9010.89.2.187 PMID:15065969

Ozuna, T. G. (2012). Examining the first-year experience and perceptions of sense of belonging among Mexican American students enrolled in a Texas HBCU. Unpublished doctoral dissertation, University of Texas, Austin, Texas.

Palloff, R. M., & Pratt, K. (2005). *Collaborating online: Learning together in community.* San Francisco, CA: Jossey-Bass.

Palloff, R. M., & Pratt, K. (2013). *Lessons from the Virtual Classroom* (2nd ed.). San Francisco, CA: Jossey-Bass.

Palmer, R., & Wood, L. (2012). *Black men in college: Implications for HBCUs and beyond*. New York: Routledge.

Pan, W., & Bai, H. (2009). A multilevel approach to assessing the interaction effects on college student retention. *Journal of College Student Retention*, *11*(2), 287–301. doi:10.2190/CS.11.2.g

Paradise, A. (2008). *Informal learning: Overlooked or overhyped?* Alexandria: American Society for Training & Development, Inc.

Parker, K., Lenhart, A., & Moore, K. (2011). *The digital revolution and higher education: College presidents, public differ on value of online learning*. Retrieved from http://www.pewinternet.org/~/media//Files/Reports/2011/PIP-Online-Learning.pdf

Park, J. H., & Choi, H. J. (2009). Factors influencing adult learners' decision to drop out or persist in online learning. *Journal of Educational Technology & Society*, *12*(4), 207–217.

Pascarella, E. T. (1982). Concluding thoughts. *New Directions for Institutional Research*, *36*(36), 89–92. doi:10.1002/ir.37019823608

Pascarella, E. T., Pierson, C. T., Wolniak, G. C., & Terenzini, P. T. (2004). First-generation college students: Additional evidence on college experiences and outcomes. *The Journal of Higher Education*, *75*(3), 249–283. doi:10.1353/jhe.2004.0016

Pascarella, E. T., & Terenzini, P. T. (2005). How college affects students (Vol. 2). San Francisco: John Wiley & Sons, Inc.

Pascarella, E. T., & Terenzini, P. T. (2005). *How college affects students: A third decade of research* (Vol. 2). San Francisco, CA: Jossey-Bass Higher & Adult Education.

Pascarella, E. T., & Terenzini, P. T. (2005). How College Affects Students: Vol. 2. *A Third. Decade of Research*. San Francisco, CA: Jossey-Bass.

Patterson, B., & McFadden, C. (2009). Attrition in online and campus degree programs. *Online Journal of Distance Learning Administration*, *12*(2). Retrieved from http://www.westga.edu/~distance/ojdla/summer122/patterson112.html

Patton, M. Q. (1990). *Qualitative Evaluation and Research Methods*. Newbury Park, CA: Sage.

Paulhus, D. L., & Vazire, S. (2007). The self-report method. In R. C. Fraley & R. F. Kruger (Eds.), *Handbook of research methods in personality psychology* (pp. 224–239). New York: Guilford.

Pearce, J. A. II, & David, F. (1987). Corporate Mission Statements: The Bottom Line. *The Academy of Management Executive*, *1*(2), 109–116. doi:10.5465/AME.1987.4275821

Perez, T., Cromley, J. G., & Kaplan, A. (2014). The role of identity development, values, and costs in college STEM retention. *Journal of Educational Psychology*, *106*(1), 315–329. doi:10.1037/a0034027

Peterson, C., & Seligman, M. E. P. (2004). *Character strengths and virtues: A handbook and classification*. Washington, DC: American Psychological Association.

Peterson, N. G., Mumford, M. D., Borman, W. C., Jeanneret, P. R., & Fleishman, E. A. (Eds.). (1999). *An occupational information system for the 21st century: The development of O*NET*. Washington, DC: American Psychological Association. doi:10.1037/10313-000

Pike, G. R. (2006). Students personality types, intended majors, and college expectations: Further evidence concerning psychological and sociological interpretations of Hollands theory. *Research in Higher Education*, *47*(7), 801–822. doi:10.1007/s11162-006-9016-5

Pintrich, P. R., & De Groot, E. V. (1990). Motivational and self-regulated learning components of classroom academic performance. *Journal of Educational Psychology*, *82*(1), 33–40. doi:10.1037/0022-0663.82.1.33

Polat, N., Mancilla, R., & Mahalingappa, L. (2013). Anonymity and motivation in asynchronous discussions and L2 vocabulary learning. *Language Learning & Technology*, *17*(2), 57–74.

Porchea, S. F., Allen, J., Robbins, S., & Phelps, R. P. (2010). Predictors of long-term enrollment and degree outcomes for community college students: Integrating academic, psychosocial, socio-demographic, and situational factors. *The Journal of Higher Education*, *81*, 750–778.

Poropat, A. E. (2009). A meta-analysis of the Five-Factor Model of personality and academic performance. *Psychological Bulletin*, *135*(2), 322–338. doi:10.1037/a0014996 PMID:19254083

Porter, K. B. (2008). Current trends in student retention: A literature review. *Teaching and Learning in Nursing*, *3*(1), 3–5. doi:10.1016/j.teln.2007.09.001

Porter, M. C. (2011). Assessing alternative breaks: Moving beyond sleeping on floors and pass-the-candle reflection. *About Campus: Enriching the Student Learning Experience*, *16*(5), 21–24. doi:10.1002/abc.20077

Porter, S. R., & Swing, R. (2006). Understanding how first-year seminars affect persistence. *Research in Higher Education*, *47*(1), 89–109. doi:10.1007/s11162-005-8153-6

Poster, E. C., Mancini, M. E., & Ganji, D. MOOCS and More: Expanding Online Access for Nurses. *Educause Review Online*. Retrieved August 5, 2013 from http://www.educause.edu/ero/article/moocs-and-more-expanding-online-access-nurses

Potgieter, M. (2010, October). Conceptual gain in chemistry: Is the gap addressed effectively? *Paper presented at the STEM, Mind the Gap Conference, Academy of Science of South Africa, Cape Town.* Retrieved January 10, 2017 from http://www.google.com.au/

Povey, H., & Angier, C. (2004). *'I can do it, but it'll be a battle': finding her place as an undergraduate mathematician.* Sheffield Hallam University.

Prevatt, F., Petscher, Y., Proctor, B. E., Hurst, A., & Adams, K. (2006). The revised Learning and Study Strategies Inventory: An evaluation of competing models. *Educational and Psychological Measurement*, *66*(3), 448–458. doi:10.1177/0013164405282454 PMID:26347055

Raab, L., & Adam, A. J. (2005). The university college model: A learning-centered approach to retention and remediation. *New Directions for Institutional Research*, (125): 86–106.

Raby, R. L. (2009). Globalization and community college model development. In R.L. Raby & E.J. Valeau (Eds.). Community college models: Globalization and high education reform (pp. 21-38). Amsterdam: Springer Press. doi:10.1007/978-1-4020-9477-4_2

Rand, K. L., & Cheavens, J. S. (2009). Hope theory. In S. J. Lopez & C. R. Snyder (Eds.), *Oxford handbook of positive psychology* (pp. 323–333). New York, NY: Oxford University Press.

Raykov, T., & Marcoulides, G. A. (2011). *Introduction to psychometric theory*. Routledge.

Reisberg, D. (2010). *Cognition: Exploring the Science of the Mind* (4th ed.). New York, NY: Norton.

Reupert, A., Maybery, D., Patrick, K., & Chittleborough, P. (2009). The importance of being human: Instructors' personal presence in distance programs. *International Journal of Teaching and Learning in Higher Education*, *21*(1), 47–56.

Reushle, S., & Loch, B. I. (2008). Conducting a trial of web conferencing software: Why, how, and perceptions from the coalface. *TOJDE/ Turkish Online Journal of Distance Education*, *9*(3), 19–28.

Rheinberg, F., Vollmeyer, R., & Rollett, W. (2000). Motivation and action in self-regulated learning. In M. Boekaerts, P. R. Pintrich, & M. Zeidner, (Eds.), Handbook of Self-Regulation (pp. 503-529). San Diego, CA: Academic Press. doi:10.1016/B978-012109890-2/50044-5

Richardson, M., Abraham, C., & Bond, R. (2012). *Psychological correlates of university students' academic performance: A systematic review and meta-analysis.* Retrieved from http://www.ncbi.nlm.nih.gov/pubmed/22352812

Richardson, J. C., & Swan, K. (2003). Examining social presence in online courses in relate to students perceived learning and satisfaction. *Journal of Asynchronous Learning Networks*, *1*, 68–98.

Richardson, M., Abraham, C., & Bond, R. (2012). Psychological correlates of university students' academic performance: A systematic review and meta-analysis. *Psychological Bulletin*, *138*(2), 353–387. doi:10.1037/a0026838 PMID:22352812

Rikoon, S. H., Brenneman, M., Kim, L. E., Khorramdel, L., MacCann, C., Burrus, J., & Roberts, R. D. (2016). Facets of conscientiousness and their differential relationships with cognitive ability factors. *Journal of Research in Personality*, *61*, 22–34. doi:10.1016/j.jrp.2016.01.002

Rikoon, S. H., Liebtag, T., Olivera-Aguilar, M., Steinberg, J., & Robbins, S. B. (2015). *Anticipating College Enrollment: Adapting SuccessNavigator for High School Students. (ETS RR-15-37).* Princeton, NJ: Educational Testing Service.

Robbins, R. (2009). Evaluation and assessment of career advising. In K.B. Hughey, D. Nelson, J. Damminger et al. (Eds.), Handbook of Career Advising (Ch. 12). San Francisco: Jossey-Bass.

Robbins, R. (2011). Assessment and accountability. In J. Joslin & N. Markee (Eds.), *Academic Advising Administration: Essential Knowledge and Skills for the 21st Century*. Manhattan, KS: National Academic Advising Association.

Robbins, S. B., Allen, J., Casillas, A., Akamigbo, A., Saltonstall, M., Campbell, R., & Gore, P. A. (2009). Associations of resource and service utilization, risk level, and college outcomes. *Research in Higher Education*, *50*(1), 101–118. doi:10.1007/s11162-008-9106-7

Robbins, S. B., Lauver, K., Le, H., Davis, D., Langley, R., & Carlstrom, A. (2004). Do psychosocial and study skill factors predict college outcomes? A meta-analysis. *Psychological Bulletin*, *130*(2), 261–288. doi:10.1037/0033-2909.130.2.261 PMID:14979772

Robbins, S. B., Oh, I. S., Le, H., & Button, C. (2009). Intervention effects on college performance and retention as mediated by motivational, emotional, and social control factors: Integrated meta-analytic path analyses. *The Journal of Applied Psychology*, *94*(5), 1163–1184. doi:10.1037/a0015738 PMID:19702363

Robbins, S., Allen, J., Casillas, A., Peterson, C., & Le, H. (2006). Unraveling the differential effects of motivational and skills, social, and self-management measures from traditional predictors of college outcomes. *Journal of Educational Psychology*, *98*(3), 598–616. doi:10.1037/0022-0663.98.3.598

Robertson, S., Canary, C. W., Orr, M., Herberg, P., & Rutledge, D. N. (2010). Factors related to progression and graduation rates for RN-to Bachelor of Science in nursing programs: Searching for realistic benchmarks. *Journal of Professional Nursing*, *6*(2), 9–107. doi:10.1016/j.profnurs.2009.09.003 PMID:20304377

Rodriguez, B. M. (1995). Creating inclusive and multicultural communities: Working through assumptions of culture, power, diversity, and equity. *Multicultural education: Strategies for implementation in colleges and universities, 4*, 33-40.

Roed, J. (2003). Language learner behaviour in a virtual environment. *Computer Assisted Language Learning*, *16*(2-3), 155–172. doi:10.1076/call.16.2.155.15880

Rollnick, M., Mphahlele, M., Ogude, N., Green, G., Huddle, P., Mahooana, P., & White, M. (1997). A team work approach to workshop delivery for tutor development. *Academic Development*, *3*(2), 91–116.

Roman, T., Kelsey, K., & Lin, H. (2010). Enhancing online education through instructor skill development in higher education. *Online Journal of Distance Learning Administration*, *8*(1). Retrieved from http://www.westga.edu/~distance/ojdla/winter134/roman_kelsey134.html

Rosen, A. (2011). *CHANGE.EDU: Rebooting for the New Talent Economy*. New York: Kaplan Publishing.

Rosenberg, L., Perraud, S., & Willis, L. (2007). The value of admission interviews in selecting accelerated second-degree baccalaureate nursing students. *The Journal of Nursing Education*, *46*(9). PMID:17912993

Roskos, K., & Boehlen, S. (2001). Enhancing teachers' awareness of their instructional talk in the teaching of reading: Differential features of self-assessment. *Action in Teacher Education*, *22*(4), 59–74. doi:10.1080/01626620.2001.10463030

Rovai, A. P. (2003). In search of higher persistence rates in distance education online programs. *The Internet and Higher Education*, *6*(1), 1–16. doi:10.1016/S1096-7516(02)00158-6

Rovai, A. P., & Jordan, H. (2004). Blended learning and sense of community: A comparative analysis with traditional and fully online graduate courses. *International Review of Research in Open and Distance Learning*, *2*(5), http://www.irrodl.org/index.php/irrodl/article/viewArticle/192/274 Retrieved December 21, 2014

Rowsell, J., Sztainbok, V., & Blaney, J. (2007). Losing strangeness: Using culture to mediate ESL teaching. *Language, Culture and Curriculum*, *20*(2), 140–154. doi:10.2167/lcc331.0

Ruberg, L. F., & Taylor, C. D. (1995). *Student responses to network resources: Formative evaluation of two classes*. Paper presented at the Annual Meeting of the American Educational Research Association, San Francisco, CA.

Rudestam, E. K. (2004). Distributed education and the role of online learning in training professional psychologists. *Professional Psychology, Research and Practice*, *23*(4), 427–432. doi:10.1037/0735-7028.35.4.427

Rudzki, R. (2000). Implementing internationalisation: The practical application of the fractal process model. *Journal of Studies in International Education*, *4*(2), 77–90. doi:10.1177/102831530000400208

Rueda, R. (2011). *The 3 dimensions of improving student performance*. New York: Teachers College Press.

Rueda, R. (2011). *The three dimensions of improving student performance: Finding the right solutions to the right problems*. New York, NY: Teachers College Press.

Rueda, R. (2011). *The three dimensions of improving student performance: finding the right solutions to the right problems*. New York: Teachers College.

Russell, S. S. (2006). An overview of adult learning processes. *Urologic Nursing*, *26*(5), 349–352, 370. Retrieved from http://www.medscape.com/viewarticle/547417_2 PMID:17078322

Ryan, M. G. (2013). Improving retention and academic achievement for first-time students at a two-year college. *Community College Journal of Research and Practice*, *37*(2), 131–134. doi:10.1080/10668926.2012.715266

Sackett, P. R., & Walmsley, P. T. (2014). Which personality attributes are most important in the workplace? *Perspectives on Psychological Science*, *9*(5), 538–551. doi:10.1177/1745691614543972 PMID:26186756

Sansone, C., & Harackiewicz, J. M. (2000). *Intrinsic and extrinsic motivation: The search for optimal motivation and performance*. San Diego, CA: Academic Press.

Sassen, S. (2001). *The global city: New York, London, Tokyo*. Princeton, NJ: Princeton University Press. doi:10.1515/9781400847488

Schacter, D. L., & Tulving, E. (1994). What are the memory systems of 1994? In D. L. Schacter & E. Tulving (Eds.), *Memory Systems 1994*. Cambridge, MA: MIT Press.

Schein, E. H. (2010). *Organization culture and leadership* (4th ed.). San Francisco, CA: Jossey-Bass.

Schlossberg, N. K. (1989). Marginality and mattering: Key issues in building community. *New Directions for Student Services*, *48*(48), 5–15. doi:10.1002/ss.37119894803

Schlossberg, N., Waters, E., & Goodman, J. (1995). *Counselling adults in transition: Linking practice with theory* (2nd ed.). New York: Spring Publishing.

Schmitt, N., Billington, A., Keeney, J., Reeder, M., Pleskac, T. J., Sinha, R., & Zorzie, M. (2011). *Development and validation of measures of noncognitive student potential* (College Board Research Report 2011-1). New York: The College Board.

Schreiner, L., & Nelson, D. (2013). The contribution of Student Satisfaction to Persistence. *Journal of College Student Retention*, *15*(1), 73–111. doi:10.2190/CS.15.1.f

Schroeder, S., Baker, M., Terras, K., Mahar, P., & Chiasson, K. (2016). Desired and experienced levels of connectivity to an asynchronous, online, distance degree program. *Online Learning*, *20*(3), 244–263.

Schuh, J. H., Kones, S. R., & Harper, S. R. (2010). *Student services: A handbook for the profession*. San Francisco, CA: Jossey-Bass Publishing.

Schunk, D. H., & Ertmer, P. A. (2000). Self-regulatory and academic learning self-efficacy enhancing interventions. In M. Boekaerts, P.R. Pintrich, & M. Zeidner, (Eds.), Handbook of Self-Regulation (pp. 631-649). San Diego, CA: Academic Press.

Schunk, D. H. (2008). Metacognition, self-regulation, and self-regulated learning: Research recommendations. *Educational Psychology Review*, *20*(4), 463–467. doi:10.1007/s10648-008-9086-3

Schwartz, H. L., & Holloway, E. L. (2012). Partners in learning: A grounded theory study of relational practice between masters students and professors. *Mentoring & Tutoring: Partnership in Learning*, *20*(1), 115–135. doi:10.1080/13611267.2012.655454

Schwartz, M. A. (2008). The importance of stupidity in scientific research. *Journal of Cell Science*, *121*(11), 1771. doi:10.1242/jcs.033340 PMID:18492790

Schwebel, D. C., Walburn, N. C., Klyce, K., & Jerrolds, K. L. (2012). Efficacy of advising outreach on student retention, academic progress and achievement, and frequency of advising contacts: A longitudinal randomized trial. *NACADA Journal*, *32*(2), 36–43. doi:10.12930/0271-9517-32.2.36 PMID:23270066

Sciarra, T. D., Seirup, J. H., & Sposato, E. (2016). High school predictors of college persistence: the significance of engagement and teacher interaction. *The Professional Counselor*, *6*(2), 189–202. doi:10.15241/ds.6.2.189

Scott, I., Yeld, N., & Hendry, J. (2007). *A case for improving teaching and learning in South African higher education. Higher Education Monitor No. 6*. Pretoria, South Africa: Council on Higher Education.

Scott, W. R. (2003). *Organizations: Rational, natural, and open systems* (5th ed.). Upper Saddle River, NJ: Pearson Education, Inc.

Scrivner, S., & Weis, M. J. (2009). More Guidance, Better Results? Three-Year Effects of an Enhanced Student Services Program at Two Community Colleges. Retrieved from http://www.mdrc.org/publication/more-guidance-better-results

Seaman, J. S. (2009). Online learning as a strategic asset. Volume II: The paradox of faculty voices -Views and experiences with online learning. Results of a National Faculty Survey, Part of the Online Education Benchmarking Study Conducted by the APLU-Sloan National Commission on Online Learning. Association of Public and Land Grant Universities Schroeder.

Seidman, A. (2012). College Student Retention Formula for Student Success (2nd ed.). Alan College Student Retention: Formula for Student Success (The ACE Series on Higher Education). Plymouth, Rowman & Littlefield Publishers.

Seidman, A. (2005). *Minority students' retention: Resources for practitioners*. New Directions for Institutional Research. Wiley Periodicals Inc.

Seidman, A. (2005b). Where we go from here: A retention formula for student success. In A. Seidman (Ed.), *College student retention* (pp. 295–316). Westport: Praeger Publishers.

Seidman, A. (2009). *College student Retention: Formula for student success*. Lanham, MD: Rowman & Littlefield Publishers.

Seldomridge, L. A., & DiBartolo, M. C. (2005). A profile of accelerated second bachelors degree nursing students. *Nurse Educator*, *30*(2), 65–68. doi:10.1097/00006223-200503000-00007 PMID:15785342

Settersten, R. A., & Ray, B. (2010). What's Going on With Young People Today? The Long and Twisting Path to Adulthood. *The Future of Children*, *20*(1), 19–37.

Shea, P., Li, C. S., Swan, K., & Pickett, A. M. (2005). Developing learning community in online asynchronous college courses: the role of teaching presence. *Journal of Asynchronous Learning Networks, 9*. Retrieved from http://www.sloan.org/publications/jaln/v9n4/pdf/v9n4_shea.pdf

Shelton, E. N. (2012). A model of nursing student retention. International Journal of Nursing Education Scholarship, 9(1). *Article*, *6*, 1–16. doi:10.1515/1548-923X.2334

Shin, S., & Koh, M. (2006). Korean education in cultural context. *Essays in Education*, *14*, 1–10.

Shi, S., & Morrow, B. V. (2006). E-Conferencing for Instruction: What Works? *EDUCAUSE Quarterly*, *29*(4), 42–49.

Simonson, M., Smaldino, S., Albright, M., & Zvacek, S. (2012). *Teaching and learning at a distance: Foundations of distance education* (5th ed.). Boston, MA: Pearson.

Sisson, G. R. (2002). *Hands-on training: A simple and effective method for on-the-job training*. San Francisco: Berrett-Koehler Publishers.

Siwatu, K. O. (2007). Preservice teachers culturally responsive teaching self-efficacy and outcome expectancy beliefs. *Teaching and Teacher Education*, *23*(7), 1086–1101. doi:10.1016/j.tate.2006.07.011

Siwatu, K. O. (2009). Student teachers' self-efficacy beliefs regarding culturally responsive teaching and their professed classroom practices. *Teacher Education and Practice*, *22*, 323–333.

Siwatu, K. O. (2011). Preservice teachers culturally responsive teaching self efficacy forming experiences: A mixed methods study. *The Journal of Educational Research*, *104*(5), 360–369. doi:10.1080/00220671.2010.487081

Skinner, B., & Austin, R. (1999). Computer conferencing: Does it motivate EFL students? *ELT Journal*, *53*(4), 270–279. doi:10.1093/eltj/53.4.270

Skipper, T. (2003). *Involvement in campus activities and the retention of first year college students*. Columbia, SC: National Resource Center.

Skylar, A. A. (2009). A comparison of asynchronous online text-based lectures and synchronous interactive web conferencing lectures. *Issues in Teacher Education*, *18*(2), 69.

Smith, G., Ferguson, D., & Caris, M. (2002). Teaching over the web versus in the classroom: Differences in the instructor experience. *International Journal of Instructional Media*, *29*(1), 61–67.

Smith-Paríolá, J., & Gòkè-Paríolá, A. (2006). Expanding the parameters of service learning: A case study. *Journal of Studies in International Education*, *10*(1), 71–86. doi:10.1177/1028315305283307

Snowmont, J., McCown, R., & Biehler, R. (2012). *Psychology applied to teaching*. Belmont, CA: Wadsworth.

Snyder, C. (2008). Grabbing hold of a moving target: Identifying and measuring the transformative learning process. *Journal of Transformative Education*, *6*(3), 159–181. doi:10.1177/1541344608327813

Snyder, C. R. (1994). *The psychology of hope: You can get there from here*. New York: Free Press.

Southern, N. L. (2007). Mentoring for transformative learning: The importance of relationship in creating learning communities of care. *Journal of Transformative Education*, *5*(4), 329–338. doi:10.1177/1541344607310576

Squire, L. R. (1994). Declarative and nondeclarative memory: Mulitiple brian systems support learning and memory. In D. Schacter & E. Tulving (Eds.), *Memory systems 1994*. Cambridge, MA: MIT Press.

Stachowiak, B. (2016, June 2). *Critical Instructional Design with Sean Michael Morris. Teaching in Higher Education Podcast*. Retrieved from http://teachinginhighered.com/podcast/critical-instructional-design/

Stastista. (2017). Percentage of the U.S. population who have completed four years of college or more from 1940 to 2015, by gender. Retrieved from https://www.statista.com/statistics/184272/educational-attainment-of-college-diploma-or-higher-by-gender/

Statista. (n. d.). Internet usage in South Korea. Retrieved April 10, 2017, from https://www.statista.com/topics/2230/internet-usage-in-south-korea/

Stebleton, M. J., Soria, K. M., & Cherney, B. (2013). The high impact of education abroad: College students' engagement in international experiences and the development of intercultural competencies. *Frontiers: The Interdisciplinary Journal of Study Abroad*, 22.

Stilborne, L., & Williams, L. (1996). Meeting the needs of adult learners in developing courses for the Internet. In *Proceedings of INET '96*.

Straker, J. (2016). International student participation in higher education: changing the focus from international students to participation. *Journal of Studies in International Education*, *20*(4), 299–318. doi:10.1177/1028315316628992

Strayhorn, T. L. (2006). *Frameworks for assessing learning and development outcomes*. Washington, DC: Council for the Advancement of Standards in Higher Education.

Strayhorn, T. L. (2009). An examination of the impact of first-year seminars on correlates of college student retention. *Journal of the First-Year Experience & Students in Transition*, *21*(1), 9–27.

Street, H. (2010, Winter). Factors influencing a learner's decision to drop-out or persist in higher education research. *Online Journal of Distance Learning Administration*, *13*(4).

Stromquist, N. P. (2002). *Education in a globalized world: the connectivity of economic power, technology and knowledge*. Lanham, MD: Rowman & Littlefield.

Sutherland, T. (2002). *Active learning toolkit*. Sarasota, FL: American Accounting Association.

Sutton, R., & Rubin, D. (2004). The GLOSSARI project: Initial findings from a system-wide research initiative on study abroad learning out- comes. *Frontiers: The Interdisciplinary Journal of Study Abroad*, *10*, 65–82.

Swales, J. M., & Rogers, P. S. (1995). Discourse and the Projection of Corporate Culture: The Mission Statement. *Discourse & Society*, *6*(2), 223–242. doi:10.1177/0957926595006002005

Tarbuck, A. F., & Paykel, E. S. (1995). Effects of major depression on the cognitive function of younger and older subjects. *Psychological Medicine*, *25*(2), 285–296. doi:10.1017/S0033291700036187 PMID:7675916

Tarrow, S., & Hall, P. A. (1998). *Globalization and area studies: When is too broad too narrow?* The Chronicle of Higher Education. Retrieved from http://chronicle.com/article/GlobalizationArea/99332/

Tauroza, S., & Luk, J. (1997). Accent and Second Language Listening Comprehension. *RELC Journal*, *28*(1), 54–71. doi:10.1177/003368829702800104

Taylor, A., & McQuiggan, C. M. (2008). Faculty development programming: If we build it, will they come? *Educause*, *3*, 28–37.

Taylor, J. M., Dunn, M., & Winn, S. K. (2015). Innovative orientation leads to improved success in online courses. *Online Learning*, *19*(4), 112–120. doi:10.24059/olj.v19i4.570

Tedx Talks. (2013, June). Designing a university for the new millennium: David Helfand at TEDx West Vancouver ED (YouTube video). Retrieved from https://www.youtube.com/watch?v=DZQe73IXZtU

Terblanché, E. (2015). Deciding to teach online: Communication, opportunities and challenges for educators in distance education. *Communicatio, 41*(4).

Terenzini, P. T., Springer, L., Yaeger, P. M., Pascarella, E. T., & Nora, A. (1996). First-generation college students: Characteristics, experiences, and cognitive development. *Research in Higher Education*, *37*(1), 1–22. doi:10.1007/BF01680039

Terjesen, S., & Elam, A. (2012). Women entrepreneurship: A force for growth. *International Trade Forum*, (2), 16.

Terrion, J., & Leonard, D. (2010). A taxonomy of the characteristics of student peer mentors in higher education: Findings from a literature review. *Mentoring & Tutoring: Partnership in Learning*, *15*(2), 149–164. doi:10.1080/13611260601086311

The Economist. (2011, May 12). *Depressed students in South Korea: We don't need quite so much education*. Retrieved from http://www.economist.com/blogs/banyan/2011/05/depressed_students_south_korea

The World Factbook. (2014). South Korea. Retrieved from https://www.cia.gov/library/publications/the-world-factbook/geos/ks.html

Thomas, L., Hill, M., O'Mahony, J., & Yorke, M. (2017) Supporting student success: strategies for institutional change What Works? Student Retention & Success programme. Retrieved from https://www.phf.org.uk/wp-content/uploads/2017/04/Full-report-Final.pdf

Thomas, L. (2002). Student retention in higher education: The role of institutional habitus. *Journal of Education Policy*, *17*(4), 423–442. doi:10.1080/02680930210140257

Thomson, W. (1883). Electrical Units of Measurement: The popular lectures and addresses. Macmillan and Co.

Tinto, V. (1993). Leaving college: Rethinking causes and cures of student attrition. Chicago, IL: University of Chicago Press.

Tinto, V. (1998). Learning Communities and the Reconstruction of Remedial Education. In *Higher Education, Replacing Remediation in Higher Education Conference*, Stamford University, January 26-27.

Tinto, V. (2004). Student retention and graduation: facing the truth, living with the consequences. The Pell Institute. Retrieved 2/9/14 from http://www.pellinstitute.org/downloads/publications-Student_Retention_and_Graduation_July_2004.pdf

Tinto, V. (1975). Dropout from higher education: A theoretical synthesis of recent research. *Review of Educational Research*, *45*(1), 89–125. doi:10.3102/00346543045001089

Tinto, V. (1982). Defining dropout: A matter of perspective. *New Directions for Institutional Research*, *36*(36), 3–15. doi:10.1002/ir.37019823603

Tinto, V. (1987). *Increasing student retention*. San Francisco: Jossey Bass.

Tinto, V. (1993). *Leaving college: Rethinking the causes and cures for student attrition* (2nd ed.). Chicago: University of Chicago Press.

Tinto, V. (1993). *Leaving College: Rethinking the Causes and Cures of Student Attrition* (2nd ed.). Chicago, IL: University of Chicago Press.

Tinto, V. (1993). *Leaving college: Rethinking the causes and cures of student attrition* (2nd ed.). Chicago: The University of Chicago Press.

Tinto, V. (2000). Linking learning and leaving: Exploring the role of the college classroom in student departure. In J. M. Braxton (Ed.), *Reworking the student departure puzzle* (pp. 81–94). Nashville, TN: Vanderbilt University Press.

Tinto, V. (2004). *Student retention and graduation: Facing the truth, living with the consequences*. Washington, DC: The Pell Institute for the Study of Opportunity in Higher Education.

Tinto, V. (2006). Research and practice of student retention: What next? *Journal of College Student Retention*, *8*(1), 1–19. doi:10.2190/4YNU-4TMB-22DJ-AN4W

Tobolowsky, B. F., & Associates. (2008). *2006 National Survey of First-Year seminars: Continuing innovations in the collegiate curriculum* (Monograph No, 51). Columbia, SC: University of South Carolina, National Resource Center for the First-Year Experience and Students in Transition.

Tough, P. (2013). *How children succeed*. Random House.

Tracey, T. J., & Sedlacek, W. E. (1982, March). Non-cognitive variables in predicting academic success by race. *Paper presented at the annual meeting of the American Educational Research Association*, New York.

Tracey, D. H. (2012). *Lenses on reading: An introduction to theories and models* (2nd ed.). New York, NY: Guilford Press.

Tresman, S. (2002). *Towards a Strategy for Improved Student Retention in Programmes of Open, Distance Education: A Case Study From the Open University UK*.

Tricot, A., & Sweller, J. (2014). Domain-specific knowledge and why teaching generic skills does not work. *Educational Psychology Review*, *26*(2), 265–283. doi:10.1007/s10648-013-9243-1

Tsourtos, G., Thompson, J. C., & Stough, C. (2002). Evidence of an early information processing speed deficit in unipolar major depression. *Psychological Medicine*, *32*(2), 259–265. doi:10.1017/S0033291701005001 PMID:11866321

Tubbs, S., & Moss, S. (2006). *Human communication: Principles and contexts*. New York, NY: McGraw Hill.

Tulving, E. (1993). What is episodic memory? *Current Directions in Psychological Science*, *2*(3), 67–70. doi:10.1111/1467-8721.ep10770899

Tuttle, K. N. (2000). Academic advising. *New Directions for Higher Education*, *111*(111), 15–24. doi:10.1002/he.11102

Twombly, S. B., Salisbury, M. H., Tumanut, S. D., & Klute, P. (2012). *Study Abroad in a New Global Century: Renewing the Promise, Refining the Purpose*. San Francisco, CA: Jossey-Bass.

U.S. Department of Education. (2009). Evaluation of evidence-based practices in online learning: A meta-analysis and review of online learning studies. Washington, D.C. Retrieved from http://www.gibill.va.gov/gi_bill_info/ch33/yellow_ribbon.htm

U.S. Department of Education. (2010). Evaluation of Evidence Based Practice in Online Learning. A Meta-Analysis Review of Online Learning Studies. Retrieved from https://www2.ed.gov/rschstat/eval/tech/evidence-based-practices/finalreport.pdf

U.S. Department of Education. (2012). Federal student financial aid handbook. Retrieved from www.ifap.ed.gove/fsahandbook/attachments/1112FSAHbkVol1Ch1.pdf

U.S. Department of Education. (2017). *Reimagining the role of technology in higher education. A supplement to the National Education Technology Plan*. Washington, DC: Office of Educational Technology.

U.S. Department of Health and Human Services, Health Resources and Services Administration (2010). The Registered Nurse Population: Findings from the 2008 National Sample Survey of Registered Nurses.

U.S. Department of Health and Human Services, Health Resources and Services Administration (2013). The U.S. Nursing Workforce: Trends in Supply and Education.

U.S. Department of Labor. (2015a). Labor Force Statistics from the Current Population Survey. Retrieved from http://www.bls.gov/cps/demographics.htm

U.S. Department of Labor. (2015b). *Earning and Unemployment Rates by Educational Attainment, 2015*. Retrieved from https://www.bls.gov/emp/ep_chart_001.htm

Umbach, P. D., & Wawrzynsli, M. R. (2005). Faculty do matter: The role of college faculty in student learning and engagement. *Research in Higher Education*, *46*(2), 153–185. doi:10.1007/s11162-004-1598-1

United States Department of Education. (2006, September). *A Test of Leadership: Charting the Future of U.S. Higher Education: A Report of the Commission Appointed by Secretary of Education Margaret Spellings*. Retrieved from https://www2.ed.gov/about/bdscomm/ list/hiedfuture/reports/pre-pub-report.pdf

United States Department of Education. (2009, October 27). *Institutional eligibility under the Higher Education Act of 1965, as amended, and the secretary's recognition of accrediting agencies; final rule* (Federal Register, Vol. 7, No. 206, Doc E9-25186). Retrieved from https://www2.ed.gov/legislation/FedRegister/finrule/2009-4/102709a.html

United States Department of Education. (2012, February). *Pathways to success, integrating learning with life and work to increase national college completion*. Retrieved from http://www2.ed.gov/about/bdscomm/list/acsfa/ptsreport2.pdf

United States Department of Education. (2014). Pathways to success, integrating learning with life and work to increase national college completion. Retrieved from http://www2.ed.gov/about/bdscomm/list/acsfa/ptsreport2.pdf

UPCEA (University Professional and Continuing Education Association). (2014, October). Federal Policy Brief. Retrieved from http://www.upcea.edu/Files/UPCEAPolicyBrief.pdf

Upcraft, L., & Gardner, J. (1990). *The freshmen year experience: Helping students survive and succeed in college*. San Francisco, CA: Jossey - Bass Publishers.

Valle-Riestra, D. M., Shealey, M. W., & Cramer, E. D. (2011). Recruiting and retaining culturally diverse special educators. *Multidisciplinary Journal of Teaching and Learning*, *1*(2), 68–87.

van der Wende, M. (2007). Internationalization of higher education in the OECD countries: Challenges and opportunities for the coming decade. *Journal of Studies in International Education*, *11*(3-4), 274–289. doi:10.1177/1028315307303543

Vande Berg, M. (2007). Intervening in the learning of U.S. students abroad. *Journal of Studies in International Education*, *11*(3-4), 392–399. doi:10.1177/1028315307303924

Vogt, E., Brown, J., & Issacs, D. (2003). *The Art of powerful questions: Catalyzing insight, innovation, and action*. Mill Valley, CA: Whole Systems Associates.

Wächter, B. (2003). An introduction: Internationalisation at home in context. *Journal of Studies in International Education*, *7*(1), 5–11. doi:10.1177/1028315302250176

Walker, C. O., Greene, B. A., & Mansell, R. A. (2006). Identification with academics, intrinsic/extrinsic motivation, and self-efficacy as predictors of cognitive engagement. *Learning and Individual Differences*, *16*(1), 1–12. doi:10.1016/j.lindif.2005.06.004

Wang, L. C. (2013). The effect of high-stakes testing on suicidal ideation of teenagers with reference dependent preferences. *Monash University*, *29*(2), 1–33.

Wankel, L. A., & Blessinger, P. (Eds.). (2013). *Increasing student engagement and retention using multimedia technologies: Video annotation, multimedia applications, videoconferencing and transmedia storytelling*. Bingley, UK: Emerald Group Publishing Limited.

Warner, J. (2014). *Fact sheet: The women's leadership gap*. Center for American Progress. Retrieved from: https://www.americanprogress.org/issues/women/reports/2014/03/07/85457/fact-sheet-the-womens-leadership-gap/

Warschauer, M. (1995). Comparing face-to-face and electronic discussion in the second language classroom. *CALIO Journal*, *13*, 7–26.

Webber, D. A., & Ehrenberg, R. G. (2010). Do expenditures other than instructional expenditures affect graduation and persistence rates in American higher education? *Economics of Education Review*, *29*(6), 947–958. doi:10.1016/j.econedurev.2010.04.006

Weekley, J. A., & Ployhart, R. E. (Eds.). (2013). *Situational judgment tests: Theory, measurement, and application*. Psychology Press.

Weidman, J. C., & Park, N. (2000). *Higher education in Korea: tradition and adaptation*. New York: Falmer Press.

Wei, F. F., Wang, Y. K., & Fass, W. (2014). Research Report: An experimental study of online chatting and notetaking techniques on college students cognitive learning from a lecture. *Computers in Human Behavior*, *34*, 148–156. doi:10.1016/j.chb.2014.01.019

Weinstein, C. E., & Palmer, D. R. (2002). *Learning and Study Strategies Inventory (LASSI):User's manual*. Clearwater, FL: H & H Publishing.

Weiss, M. J., Bloom, H. S., & Brock, T. (2014). A conceptual framework for studying the sources of variation in program effects. *Journal of Policy Analysis and Management*, *33*(3), 778–808. doi:10.1002/pam.21760

Wickersham, L., & Dooley, K. (2006). A content analysis of critical thinking skills as an indicator of quality online discussion in virtual learning communities. *Quarterly Review of Distance Education, 7*(2), 185–193.

Wiggers, R., & Arnold, C. (2011). Defining, Measuring and Achieving 'Student Success' in Ontario Colleges and Universities. The Higher Education Quality Council of Ontario. Retrieved January 19, 2017 from http://www.heqco.ca/

Wiggins, G., & McTighe, J. (2005). Understanding By Design. The Association for Curriculum and Development. https://www.amazon.com/Understanding-Design-Grant-Wiggins/dp/1416600353/ref=sr_1_1?ie=UTF8&qid=1488596859&sr=8-1&keywords=wiggins+understanding+by+design#reader_1416600353

Wighting, M. J., Liu, J., & Rovai, A. P. (2008). Distinguishing sense of community and motivation characteristics between online and traditional college students. *Quarterly Review of Distance Education, 9*(3), 285–295.

Williams, I. L., Glenn, P. W., & Wider, F. (2008). Nurtured advising: An essential approach to advising students at historically black colleges and universities. *Academic Advising Today, 31*(1). Retrieved from January 10, 2017 from http://www.nacada.ksu.edu/

Williams, D., Howell, S., & Hricko, M. (2006). *Online assessment, measurement, and evaluation: Emerging practices.* Hershey, PA: Information Science Publishing. doi:10.4018/978-1-59140-747-8

Williamson, L. (2014, February 4). Kimchi: South Korea's efforts to boost its national dish. *BBC.* Retrieved from http://www.bbc.com/news/world-asia-25840493

Willingham, D. T. (2009). *Why don't students like school?: A cognitive scientist answers questions about how the mind works and what it means for the classroom.* San Francisco, CA: Jossey-Bass.

Witherspoon, R., Goldsmith, M., Lyons, L., & Freas, A. (2000). *Coaching for leadership: How the world's greatest coaches help leaders learn.* San Francisco, CA: Jossey-Bass/Pfeiffer.

Wlodkowski, R. J. (2008). *Enhancing adult motivation to learn: A comprehensive guide for teaching all adults.* San Francisco, CA: Jossey-Bass.

Wolf, P. (2006). Best practices in the training of faculty to teach online. *Journal of Computing in Higher Education, 17*(2), 47–78. doi:10.1007/BF03032698

Wong, A. (2016). The Atlantic. Where are all the high school grads going? Retrieved from https://www.theatlantic.com/education/archive/2016/01/where-are-all-the-high-school-grads-going/423285/

Wong, M. L., & Faller, P. (2104). Creating time and space for transformative learning in a cohort-based adult education doctoral program. In *Proceedings of the XI International Transformative Learning Conference.* Teachers College, Columbia University, New York, NY.

Woolf, M. (2002). Harmony and dissonance in international education: The limits of globalisation. *Journal of Studies in International Education, 6*(1), 5–15. doi:10.1177/1028315302006001002

Worldatlas.com. (2016, November 24). Countries Where People Spend the Most Time Online. Retrieved March 20, 2017 from http://www.worldatlas.com/articles/top-countries-which-spend-the-greatest-amount-of-time-online.htm

Worldatlas.com. (2017, February 21). Suicide rates by country. Retrieved March 27, 2017 from http://www.worldatlas.com/articles/countries-with-the-most-suicides-in-the-world.html

Yang, R. (2002). University internationalization: Its meanings, rationales and implications. *Intercultural Education, 13*(1), 81–95. doi:10.1080/14675980120112968

Yarnell, M., & Yarnell, R. R. (2010). *Your first year in network marketing: Overcome your fears, experience success, and achieve your dreams!* Three Rivers Press.

Yeager, D. S., & Dweck, C. S. (2012). Mindsets that promote resilience: When students believe that personal characteristics can be developed. *Educational Psychologist*, *47*(4), 302–314. doi:10.1080/00461520.2012.722805

Yeager, D. S., Romero, C., Paunesku, D., Hulleman, C. S., Schneider, B., Hinojosa, C., & Trott, J. et al. (2016). Using design thinking to improve psychological interventions: The case of the growth mindset during the transition to high school. *Journal of Educational Psychology*, *108*(3), 374–391. doi:10.1037/edu0000098 PMID:27524832

Yeager, D. S., & Walton, G. M. (2011). Social-psychological interventions in education: Theyre not magic. *Review of Educational Research*, *81*(2), 267–301. doi:10.3102/0034654311405999

Yeager, D., Walton, G., & Cohen, G. L. (2013). R&D: Addressing achievement gaps with psychological interventions. *Phi Delta Kappan*, *94*(5), 62–65. doi:10.1177/003172171309400514

Yeakey, C., & Henderson, R. (2000). *Surmounting all odds: Education opportunity, and society in the new millennium*. Greenwich, CT: Information Age Publishing.

Yeaple, R. (2014). MBA Pay. *Forbes*. Retrieved from http://www.forbes.com/sites/ronaldyeaple/2014/03/17/mba-alumni-profile-2/#7f92ac6e503e

Yorke, M. (2008). Is the first year experience different for disadvantaged students? In F. Ferrier & M. Heagney (Eds.), *Higher Education in Diverse Communities: Global Perspectives Local Initiatives* (pp. 113–120). Dublin: Higher Education Authority.

Yorke, M., & Longden, B. (2004). *Retention and student success in higher education*. Berkshire, England: Open University Press.

Yorks, L., & Kasl, E. (2002). Toward a theory and practice for whole-person learning: Re-conceptualizing experience and the role of affect. *Adult Education Quarterly*, *52*(3), 176–192. doi:10.1177/07417136020523002

Yun, C., Kim, K., & Kim, J. (2009). Praise types perceived by pre-service secondary teachers. *The Journal of Korean Teacher Education*, *26*(4), 55–78. doi:10.24211/tjkte.2009.26.4.55

Zak, P. (2013). Measurement Myopia. Retrieved from http://www.druckerinstitute.com/2013/07/measurement-myopia/

Zaleski, Z. (1988). Close relationships and acting for self-set goals. *European Journal of Social Psychology*, *18*(2), 191–194. doi:10.1002/ejsp.2420180210

Zhang, Y. B., Lin, M., Nonaka, A., & Beom, K. (2005). Harmony, Hierarchy and Conservatism: A Cross-Cultural Comparison of Confucian Values in China, Korea, Japan, and Taiwan. *Communication Research Reports*, *22*(2), 107–115. doi:10.1080/00036810500130539

Zheng, B., & Warschauer, M. (2015). Participation, interaction, and academic achievement in an online discussion environment. *Computers & Education*, *84*, 78–89. doi:10.1016/j.compedu.2015.01.008

Ziegler, M., MacCann, C., & Roberts, R. (Eds.). (2011). *New perspectives on faking in personality assessment*. Oxford University Press. doi:10.1093/acprof:oso/9780195387476.001.0001

Zimbardo, P. G. (1969). The human choice: Individuation, reason, and order vs. deindividuation, impulse, and chaos. In W. J. Arnold & D. Levine (Eds.), *Nebraska symposium on motivation* (Vol. 17, pp. 237–307). Lincoln: University of Nebraska Press.

About the Contributors

Ruth Claire Black is an experienced university administrator and faculty member with a history of success building outstanding online, blended, military, vocational, international and offsite education programs within large, distributed public and private university systems. In her current role as a visiting scholar at Namseoul University, Ruth is expanding her international experience while working on several national education projects within the Republic of Korea. Ruth is also serving as one of the founding faculty members of the new College of International Education at Namseoul University. Ruth's academic and research interests include, coaching to improve retention and student success and embedded digital content options for online courses as well as comparative analysis of outcomes and student satisfaction in technology delivered courses versus traditional face to face courses. Ruth has a BA in Political Science, with minors in English and History from the University of California, Irvine, a Juris Doctorate (JD) from Vanderbilt University and a Doctorate in Education (EdD) from the University of Southern California.

* * *

Neeta Baporikar is currently Director/Professor (Business Management) at Harold Pupkewitz Graduate School of Business (HP-GSB), Namibia University of Science and Technology, Namibia. Prior to this she was Head-Scientific Research, with Ministry of Higher Education CAS-Salalah, Sultanate of Oman, Professor (Strategic Management and Entrepreneurship) at IIIT Pune and BITS India. With more than a decade of experience in industry, consultancy and training, she made a lateral switch to research and academics in 1995. Dr. Baporikar holds D.Sc. (Management Studies) USA, PhD in Management, University of Pune INDIA with MBA (Distinction) and Law (Hons.) degrees. Apart from this, she is also an External Reviewer, Oman Academic Accreditation Authority, Accredited Management Teacher, Qualified Trainer, Doctoral Guide and Board Member of Academics and Advisory Committee in accredited B-Schools. Reviewer for international journals, she has to her credit 5 conferred doctorates, several refereed research papers, and authored books in the area of Entrepreneurship, Strategy, Management and Higher Education.

Eric Canny serves as Vice President for Program Management at CEA Study Abroad, a newly created role, responsible for globally implementing a new division to ensure a steady two-way communication between program sites and headquarters concerning all aspects of daily operations. This communication, and new organizational structure, helps CEA perform as a more integrated organization with Program Management as the conduit to all other operational and academic operations with the goal of ensuring best practices are adopted across all locations, and innovation is incorporated into CEA's practices.

Eric is responsible for all 21 program sites functioning at a consistently high level across a wide range of geographies, cultures, and program types. Programs serve approximately 4,000 students per year in standard, hybrid and customized programs spanning time-frames from 10 days to an entire academic year; with annual growth goals across the organization ranging from 12-50% depending upon area and program type. Previously, Eric served as Dean of Global Education at Augsburg College, overseeing the nationally award-winning Center for Global Education and Experience. He led an expanding holistic model of globalization for Augsburg College based upon strengthening as well as expanding, the institution's unique approach and commitment to programming globally, aligned with the 35 years of operations of the Center for Global Education and Experience. Prior positions held include: Executive Director of International Learning (Stetson University); Director International Initiatives (University of North Texas); and Director Global Student Services (New York University) where he was part of the team that expanded NYU's study abroad presence from 4 to 10 sites, and 400 to 4,000 students annually in a span of 8 years. He has presented nationally, and internationally, as well as authored chapters on preparing students to study abroad, working with non-traditional students, and student success models in foreign operations of U.S. higher educational institutions (forthcoming). He is currently focused on the development of new cutting-edge models of student learning assessment in, and outside of, the classroom setting. This model, would utilize databases linked to student data, to track learning outcomes in a multi-dimensional manner. Eric has 22 years' experience in educational policy and programming, academic curriculum development and accreditation, admissions, risk assessment and crisis management, student and staff development, and non-profit event management. He was team-lead for opening US higher educational institutions in: Argentina, China, Costa Rica, France, Ghana and the United Kingdom as well as expansion of existing programs in Chile, Cuba, Czech Republic, France, Germany, Mexico, Namibia, Nicaragua, Spain, and Thailand as well as other countries. Eric has traveled to more than 55 countries. He received his B.F.A and M.A. from New York University, and his Doctorate from the University of Southern California.

Daisha J. Cipher is Associate Professor with Tenure at The University of Texas at Arlington College of Nursing and Health Innovation. Prior to serving in this capacity, Dr. Cipher was the Associate Director for the UT Arlington College of Nursing and Health Innovation Center for Research and Scholarship. Prior to serving UT Arlington, Dr. Cipher served as Program Director for the University of North Texas Health Science Center School of Public Health Master's of Public Health Clinical Research Program, which also serving as Director of Biostatistics for the University of North Texas Health Science Center Osteopathic Research Center. Dr. Cipher received her Ph.D. from Southern Methodist University in 1998, and spent the next year at the University of North Texas in a post-doctoral fellowship in advanced statistical methods with a special focus on structural equation modeling and path analysis. Since that time, she has developed research interests in multivariate statistics and statistical risk modeling, including the application of structural equation modeling, hierarchical linear mixed models, path analysis, generalized estimating equations, receiver operator curves, logistic regression, multiple regression, ordinal regression, and sensitivity analysis to identify risks and protective factors of outcomes in clinical, medical and educational settings. Dr. Cipher is also dedicated to increasing the number of analytic projects focusing on national population data, particularly the analysis of large national databases to investigate specific diseases, health determinants, health policy issues, epidemiologic issues, and public and clinical health practices. Finally, she has developed a niche in the development of statistical instruction tools for academic programs across health disciplines, including the development and publication of concise practical instruction materials that help graduate students, undergraduate students, and instructors to develop confidence in interpreting the statistical data found in health sciences research articles. Dr. Cipher has published over 70 manuscripts in databased, refereed journals. Moreover, she has authored 11 book chapters and a statistics textbook. She has created and taught over 65 courses in statistics, research design, epidemiology, and information technology. She has received federal grant funding continuously since 2005. She has served as co-investigator on three grants funded by the National Institutes of Health and three grants funded by the Department of Veterans Affairs.

Alan Cromlish is an Assistant Professor in the new College of International Teaching at Namseoul University, in Seoul, South Korea. Alan is developing master's programs that create and support global education leaders. Alan's academic and research interests include, well-being and hope theory in the Korean education system. Alan has a BS in Psychology from the Georgia College and State University and a Masters of Education from the University of Missouri. Alan is a doctoral candidate in the Global Executive Doctorate of Education Program at the Rossier School of Education at the University of Southern California.

Helen Eckmann is a full-time faculty with Brandman University Business School. She was elected by staff and students as "Teacher of the Year" in 2011 and "Teacher of the Year" for the Business School in 2012. She is responsible for the Management program which includes Strategy Sustainability, Innovation and Supply Chain for both the BBA and the MBA degrees. She also designed and teaches Strategic Thinking for Brandman University's new doctoral program. Helen has designed or helped to design over 30 courses for Brandman in the BBA and MBA programs, both blended and online. She has also taught as an Adjunct at the University of San Diego Supply Chain Management Master's Program for the past ten years. This program is approved by the Institute for Supply Management. She has received "Teacher of Impact" Awards from students within that program eight times, the most recent being April of 2012.

Judi Simmons Estes received a Ph.D. in Behavioral Psychology from the University of Kansas. She currently serves as an Associate Professor and Associate Dean for the School of Education, Park University in Parkville, Missouri. Dr. Estes has developed and delivered online courses for the past eight years at the undergraduate and graduate levels; in addition, she has supported the development of a fully online degree program. Research interests include identifying strategies for increasing student retention.

Scott Gaier earned his PhD in Educational Studies from Purdue University. Dr. Gaier is the Director of Academic Enrichment and also Associate Professor of Higher Education at Taylor University. He oversees Taylor's Academic Enrichment Center which includes first year transition initiatives, retention efforts, disability support services, and interventions for academically at-risk students. Dr. Gaier also teaches undergraduate courses on reading, learning, study skills, and transition to college and also teaches graduate courses on research methods. He is the recipient of multiple teaching awards. His research interests include student learning, academic success, and college teaching.

Laura Galloway is currently an Associate Professor in the School of Business and Professional Studies at Brandman University. Dr. Galloway has over 13 years' experience in higher education online instruction. Laura holds two Master's degrees, the first she earned from Chapman University in Organizational Leadership with concentration in Human Resource Management. She received her second Master's degree in Human Development from Fielding Graduate University, where she also received her PhD in Organizational Development. Her dissertation topic was on instructional design to improve student learning and student transformation. In addition to being a full-time faculty member, Laura is a certified Executive Coach, an Instructional Designer, and curriculum developer. She has also served as Director of Human Resources for a school district. Laura has over 15 years of experience in adult education in both academic and corporate environments with a strong background and extensive experience in providing innovative academic and professional support services and programs to diverse adult student populations. Her expertise includes strategic planning, program development, organizational behavior, systems, organizational development, course design, instructional design methodologies, talent management systems and adult learning theory. She has taught college and graduate level courses in Supervision and Leadership, Human Resources Management Systems, Teams, Organizational Development and Change, Contemporary Issues in Organizational Leadership and Organizational Analysis. Her consulting work includes specialization in Human Resources Systems and Development for mid-size, non-profit organizations.

Darab Ganji ("Rob") is vice chairman of Academic Partnerships (AP), a higher education service provider that helps public and private not-for-profit universities expand access to higher education by scaling their online programs. As a leader in Online Program Management, AP assists universities by converting their traditional high demand degree programs into an online format, recruiting qualified students and retaining them through graduation. Darab has extensive experience in the edtech and healthcare sectors with strategic and operational leadership roles in higher education services. As vice chairman, he provides strategic guidance to the Academic Partnerships executive team and specifically focuses on growth and strategic initiatives. Before becoming vice chairman of AP in 2012, he served as the company's president and, prior to that, successfully launched and led its healthcare division which is the largest vertical of the company. Darab is a member of the board of directors and guest lecturer at Southern Methodist University's John Tower Center for Political Studies. He is also a board member of the Dallas Holocaust and Human Rights Museum and is a member of the Advisory Board of Ilumno, a network of higher education institutions across Latin America serving over 250,000 students. Darab's articles on education, healthcare, political economy and foreign policy have appeared in numerous publications including EDUCAUSE, the Los Angeles Times, the Houston Chronicle and The Dallas Morning News. He holds a bachelor's degree in political science from Southern Methodist University and a master's degree in political economy from the University of Texas at Dallas.

Samuel L. Hinton is a native of Rock Hill, South Carolina. Dr. Hinton received his Bachelor of Arts degree in History from South Carolina State University, Orangeburg, SC; Master of Science in Post-Secondary Education with a concentration in Higher Education Administration from Troy University, Troy, Alabama and a Doctor of Education in Educational Leadership with a concentration in Higher Education Administration from Argosy University in Atlanta, Georgia. Dr. Hinton is the co-founder of the Crawford-Hinton History scholarship along with Dr. Fashaad Crawford for History, History Education and Social Studies majors in the department of Political Science and History established on May 21, 2001 in the division of Institutional Advancement at South Carolina State University where he served in the capacity of Development Associate. Dr. Hinton is the 2006 Minority Scholarship recipient from the Southern Association of Collegiate and Admission Officers to attend the annual conference for Registrars and Admissions officers in Lexington, Kentucky where he served in the capacity of Sr. Admissions Counselor and Recruiter at Voorhees College. Dr. Hinton is also an abstract reviewer for "Call for Papers" for the annual Black Doctoral Network Conference. Dr. Hinton has served as a former member of the Quality Enhancement Plan (QEP) at Benedict College, which is a course of action, designed to address a specific issue critical to student learning which supports reaffirmation related to SACS (Southern Accreditation of Colleges and Schools) where he served in the capacity of Retention Specialist and Seminar Instructor. Dr. Hinton is a contributing author for the book "Setting a New Agenda for Student Engagement and Retention in Historically Black Colleges and Universities edited by Mr. Charles Prince, Howard University in Washington, DC and Dr. Rochelle Ford, Syracuse University, Syracuse, New York. Dr. Hinton is also a published research writer with ProQuest Dissertations and Thesis. Dr. Hinton dissertation title is "Factors That Affect Retention Among Freshman Students at Historically Black Colleges and Universities. Dr. Hinton formerly served as a Board member for the Garden City Preparatory Academy for Boys (2014-2016) in Orangeburg, SC and currently serves as a Board member for the Armstrong Achievement Academy for Boys in Columbia, SC. Dr. Hinton is a member of Alpha Phi Alpha Fraternity, Incorporated and the South Carolina State University Alumni Association as a Life Member. Dr. Hinton was also selected as 2017 fellow into the Epsilon Cohort by the Higher Education Leadership Foundation Institute. Dr. Hinton has over a decade of experiences in Higher Education and has served in various leadership capacities throughout his career. Dr. Hinton is highly motivated and result oriented professional with exceptional leadership and communication skills; Strong background in the development and delivery of high-impact presentations utilizing verbal acuity and technical proficiencies; experienced in leading and motivating both students and associates; dedicated to the advancement of students through the development and implementation of organized, engaging, and innovative programs with meaningful involvement. Dr. Hinton is skilled in recognizing individual student needs through assessing overall goals while enhancing success for the advancement of Higher Education.

Ryan Korstange is an Instructor in the University College at Middle Tennessee State University, where he teaches University 1010, MTSU's FYE course, and coordinates the UNIV 1010 program. Ryan's research time is split between determining effective ways to help promote student learning and describing the way ancient philosophers and theologians understood the development of the human soul.

Kevin S. Krahenbuhl is Assistant Professor of Education at Middle Tennessee State University and teaches in the Assessment, Learning, and School Improvement Ed.D. program and the Department of Educational Leadership. His dissertation analyzed a college-specific iteration of the FYE seminar and since joining higher education he has played an active role in student success including being awarded for his efforts towards improving student learning outcomes. His research interests place an emphasis on cognition and learning history, increasing student success, and integration of technology to enhance learning.

Julie Little is an Associate Professor of Management in the Business Department at Taylor University in Upland, Indiana. She teaches courses in entrepreneurship, management, basic business and human resources. She is the faculty advisor to the Taylor University SHRM Student Chapter. Her research interests include leadership, student engagement and organizational socialization.

Mary (Beth) Mancini is Professor, Senior Associate Dean for Education Innovation and Chair for Undergraduate Nursing Programs at The University of Texas at Arlington College of Nursing and Health Innovation where she holds the Baylor Professorship for Healthcare Research. Prior to moving to an academic role in 2004, Dr. Mancini was Senior Vice President for Nursing Administration and Chief Nursing Officer at Parkland Health & Hospital System in Dallas, Texas, a position she held for 18 years. In 2013, Dr. Mancini was appointed a Visiting Scholar in Innovation and Simulation at The University of Pennsylvania School of Nursing. In 2014, she was reappointed as a Visiting Scholar in Simulation and Curriculum. Dr. Mancini received a BSN from Rhode Island College, a Masters in Nursing Administration from The University of Rhode Island and a PhD in Public and Urban Affairs from The University of Texas at Arlington. She completed a Johnson & Johnson Wharton Nurse Executive Fellowship at the Wharton School of Business of the University of Pennsylvania and a National Association of Public Hospitals Management Fellowship program through the Robert F. Wagner Graduate School of Public Service at New York University. In recognition for her contributions in the fields, Dr. Mancini was inducted as a Fellow in the American Academy of Nursing, a Fellow in the National League for Nursing's Academy of Nurse Educators, and a Fellow of the American Heart Association. In 2013, Dr. Mancini was recognized with a Regent's Outstanding Teaching Award from the University of Texas System. Her groundbreaking work in high quality, high volume, accelerated online education, led to the UT Arlington's College of Nursing receiving the Texas Higher Education Coordinating Board's prestigious Star Award in 2012. Dr. Mancini is an active volunteer with numerous professional organizations. She currently serves as a member of the American Heart Association's Educational Science and Program committee as well as its Get With The Guidelines – Resuscitation Work Group, member of the National Academies of Science Global Task Force on Innovations in Health Professions Education, and co-chair of the Basic Life Support Task Force for the International Liaison Committee on Resuscitation. She has served as President of the international Society for Simulation in Healthcare as well as having served as a member of the Royal College of Physicians and Surgeons of Canada's Simulation Task Force and the World Health Organization's Initiative on Training, Simulation and Patient Safety. Dr. Mancini's research interests include innovations in education, interprofessional collaborative practice, and the development of high performing healthcare teams through the use of simulation. She has received over $6.5 million in competitive grants, has more than 90 publications to her credit, and is a sought-after speaker at local, national and international conferences on topics such as innovations in online education, development of high-volume, high-quality educational programs, simulation in healthcare, health professions education, patient safety, outcomes related to basic and advanced life support education, and work redesign.

Ross Markle is currently a Higher Education Assessment Advisor at Educational Testing Service. In his role, he supports ETS' thought leadership efforts in higher education by collaborating with operational and research areas, as well as the higher education community. Ross also works directly with colleges and universities to promote the effective use of assessments and data in student success efforts, focusing on the areas of holistic advising, course placement, institutional planning, predictive modeling, and supporting traditionally underserved populations. He has also worked in ETS' Research and Development Division, researching the assessment of noncognitive and 21st century skills, student success, and student learning outcomes assessment in higher education. Prior to joining ETS' Higher Education Division, Ross obtained his Ph.D. in Assessment & Measurement Psychology from James Madison University where he assisted in evaluating the quality of program level assessment activities. He also served as the Director of Co-curricular Assessment and Research at Northern Kentucky University.

Meagan Moreland joined the faculty at Northeastern State University in 2008. Dr. Moreland is an Associate professor of Literacy Education and the department chair for Curriculum and Instruction. She obtained her Ph.D. from the University of Oklahoma in Instructional Leadership and Academic Curriculum with an emphasis in reading. Her M.Ed is in Reading from Northeastern State University and her BS is in Elementary Education also from Northeastern State University. Dr. Moreland's research interests include engagement and motivation in reading, clinical work and reading assessment. She has a passion for the reciprocal nature of reading clinics which create supportive experiences for NSU students and children and their parents.

Sean Nemeth is the Associate Vice Chancellor of Enrollment Services, Retention, and Advising at Brandman University. Sean completed his B.S. in Management at Georgia Tech and his M.B.A. at East Carolina University and his Doctorate at Brandman Unviersity. Sean has worked primarily in higher education and businesses that support higher education. He helped to develop InsideTrack's executive coaching program while working in collaboration with several institutions, including Chapman University. In 2007, Sean began his work with Brandman University overseeing the redesign of its academic advising structure and retention efforts. Since his arrival, student satisfaction with academic advising has improved by 15% and term to term retention has increased by 5.5%.

Sam Rikoon is a research scientist in the Student and Teacher Research area at Educational Testing Service. His work focuses primarily on building and establishing the psychometric integrity of student assessments of "noncognitive" or social and emotional learning constructs for higher education and high school applications. He also works directly with institutions to help them make appropriate use of the data and reports emerging from such assessments, and conducts research on student learning outcomes. His research has been published in the Journal of Higher Education, Journal of Educational Psychology, and Journal of Personality Research, among others.

Gokhan Savas is currently an Assistant Professor of Sociology and also serves as the Director of Center for Survey Research at Social Sciences University of Ankara. He holds a B.S. in sociology from Hacettepe University, an M.A. in sociology and a certificate of advanced studies in women's and gender studies from Syracuse University, USA. He earned his PhD in sociology from the same institution. As a quantitative sociologist, he is broadly interested in social inequality particularly pertaining to the intersections of gender, race, and social class; achievement and attainment in postsecondary education; international immigration and the children of immigrants; and critical race theory. His works have been published in several journals including Research in Higher Education, British Journal of Sociology of Education; European Journal of Social Sciences; American Journal of Educational Research, and Race, Gender & Class Journal. Prior to joining Social Sciences University of Ankara, he was a tenure-track

Assistant Professor of Sociology at Luther College from 2013-2014; and a Visiting Assistant Professor of Sociology at the State University of New York – Cortland in 2013. He has previously taught several courses such as Introduction to Sociology, Research Methods for Social Sciences, and Sociology of Family. He received an NSF Dissertation grant, and also received an Outstanding Teaching Award from Syracuse University.

Melinda L. Smith is a Professor of Literacy Education at Northeastern State University. Dr. Smith obtained her Ed.D from Oklahoma State University in Curriculum Studies and Educational Leadership with a specialization in Literacy. Her M.Ed is from Northeastern State University in Reading Education and her BS is from the University of Tulsa in Elementary Education. During her career Dr. Smith has served in a variety of leadership positions and has been recognized for outstanding Research and Teaching. She is passionate about teaching, research and service initiatives aimed at advancing literacy instruction and increasing students' literacy achievement outcomes. Her over 30 years in education also includes public school teaching and extensive experience working with primary grade struggling readers in clinical settings. Her research interests included reading acquisition, struggling readers and comprehension.

Danielle Spoutz is pursuing a Masters of Arts in Higher Education and Student Development at Taylor University. Ms. Spoutz serves as the Graduate Assistant to Taylor's Business Department. Her research interests include college traditions and culture.

Allyson L. Watson is assistant dean and Northeastern State's first named endowed chair. Her educational background includes master's (M.Ed.) and doctoral (PhD) degrees in educational administration, curriculum and supervision from the University of Oklahoma. Her baccalaureate degree is from Bethune Cookman University in Elementary Education. Dr. Watson has focused her research on urban education, faculty of color in higher education, and urban school and university partnerships. Dr. Watson founded Teaching & Urban Reform Network (TURN), a program to prepare pre-service teachers in urban education and encourage effective pedagogical practices. In 2015, Dr. Watson coordinated a team of educators from across the US to help establish the first ever robotics lab for girls in north Haiti. The sustainable work has empowered girls and women in STEM. Through this innovative initiative, Haiti sent the first competitive girls team in the history of the nation to compete in the VEX World Championship Robotics Competition. Dr. Watson believes that her job as an educator is to empower others to believe in themselves and their capacity to reach their true potential.

Alexandra McDermott Wilcox is an innovative educator, community advocate, and writer. Dr. Wilcox is currently an Assistant Adjunct Professor at the University of Southern California and an Adjunct Faculty Member at American Public University System. Dr. Wilcox has also served in the professional sector as a litigation/trial attorney, franchise education business owner, and college academic advisor and counselor. Dr. Wilcox is also a published author and her work appears in various literary journals. Dr. Wilcox has a Bachelor of Arts in English Literature from Washington University, Juris Doctor from Chapman University School of Law, Master of Fine Arts in Creative Writing from the University of California at Riverside, and a Global Executive Doctor of Education from the University of Southern California.

Index

Q

R

S

T

U

V

W

CPSIA information can be obtained
at www.ICGtesting.com
Printed in the USA
BVOW07*1138161217
502906BV00002B/5/P

9 781522 529989